Metal-Organic Framework Composites

Volume I

Edited by

**Anish Khan[1,2], Bahaa M. Abu-Zied[1,2,3], Mahmoud A. Hussein[2,3],
Abdullah M. Asiri[1,2], Mohammad Azam[4]**

[1]Center of Excellence for Advanced Materials Research, King Abdulaziz University, Jeddah, 21589, Saudi Arabia

[2]Chemistry Department, Faculty of Science, King Abdulaziz University, P.O. Box 80203, Jeddah, 21589, Saudi Arabia

[3]Chemistry Department, Faculty of Science, Assiut University, 71516, Assiut, Egypt

[4]Department of Chemistry, College of Science, King Saud University, Riyadh 11451, Saudi Arabia

Copyright © 2019 by the authors

Published by **Materials Research Forum LLC**
Millersville, PA 17551, USA

Published as part of the book series
Materials Research Foundations
Volume 53 (2019)
ISSN 2471-8890 (Print)
ISSN 2471-8904 (Online)

Print ISBN 978-1-64490-028-4
eBook ISBN 978-1-64490-029-1

This book contains information obtained from authentic and highly regarded sources. Reasonable efforts have been made to publish reliable data and information, but the author and publisher cannot assume responsibility for the validity of all materials or the consequences of their use. The authors and publishers have attempted to trace the copyright holders of all material reproduced in this publication and apologize to copyright holders if permission to publish in this form has not been obtained. If any copyright material has not been acknowledged please write and let us know so we may rectify this in any future reprints.

Distributed worldwide by

Materials Research Forum LLC
105 Springdale Lane
Millersville, PA 17551
USA
http://www.mrforum.com

Manufactured in the United States of America
10 9 8 7 6 5 4 3 2 1

Table of Contents

Preface

Metal-organic frameworks (MOFs) have gotten extraordinary consideration because of their exceptional physical and chemical properties that make them pertinent in different fields, for example, designing, medicine, and environment. It is a new option for miniature solutions for environment safety. Since combination conditions incredibly affect the properties of these compounds, it is especially important to choose an appropriate synthesis technique that produces the product with homogenous morphology, small size dispersion, and high thermal stability and also with a surface area so the properties could prompt broad applications in different fields. This book covers different aspects of composites fabrication from energy storage and catalysts including its preparation, design, characterization techniques of Metal Organic Frameworks (MOFs) material and its application. The special features in this book comprise of illustrations and tables summarize upto date information on research carried out on manufacturing, design, and characterization and water treatment applications of MOFs composites.

Chapter 1 deals with the different methods of preparation of metal-organic frameworks (MOFs) and their application in industrial wastewater treatment. Current approaches in the extraction and degradation of water pollutants by MOFs are highlighted. In chapter 2, advantages of MOFs and their composites in the construction of electrochemical sensors as well as their applications in the determination of several environmental contaminants are discussed. Chapter 3 summarizes the results of a large number of literatures in recent years on the adsorption of pollutants in wastewater by MOFs, including decontamination strategies and practical applications. While in chapter 4, it is discussed the development and comparison of traditional materials based on metal organic frameworks based wastewater treatment techniques. The process of pollutants in water and the performance and details of MOFs in different treatment areas have been emphasized. Chapter 5 presents techniques of metallizing the high-performance polymer fibre surfaces for compatibility with the metallic matrices, using techniques such as electroless coatings, Radio Frequency (RF) ion sputtering and High Velocity Oxygen Fuel (HVOF) thermal spray processes. This chapter also presents metal-polymer composites made by the HVOF thermal spray technique with low melting metallic matrices. Chapter 6 describes the condensation of WO_4^{2-} Polyhedra units on layered rare earth hydroxides nanosheets: hierarchical channels and heavy metal adsorption. Chapter 7 discusses the design approach of MOFs based on green applications in four categories. It starts with an overview of the MOFs approach for the giant problem of clean energy. It then explains how the MOFs material is a strong candidate to overcome this problem by comparing the different design approaches. By the end of the chapter, the challenges facing to commercialize this technique is discussed. Chapter 8 reviewed the recent progress in MOFs for water adsorption, including adsorption isotherm, mechanisms, water harvesting processes, criteria for MOFs, and affecting factors. Chapter 9 focuses on the origin of flexibility, and the substantial geometrical changes that can occur due to external stimuli, such as temperature, pressure, light, gas or solvent adsorption. Flexibility control methods have also been discussed along with possible characterization techniques to help to identify the source of flexibility. Practical applications of flexible MOFs are also discussed. In this respect, several prized examples covered by literature are present to help in a comprehensive understanding in terms of design and structure tunability of flexible MOFs. Recent developments on the preparative strategies and applications of MOF encapsulated enzyme with special focus on catalysis are summarized in this chapter 10. The enhancement/retention of enzymatic activity of the composite material compared to free enzymes in denaturation conditions and advantages of encapsulation of the enzymes has been reviewed. Herein in chapter 11 the nanocarbons, derived from the metal-organic frameworks (MOFs), are reviewed with different aspects systematically. In chapter 12 a study on polyoxometalate-based Metal-Organic Framework composites was described.

We are highly thankful to contributors of different book chapters who provided us their valuable innovative ideas and knowledge in this edited book. We attempt to gather information related to MOF composites for environmental application from diverse fields around the world and finally complete this venture in a fruitful way. We greatly appreciate contributor's commitment to their support to compile our ideas in reality. We are highly thankful to the MRF team for their generous cooperation at every stage of the book production.

Anish Khan, Bahaa M. Abu-Zied, Mahmoud A. Hussein,
Abdullah M. Asiri,, Mohammad Azam

Metal-Organic Framework Composites - Volume I
Materials Research Foundations **53** (2019) 1-28

Materials Research Forum LLC
https://doi.org/10.21741/9781644900291-1

Chapter 1

Metal-Organic-Frameworks (MOFs) for Industrial Wastewater Treatment

Afzal Ansari[1]*, Vasi Uddin Siddiqui[1], Imran Khan[2], M. Khursheed Akram[3], Weqar Ahmad Siddiqi[1], Anish Khan[4,5], Abdullah Mohamed Asiri[4,5]

[1]Department of Applied Sciences and Humanities, Faculty of Engineering and Technology, Jamia Millia Islamia, New Delhi - 110025, India

[2]Applied Sciences and Humanities Section, University Polytechnic, Faculty of Engineering and Technology, Aligarh Muslim University, Aligarh - 202002, India

[3]Applied Sciences and Humanities Section, University Polytechnic, Faculty of Engineering and Technology, Jamia Millia Islamia, New Delhi - 110025, India

[4]Chemistry Department, Faculty of Science, King Abdulaziz University, Jeddah, Saudi Arabia

[5]Center of Excellence for Advanced Materials Research, King Abdulaziz University, Jeddah, Saudi Arabia

afzal.evs@gmail.com
ORCID ID: https://orcid.org/0000-0002-8199-5243

Abstract

Water pollution is one of the most severe problems in the world, which puts the survival and development of human society at risk. Therefore, developing efficient and cost-effective technologies for the removal of water pollutants has become a hot topic. In this chapter, we deal with the different methods of preparation of metal-organic frameworks (MOFs) and their application in wastewater treatment. The modular structure with a wide variety of MOFs with different active metal sites and organic linkers must prove to be ideal adsorbents or photocatalysts for water purification. In this chapter, current approaches in the extraction and degradation of water pollutants by MOFs are highlighted.

Keywords

Water Pollutants, Metal-Organic Frameworks, Adsorption, Treatment, Composites

Materials Research Forum LLC
https://doi.org/10.21741/9781644900291-1

Contents

1.   Introduction

Currently, water pollution has become one of the top most problems in urban aquatic ecosystems throughout the world generated from industrial effluents or anthropogenic sources [1,2]. Pollutants are released in the water without any treatment, which leads to severe water pollution and health menace [3–5]. Up to the present time, the origin of water pollution mainly includes industrial and domestic effluents, resource mining, oil spills, nuclear waste leakage and pesticides [6,7]. Recent scientific knowledge suggests that environmental pollution is already at a dangerous level, especially in a water environment [8,9]. Water is a major source of survival on this planet. Therefore, its conversation is a priority. The consumption of water resource is increasing at more than double the rate of the world's population growth [10]. The demand for fresh water has increased but also increased in water treatment loading [6]. Therefore, cost-effective water treatment technologies are in high demand. Several methods have been used commonly for wastewater treatment, such as advanced oxidation process [2], adsorption [11,12], oxidation-reduction [13,14], membrane filtration [15–17], chemical treatment [18–20], mechanical [21,22] and incineration [2]. However, these treatment techniques are not superlative because of the need for space-consumption facilities with high maintenance cost and facilities with complex equipment [23]. Among them, adsorption is one of the most effective and promising methods used for the water treatment process [24]. Commonly, porous nanomaterials are considered good adsorbents which play an important role in adsorptive separations or purification of wastewater [25]. Therefore, in the past few decades, there has been great attention to the investigation of advanced porous nanomaterial [26]. Presently, the porous nano adsorbents such as zeolites, activated carbon, clays and aluminophosphates are widely used for the treatment of wastewater [27–29]. However, some important drawbacks of these porous nanomaterials are low surface areas with lack of reduction capability which have a negative effect on their performance [4,24]. Consequently, there is still a great demand to develop newly improved adsorbents for the treatment of industrial wastewater. About two decades ago, a complex and relatively new class of highly porous nanomaterials, metal organic framework (MOFs) developed and it proved versatile performance in adsorption removal/isolation and treatment of wastewater [30–32]. MOFs has a better effect on wastewater treatment through various treatment mechanisms such as electrostatic adsorption, π-π bonding, hydrogen bonding open metal sites, and acid-base adsorption [6]. In this chapter, the potential application MOFs in wastewater treatment is discussed in terms of treatment/removal of pollutants and typical fabrication of MOFs, including their structure and applications.

1.1 Background of MOFs

Over the past few decades, MOFs is considered the most effective porous materials for wastewater treatment. MOFs are defined as a type of porous, network-structural material [33], which can be produced by the self-assembly through organic ligands and metal ions to form a complex metal-ligand [30,34]. MOFs are also known as porous coordination polymers (PCPs), prepared by combined metal ions (or cluster), organic and inorganic ligands via coordination linkages [35], have developed rapidly as one of the most active areas in coordination chemistry [33,36]. Usually, inorganic units are metal ions or metal groups, and organic units [30] (linkers or bridging-ligands) are di-, tri-, or tetradentate organic ligands such as carboxylate or other organic ions (phosphonate, sulfonate, and heterocyclic compounds) [37]. The foremost MOF with permanent porosity was stated in 1995 [35]. Since, there was no standard nomenclature accepted during the development of this new class of crystalline hybrid solids, it has been proposed several names and are in use including zeolite MOFs [27,38], porous (microporous and mesoporous) coordination polymers [39–41], porous coordination networks [42–44] and isoreticular MOFs [45]. MOFs represent a new class of porous materials that have a modular structure, which provides huge structural diversity and wide potential for the invention of materials with tailored properties and broad prospects [46]. The special interest in MOFs materials due to their easily adjustable and microporous to a mesoporous size and shape scale, by the changing of inorganic component and nature of the organic ligands [42,47]. As mentioned, the potentials of MOFs increase due to their physical and chemical properties and to the simply modified which make them a notable group of materials [31]. Furthermore, due to their notable properties and their vast range of applications such as adsorption/storage of organic molecules and gas, films, catalysis, separation, drug delivery, polymerization, luminescence, electrode materials, magnetism, carriers for nanomaterials and imaging [48–51].

MOFs are produced traditionally as well as specific ones with the use of frequent methods such as mechanochemical, electrochemical processing, microwave and ultrasonic [46]. One of the attractive features of synthesized MOFs materials is that they are topological diverse with neutral skeleton and aesthetically pleasing structures [1,33]. The solvent acts as the main template during the preparation of MOFs, usually organic or inorganic template is required, which is an advantageous comparison with the synthesized zeolite-type materials [25,26]. Furthermore, usually weak interactions with the solvents of MOFs and it is important to significantly obtain the final product remained at low temperature with a neutral framework and existing pores [31,32]. Generally, MOFs have unique properties to the treatment of wastewater with a wide development prospect due to the large porous structure, high adsorptive capacity, low density and high specific area [52]. These factors will determine the structure and performance of the fabricated MOFs, therefore, to find a solution to formulate the MOFs for better wastewater treatment.

2.   Structure and types of Metal-Organic-Frameworks

MOFs have different types and subtypes of products which precise structures are shown in Table 1. MOFs can be distinguished by two components: metal cluster/ ions (secondary buildings units) (Figure 1 A) and organic molecules (Figure 1 B) ligands, basically to give periodic porous structures connecting to the original [33,46]. These structures have an enormous number of MOFs with different combinations of elements. Moreover, the central metal atom can be replaced by the secondary building units in each geometry version of MOFs while the linkers are the same [46]. Likewise, the structure of different MOFs with the terephthalate dianion (BDC = Benzenedicarboxylate) as a linker can lead to differences (Figure 2) [53]. First synthesized MOFs (MOF-5) are contained with tetrahedral ZnO_4 component that is located on the sites of cubic lattice and connected to each other by terephthalate dianions linkers [46]. Increasing the carbon chain linker, being sustained the initial topology, this makes it possible to synthesize those materials with the same structure and uniformity, but different in pore sizes [33,54].

Table 1 - Different types and shapes of Metal-Organic Framework in previous research studies [6].

No.	Types of MOFs	Subtypes	Structures
1	UiOs (University of Oslo)	UiO-66	
		UiO-67	
		UiO-68	
2	MILs (Materialsof Institut Lavoisier)	MIL-101 (Cr)	
		MIL-100 (Al)	-----
		MIL-43 (Fe)	

3	ZIFs (Zeolitic Imidazolate Frameworks)	ZIF-8	
		ZIF67	
		ZIF-L	
4	HKUSTs (Hong Kong University of Science and Technology)	HKUST-1	
		Co-HKUST	
5	JUCs (Jilin University China)	JUC-165	
		JUC-167	
		JUC-199	

		MOF-5	
6.	MOF-N (Metal-Organic Frameworks)	MOF177	
		MOF-199	
7	MbioFs (Metal-Biomolecule Frameworks)	Zeolitic MBioF	---

The lack of control over the structural integrity of conventional solid-state materials such as zeolites and activated carbon is due to the poor correlation between the reactants and products [55]. In contrast, the MOFs can be produced under reactive conditions with a rigid molecular building block that maintains their structural integrity during the synthesis process. The metallic building units generally produced polygonal or polyhedral during the synthesis process [33]. The geometry and shape of the organic building units are predetermined but their flexibility often determines the final architecture. Therefore, it is important to identify all the corners and edges rather than topology of structure [56]. All types of a structure such as organic, inorganic and hybrid materials have an inherent topology and temporarily construct MOFs network based on the rational assembly of building units in different dimensions by the researchers [55]. For example, a prototypical metal-organic framework, MOF-5, possess ZnO_4 tetrahedra $[(Zn_4O)O_{12}$ octahedral SBU] joined by benzene dicarboxylate (BDC) ligands, resulting in a three-dimensional cubic network structure with interconnected pores of 8 Å aperture and 12 Å pore diameter [57].

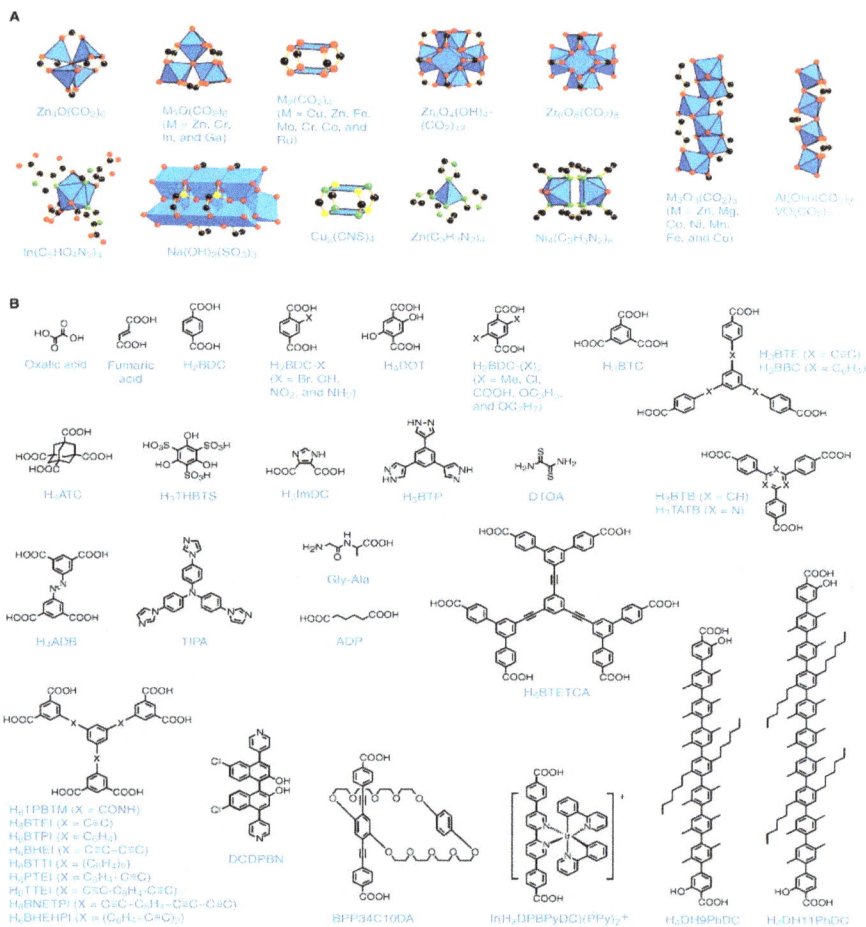

Figure 1. Constituents of MOFs (A) Metal clusters (B) Organic molecules with linkers [46].

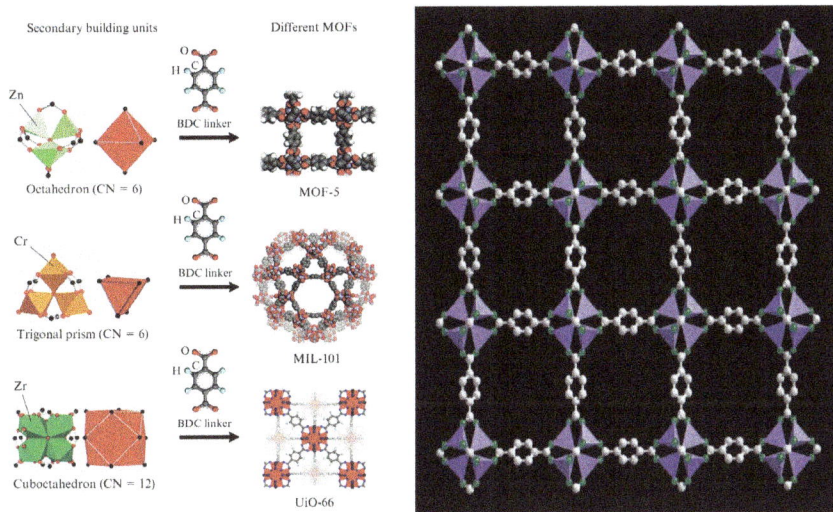

Figure 2. Structure of different MOF with BDC linker (left) and Representation of layer of the MOF-5 framework (right) [53,57]

3. Fabrication/preparation of Metal-Organic-Frameworks

The MOFs have been synthesized by various methods including traditional synthesis (solvothermal and non-solvothermal) and alternative synthesis (microwave, electrochemical, mechanochemical and sonochemical) (Figure 3). Generally, the MOFs are crystallized with a solution. Water and especially organic solvents fill the pore with the unused residue of the initial compound. Thermodynamically, dense structures are more stable [46]. Therefore, there is an important role of encapsulation of foreign molecule in porous structure and inorganic units in the formation of MOFs.

3.1 Conventional synthesis

Conventional synthesis can be divided into solvothermal and non-solvothermal. The term solvothermal means the use of any solvent and it is more common than that hydrothermal that has solvent water. Non-solvothermal synthesis is below the solvent boiling point in the open flask at atmospheric pressure, while solvothermal synthesis is done at the boiling temperature of the solvent or in special closed chemical reactors above this boiling point, which is due to solvent vapor. Non-solvothermal synthesis can be completed both at room temperature and by heating, it does not require complex instruments. A general plan of this type of synthesis means the adjustment of pH and temperature as well as the choice of a metal salt, organic linker and solvent to provide the maximum yield of the target MOFs. The reagent concentrations should be chosen in such a way that the nucleation conditions for precipitation is achieved [46]. It is usually promoted by rising temperatures and evaporation of the solvent. Moreover, the concentration gradient can be prepared by slower cooling of the solution or by slow propagation of one of the solvent reactants [58]. Majority of known MOFs have been synthesized by this method. Specifically, the simple mixing of solutions without heating gave MOF-5 [57], MOF-74, MOF-177 [33] and ZIF-8 [59], respectively.

3.2   Alternative synthesis

Generally, MOFs synthesis takes place with the chemical reaction of solvent which requires temperatures ranging from ambient temperature to 250 °C approximately. The energy is typically used as conventional electric heating, i.e., the heat is transferred through convection, from the oven and heat sources. On the other hand, the energy can be accessed through other ways, such as electromagnetic radiation, mechanical waves (ultrasound)/ mechanically and an electrical potential. The energy source is strictly correlated with the pressure, time, and energy per molecule which is presented in a system and each of these parameters can have a strong effect on the formation of the product and its morphology [58]. In this section, the alternative routes of the MOFs synthesis are described where energy is presented through microwave, electrochemical, mechanochemical and sonochemical.

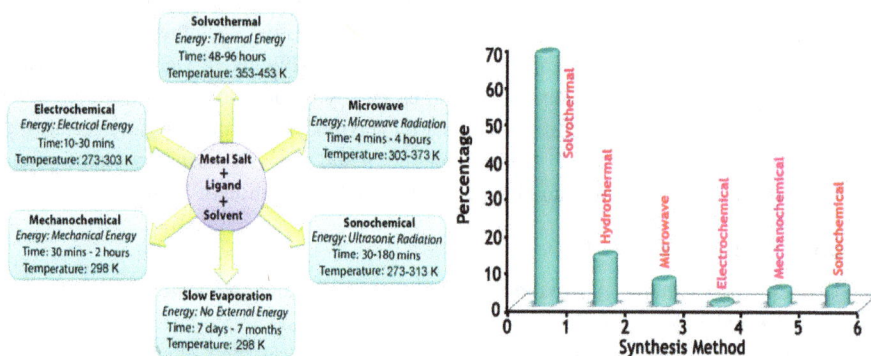

Figure 3. Synthesis conditions commonly used for MOF preparation and indicative summary of the percentage of MOFs synthesized using the various preparation routes [55]

3.2.1 Microwave synthesis

Microwaves (MW) are an array of electromagnetic radiation with two components of microwave radiation (electrical and magnetic) with frequencies between 300 and 300,000 MHz, only in the previous, there is normally an influence on the synthesis of compounds. There are two types of MW effects on a substance: action on polar molecules and free ions that lead to heating. Try to align with the alternate field in both cases, polar molecules and free ions. Electric current is produced in an electrolyte solution and hence due to resistance the heating occurs. There is resistance due to the presence of hydrogen bonds in polar molecules (water) which prevent the dipole from easily aligning with the variable area. There are several factors that affect the heating efficiency such as dipole moment of polar molecules and its free rotation [46]. The effects of MW radiation in different physical states of water shows that strongest in a liquid state while weak in vapor (absence of hydrogen bonds) and ineffective in the solid state (molecules firmly bound in the crystal lattice of ice). The solvent plays an important role in the synthesis of MOFs and the use of MW radiation implements a new specific requirement on the solvent.

Microwave radiation was first used for the synthesis of MOF in 2005 [60]. This process enables to decrease the time of the MIL-100 framework synthesis from 96 to 4 h. The MIL-100 framework was synthesized with an aqueous solution of chromium metal, H_3BTC (benzene-1,3,5-tricarboxylic acid) and hydrofluoric acid. The reactants were stirred well by using conventional electric heating or MW radiation, placed in an autoclave and gradually heated to 220 °C [60,61]. The MIL-100 was synthesized within 1 h while the impurities of chromium metal dispersed only after 4 h heating. It should be considered that, after 24-fold reduction in synthesis time, the yield of the product remains the same as traditional synthesis. Instead of chromium, an isostructural analog with iron was also obtained within 30 minutes of microwave radiation [62]. MIL-101 was synthesized using microwave radiation with the terephthalate linker in 2007. The author found that when temperature time varying from 1 to 60 min or longer than 60 min led to the appearance of impurities in the product [32,60]. It has been shown that microwave radiation for 15 min provides good yield with a 50 nm crystal size. Instead of chromium, the isostructural compound with iron was synthesized first time with microwave radiation, solutions of iron chloride $FeCl_3$ and terephthalic acid in DMF for 10 minutes [63]. The aminoterephthalic acid act as a linker in the synthesis of Fe-MIL-101-NH_2 compound and reaction was carried out for 5 min in microwave radiation and produce ~ 200 nm crystal size [62].

3.2.2 Electrochemical synthesis

The main theme of this method of MOFs synthesis is metals ions which are introduced as a result of the electrochemical process, not with the formation of these ions during the reaction of the solution of the associated salt or with the reaction of the acid. Specifically, the metal ions are supplied through the dissolution of anode into the reaction mixture containing linker molecules and the electrolytes. Make it possible to start a continuous process and to avoid the creation of ions during the reaction, which is necessary to produce a relatively huge and good yield of MOFs [58].

Figure 4. Setup for electrochemical synthesis of HKUST-1 and SEM image of the product [35,46].

An Electrochemical method was first used for the synthesis of HKUST-1 MOF in 2005 (figure 4) [64]. Copper sheets of 5 mm thick were used in the form of dissolving anode, sheets with copper cathode are immersed in the solution of H_3BTC in methanol. After applying voltage, a green-blue precipitate was collected within 150 minutes. Protic solvents are used to prevent metal cations from being stored on the cathode; however, under those circumstances, hydrogen is released in the electrochemical process. Simultaneously, some solvents are initially reduced such as acrylonitrile, acrylates and maleates, and the problem is avoided with a small amount of these solvents. A certain modified method was suggested by Hartmann et al. comparing the methods (hydrothermal, conventional non-hydrothermal and electrochemical) of synthesis of HKUST-1 [65,66]. All the synthesis methods of MOFs gave the $Cu_3(BTC)_2$ phase with

Metal-Organic Framework Composites - Volume I   Materials Research Forum LLC
Materials Research Foundations **53** (2019) 1-28   https://doi.org/10.21741/9781644900291-1

homologous surface areas and pore sizes. A thin layer of HKUST-1 was electrically deposited on a copper mesh, which was used as the anode. It should be clearly seen that this method allows both powders and films [46,64].

3.2.3 Mechanochemical synthesis

Typically, the studies of mechanochemical mechanism between solids are introduced through mechanical energy such as ball milling. This technique has become popular for high yield and fast responses without or with a small amount of solvent. Prior models consider on contact of two solids surfaces with mechanical treatment. The most common methods are the hot-spot and magma-plasma model. With the area of ~ 1 m^2, the local deformation of the surface of the tensile material increases temperature rapidly (> 1000 °C) in this place, which remains for a short time (10^{-3}- 10^{-4} sec). For hard materials, slightly different methods are considered for local heating, i.e., the formation of hot-spot on the edge of the growing cracks [46,58].

The magma-plasma model puts attention directly on the methods that occur at the interaction spot of colliding particles. According to this model, as such spots, the local temperature may be more than 10 °C due to the release of large energy, including the formation of plasma and emission of free electrons [67]. It seems that these principles are inappropriate for the mechanochemical synthesis of the MOFs because such high temperatures will lead towards the disintegration of early biological components and products. However, this type of temperature breaks the local character strongly and within a very short period can start the reaction between the milled components rather than their degeneration. The presence of liquid components during mechanochemical reaction provides additional benefits: easier crystallization, due to the greater mobility of the reactants higher the degree of crystallinity of reaction product and yield of the target product [68]. Furthermore, the constant milling of the components increases dispersion and creates new surfaces by improved reactivity. This method was first used for the synthesis of MOF in 2006 [67,68]. According to this method, copper acetate and isonicotinic acid (HINA) mixed in a ball mill for few minutes and due to this process led to a well-crystallized product with the formula Cu(INA)$_2$. x H$_2$O . y AcOH. HKUST-1 is synthesized in 2010 by mechanochemical method [69]. In both studies, copper acetate and H$_3$BTC have been used in solvent-free situations. Yuan et al focus on the fact that the Cu(INA)$_2$ is built almost naturally (1-minute milling is sufficient) while the formation of HKUST-1 requires more time for mixing. Moreover, a small amount of acetic acid (added during the reaction) increases rapidly the synthesis rate of Cu(INA)$_2$ instead of HKUST-1[69,70].

3.2.4 Sonochemical synthesis

The effect of ultrasound on the liquid and colloid systems is mainly due to cavitation. This is the occurrence of vapor formation and the release of air which is caused by a decrease in pressure in the liquid because high-intensity acoustic wave spreads through it. Though, in fact, the fall in pressure leads to the release of air dissolved from the liquid and the formation of gas voids, cavities. The pressure in the cavities is more than the pressure of saturated vapor, therefore, the cavity starts to form from a nucleus, increases to a finite shape and then collapses. The entire method only takes a few milliseconds. For cavity formation, the nuclei are microscopic bubbles. They form small cracks in the container surface and suspended particles on the edges of the vessels. When the bubbles fall, a faint glow is observed due to the heating of the gas in high pressure-induced bubbles. Thus, the bubble fall leads to a significant increase in temperature and the high-pressure difference in the surrounding liquid of the bubble [71–73].

The use of ultrasonic waves and cavitation has many important effects as it applies to chemical processes. The resonant bubbles act as the instigator, increasing the area of contact among the reagents. In addition, thermal effects and pressure are due to the disruption of differential particle aggregates, which also increase the contact area. The Sonochemical method was first used for the synthesis of MOF ($Zn_3(BTC)_2$) in 2008 [74]: Zinc acetate and H_3BTC were mixed in 20% ethanol and subjected to sonication for 90 minutes, however, high yield (75.3%) of the product was achieved even after 5-minute sonication. Schlesinger et al compared the six different methods of synthesis of HKUST-1: a synthesis on atmospheric pressure with heating under reflex, solvothermal synthesis, electrochemical synthesis, microwave, mechanical synthesis and sonication [75]. All procedures were a pure phase of the product. The maximum yield during the minimum synthesis was achieved through microwave radiation.

4. Application of MOFs for wastewater treatment

Typical pollutants in industrial/domestic wastewater include heavy metal ions, radioactive elements, polyaromatic hydrocarbons (PAH), pesticides, pharmaceuticals, and dyes [4]. According to their chemical composition, these contaminated substances can be divided into organic and inorganic categories Table 2 [6]. In the last twenty-five years, massive efforts have been made to limit the type of pollution, improve the processes, recycling products, and controlling waste treatment at the production level. Traditional discharge methods, for their part, problems of corrosion and, more seriously, emissions, if the condition of treatment is not fully controlled.

Table 2. Some common contaminants in water [6]

Contaminants	Maximum contaminants level (mg/L)	Major sources	Health menace
Antimony	0.006	Petroleum refining; fire retardants; ceramics; solder; electronics	Increase in blood cholesterol and decrease in blood sugar
Arsenic	0.05	Erosion of natural deposits; glass and electronics production wastes.	Skin problems with the circulatory system; increase risk of cancer
Barium	2	Drilling wastes; metal refining and erosion of natural deposits	Increase in blood pressure
Beryllium	0.004	Metal refining and coal burning; discharge from electrical, aerospace and defense industries	Intestinal lesions
Bromate	0.01	By-product of water chlorination	Increased risk of cancer
Cadmium	0.005	Corrosion of galvanized pipes; erosion of natural deposits; metal refining; waste batteries & paints	Kidney damage
Chromium	0.1	Steel and pulp mills; erosion of natural deposits	Allergic dermatitis
Copper	1.3	Corrosion of plumbing systems; erosion of natural deposits; wood preservatives	Gastrointestinal distress (short term exposure); liver or kidney damage
Cyanide	0.2	Discharge from metal factories or plastic and fertilizer factories	Nerve damage or problems with the thyroid gland
Fluoride	4	Erosion of natural deposits; water additive in toothpaste; fertilizer and aluminum factories	Bone disease; children's teeth mottling
Lead	0.015	Corrosion of plumbing system; erosion of natural deposits	Delay in physical or mental development, kidney problems, high blood pressure
Mercury	0.002	Erosion of natural deposits; discharge from refineries and factories; runoff from landfills or cropland	Kidney damage
Nitrate/Nitrite	10/1	Fertilizer use; septic tanks; erosion of natural deposits	Seriously ill even die, Symptoms include shortness of breath and blue-baby syndrome
Selenium	0.05	Petroleum and metal refining; erosion of natural deposits; discharge from mines	Hair or fingernail losses, numbness in fingers or toes or circulation problems
Thallium	0.002	Ore-processing sites; discharge from electronics, glass and drug factories	Hair loss, blood change, problems with kidneys, intestines or liver
Radioactive contaminants	--	Decay of natural and man-made deposits	Increase risk of getting cancer
2,4-dinitrophenol	0.07	Herbicide	Kidneys, liver or adrenal glands problems
Atrazine	0.003	Herbicide	Cardiovascular system problems or reproductive difficulties
Dalapon	0.2	Herbicide	Minor kidney changes

Dibromo-chloropropane	0.0002	Soil fumigant	Reproductive problems, increased risk of getting cancer
Diquat	0.02	Herbicide use	Cataracts
Hexachlorobenzene	0.001	Metal refining or pesticide use	Liver or kidneys problems; adverse reproductive effects; increased risk of getting cancer
Benzene	0.005	Discharge from factories; leaching from gas storage tanks and landfills	Anemia; increased risk of getting cancer

4.1   Effect of MOFs on water pollutants

Generally, pollutants in wastewater generated from different sources have a huge difference in species. The inorganic pollutants are mostly toxic heavy metal ions, excessive halide ions and radioactive substances. Due to the inefficiency of most inorganic substances, selective adsorption seems to be the only way to remove them from water and this process has been done intensively. Compared to inorganic pollutants, the types of organic pollutants in wastewater are richer and more varied, including dye, insecticide, polyaromatic hydrocarbon (PAH), agrochemicals, pharmaceuticals and so forth, most of which are highly toxic, causing serious problem or environmental health hazards [4]. To remove these organic pollutants, the adsorption through porous materials or degradation by advanced oxidation process (AOPs) has proven to be low cost and effective process.

MOFs have been also known as porous coordination polymers (PCPs) due to the large surface area and high porosity which are beneficial for their application in separation, drug delivery, gas storage and catalysis, as well as removal of heavy metals from industrial wastewater [6,76]. Heavy metals (hazardous metals) have prolonged been a serious menace to human health and by using conventional adsorbents, treatment capacity will not be enough to certify safe water. The structural features of surface holes are usually certain factors influencing the adsorbent's performance. It has been noticed that MOFs are particularly suitable for the removal and dissemination of pollutants due to their good structure, specifically with their huge distinct surface area and aperture [33]. The heavy metal ions which can be efficiently removed by the MOFs include chromium, lead, cadmium, arsenic and mercury [9,77]. Through the precise modifications in MOFs, adsorption performance can be greatly increased for heavy metal ions. The removal of various heavy metal ions by several MOFs is shown in Table 3. Ability to remove heavy metal ions by MOF can reach 99%; the maximum absorption capacity of Cd^{2+}, Cr^{3+}, Pb^{2+}, and Hg^{2+} was 49 mg/g, 117 mg/g, 232 mg/g and 769 mg/g, respectively [78]. According to Tokalıoğlu et al zirconium-based MOF has been fabricated for the removal of Pb (II) from cereals, drinks and water sample and designated as MOF-545. MOF-545 has a great adsorption potential and pre-enhancement coefficient, from which precise surface area was up to 2192 m^2/g, resulting in adsorption capability of 73 mg/g [79]. Moreover, MOF can be fabricated in the fluorescent probes for the selective exposure of metal ions [80]. According to El-Sewify et al fabricated a zirconium-based MOF using the fluorescent sensor to identify heavy metal ions in water [81].

Table 3. The effects of various types of MOFs on the removal of heavy metals ions [6].

MOFs	Preparation Methods	Modification Methods	Modifiers	Contaminants	Adsorption Capacities (mg/g)	pH	Removal Mechanisms
UiO-66	Solvothermal method	Oven-promoted method	Melamine	Pb (II)	205	6	The coordination interaction between the amino groups (-NH$_2$) and Pb (II).
MOF-74-Zn	Solvothermal method	--	--	Hg (II)	63	6	Weak interactions between the MOF skeleton (carboxylate and hydroxy group) and Hg^{2+} ions.
ZIF-8@CA	Solution method	Pre-synthetic modification	Flexible and porous cellulose aerogels	Cr (IV)	41.8	--	N-pyridine densely populated on the inner surface of porous cellulose aerogels (CA) with unique large specific surface areas and a high density of adsorption sites.
MIL-68 (W)	Microwave synthesis	Liquid phase impregnation post-synthetic modification (LP-PSM)	Na$_2$S	Cd (II)	0.139	--	Cd (II) was expected to bind to individual S atoms rather than polymers, with dense hydration spheres inhibiting sorption on adjacent binding sites.
Cu-MOFs	Solvothermal method	Load modification	Fe$_3$O$_4$	Pb (II)	219.00	--	The adsorption was chemical adsorption, which is dominated by the coordination of Pb^{2+} with amino from Cu-MOFs.

HKUST-1	Solvothermal method	LP-PSM	Potassium nickel hexacyanoferrate	Cs (I)	153	--	Physical adsorption onto the crystal lattice followed by strong electrostatic forces could lead to chemical adsorption.
UiO-66	Solvothermal method	LP-PSM	Schiff base-derived material	Co (II)	256	8.4	The carboxyl oxygen and Schiff base nitrogen in the ligand had a stronger bonding ability towards Co (II) ions

In addition, MOFs are used to remove heavy metal ions as well as adsorption and photocatalytic degradation of organic pollutants are also effective [1,82]. In fact, the effect of adsorption capacity of tested MOFs in prior studies was found to be better than commercial active carbon [83]. According to Ramezanalizadeh et al. treated methylene blue and 4-nitrophenol through photo-catalytic degradation with MOFs/CuWO4 and noticed that the photo-catalytic efficiency of MOFs/CuWO4 was significantly increased in the presence of MOFs [84]. Zhou et al. fabricated a zirconium-based MOF, which has high adsorption potential for antibiotic tetracycline [85]. Chen et al. developed a magnetic porous carbon-based adsorption using a Fe (III) based modified MOF-5, which had a large impact on the adsorption of organic contaminants including bisphenol-a, carbamazepine, 4-nitrophenol, Noroxin and atrazine. Among them, the adsorption of Norexin was most effective through the primary π-electron donor-acceptor interactions [86]. Azhar et al. synthesized HKUST-1 and UiO-66 through the solvothermal method and compared their treatment efficiency to the removal of methylene blue in wastewater. The formulation of MOFs can be used across a broad pH range and with the HKUST-1 the adsorption efficiency was higher than the UiO-66 [83].

The MOFs materials have been linked for improving water treatment efficiency with water treatment units such as membrane filtration, magnetic nanoparticles and with graphene. Liu et al. synthesized a pure zirconium-based MOF polycrystalline film using an in-situ solvothermal method on alumina hollow fiber [87]. A simple flow diagram showing the major synthesis steps is shown in Figure 5. Its performance was evaluated by measuring gas diffusion and ion separation. It has shown excellent polyvalent ion removals such as 86.3% for Ca^{2+}, 98% for Mg^{2+}, 99.3% for Al^{3+} and good permeability also. In addition, Ingole et al. also increased the water vapor transport performance of mixed gas from the combination of MOF and thin-film nanocomposite, which was described as MOF@TFN4 [88]. The results showed that the maximum selectivity of the

Materials Research Forum LLC
https://doi.org/10.21741/9781644900291-1

fabricated MOF@TFN4 membrane was up to 543 and the water vapor transmission rate was 2244 GPU, respectively. The coupling of the MOFs and membrane systems can be further improved under good operating conditions or coupling conditions.

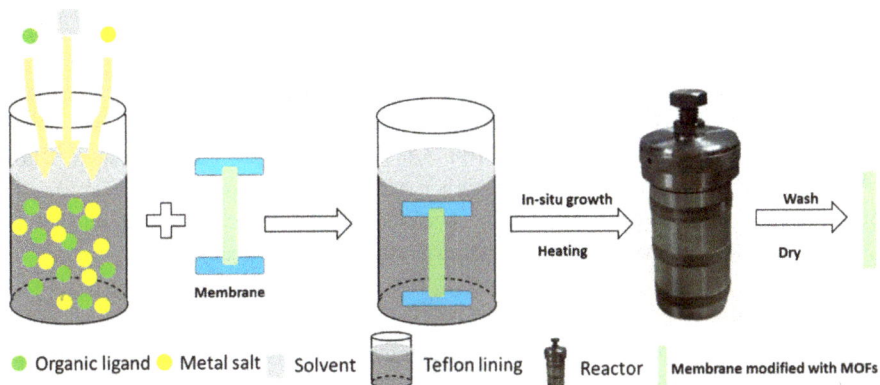

Figure 5- Schematic diagram for the solvothermal synthesis of an MOF-based membrane [6].

MOFs and magnetic nanoparticles are hybridized to develop nanocomposites, which is favorable to the removal of those fine MOFs particles from water when placed in an external magnetic field. It has been reported that hybridizing of superparamagnetic nano-Fe_3O_4 and MOFs can increase the range of MOFs applications [89]. Similarly, the hybridization of MOFs with graphene oxide also has the capacity to improve the functions of MOFs. Graphene oxide is defined as a soft two-dimensional carbon nanomaterial, which contains several functional groups of oxygen such as epoxy groups and carboxyl groups. It is characterized by a large specific surface area, high mechanical strength and good electrical conductivity. Graphene oxide can significantly increase the specific surface area of MOFs and show a good adsorption effect [90].

Conclusion

Water pollution has become a serious environmental problem, with the growth of industry and agriculture, particularly in developing countries. Cost-effective techniques that address this issue have an urgent need. From this outlook, MOFs have plenty of potential for the removal of pollutants from wastewater. In this chapter, we deal with the different methods for the synthesis of MOFs and its application in the removal of various pollutants from wastewater. In these techniques, the relationship between MOFs selection and design is closely related to their structural distribution, purity, yield, energy consumption and efficiency to remove pollutants from wastewater. Luckily, there are several types of MOFs that have different structures that can be adjusted according to pre-defined situations. Although the MOFs has less water stability and even because of its small size it is not easily separated from water, but it is possible to combine them with other techniques such as coupling with magnetic nanoparticles, graphene oxide and membrane systems. Water purification by filtration through a membrane is the most promising method for future development. Other materials that can be coupled with MOFs such as nanoparticles, graphene, carbon nanotube and the organic polymer are all possible materials. Overall, MOF-based composite materials are expected to play a significant role in improving further performance in wastewater treatment.

References

[1]   Z. Hasan, S.H. Jhung, Removal of hazardous organics from water using metal-organic frameworks (MOFs): Plausible mechanisms for selective adsorptions, Journal of Hazardous Materials. 283 (2015) 329–339. https//doi.org/10.1016/j.jhazmat.2014.09.046

[2]   M.M. Khin, A.S. Nair, V.J. Babu, R. Murugan, S. Ramakrishna, A review on nanomaterials for environmental remediation, Energy & Environmental Science. 5 (2012) 8075. https//doi.org/10.1039/c2ee21818f

[3]   G. Zhu, Y. Bian, A.S. Hursthouse, P. Wan, K. Szymanska, J. Ma, X. Wang, Z. Zhao, Application of 3-D Fluorescence: Characterization of Natural Organic Matter in Natural Water and Water Purification Systems, Journal of Fluorescence. 27 (2017) 2069–2094. https//doi.org/10.1007/s10895-017-2146-7

[4]   Q. Gao, J. Xu, J.-H. Bu, Recent advances about metal–organic frameworks in the removal of pollutants from wastewater, Coordination Chemistry Reviews. 378 (2019) 17–31. https//doi.org/10.1016/j.ccr.2018.03.015

[5]   C. Bao, C. Fang, Water Resources Flows Related to Urbanization in China: Challenges and Perspectives for Water Management and Urban Development, Water Resources Management. 26 (2012) 531–552. https//doi.org/10.1007/s11269-011-9930-y

[6]   Y. Bian, N. Xiong, G. Zhu, Technology for the Remediation of Water Pollution: A Review on the Fabrication of Metal Organic Frameworks, Processes. 6 (2018) 122. https//doi.org/10.3390/pr6080122

Materials Research Forum LLC
https://doi.org/10.21741/9781644900291-1

[7]   C.A. Martínez-Huitle, S. Ferro, Electrochemical oxidation of organic pollutants for the wastewater treatment: direct and indirect processes, Chem. Soc. Rev. 35 (2006) 1324–1340. https//doi.org/10.1039/B517632H

[8]   C. Moreno-Castilla, Adsorption of organic molecules from aqueous solutions on carbon materials, Carbon. 42 (2004) 83–94. https//doi.org/10.1016/j.carbon.2003.09.022

[9]   P. Kumar, V. Bansal, K.-H. Kim, E.E. Kwon, Metal-organic frameworks (MOFs) as futuristic options for wastewater treatment, Journal of Industrial and Engineering Chemistry. 62 (2018) 130–145. https//doi.org/10.1016/j.jiec.2017.12.051

[10]   S. Kar, P.K. Tewari, Nanotechnology for domestic water purification, in: Nanotechnology in Eco-Efficient Construction, Elsevier, 2013: pp. 364–427. https//doi.org/10.1533/9780857098832.3.364

[11]   J. Ma, J. Shi, H. Ding, G. Zhu, K. Fu, X. Fu, Synthesis of cationic polyacrylamide by low-pressure UV initiation for turbidity water flocculation, Chemical Engineering Journal. 312 (2017) 20–29. https//doi.org/10.1016/j.cej.2016.11.114

[12]   M. Kramer, U. Schwarz, S. Kaskel, Synthesis and properties of the metal-organic framework Mo3(BTC)2 (TUDMOF-1), Journal of Materials Chemistry. 16 (2006) 2245. https//doi.org/10.1039/b601811d

[13]   G. Zhu, J. Liu, J. Yin, Z. Li, B. Ren, Y. Sun, P. Wan, Y. Liu, Functionalized polyacrylamide by xanthate for Cr (VI) removal from aqueous solution, Chemical Engineering Journal. 288 (2016) 390–398. https//doi.org/10.1016/j.cej.2015.12.043

[14]   A. Gianico, G. Bertanza, C.M. Braguglia, M. Canato, G. Laera, S. Heimersson, M. Svanström, G. Mininni, Upgrading a wastewater treatment plant with thermophilic digestion of thermally pre-treated secondary sludge: techno-economic and environmental assessment, Journal of Cleaner Production. 102 (2015) 353–361. https//doi.org/10.1016/j.jclepro.2015.04.051

[15]   J. Yin, G. Zhu, B. Deng, Graphene oxide (GO) enhanced polyamide (PA) thin-film nanocomposite (TFN) membrane for water purification, Desalination. 379 (2016) 93–101. https//doi.org/10.1016/j.desal.2015.11.001

[16]   S. Wang, C.M. McGuirk, A. D'Aquino, J.A. Mason, C.A. Mirkin, Metal-Organic Framework Nanoparticles, Advanced Materials. 30 (2018) 1800202. https//doi.org/10.1002/adma.201800202

[17]   L. Wang, M.S.H. Boutilier, P.R. Kidambi, D. Jang, N.G. Hadjiconstantinou, R. Karnik, Fundamental transport mechanisms, fabrication and potential applications of nanoporous atomically thin membranes, Nature Nanotechnology. 12 (2017) 509–522. https//doi.org/10.1038/nnano.2017.72

[18] G. Zhu, Q. Wang, J. Yin, Z. Li, P. Zhang, B. Ren, G. Fan, P. Wan, Toward a better understanding of coagulation for dissolved organic nitrogen using polymeric zinc-iron-phosphate coagulant, Water Research. 100 (2016) 201–210. https//doi.org/10.1016/j.watres.2016.05.035

[19] G. Zhu, J. Liu, Y. Bian, Evaluation of cationic polyacrylamide-based hybrid coagulation for the removal of dissolved organic nitrogen, Environmental Science and Pollution Research. 25 (2018) 14447–14459. https//doi.org/10.1007/s11356-018-1630-1

[20] M. Hartmann, S. Kullmann, H. Keller, Wastewater treatment with heterogeneous Fenton-type catalysts based on porous materials, Journal of Materials Chemistry. 20 (2010) 9002. https//doi.org/10.1039/c0jm00577k

[21] E. Paul, P. Camacho, M. Sperandio, P. Ginestet, Technical and Economical Evaluation of a Thermal, and Two Oxidative Techniques for the Reduction of Excess Sludge Production, Process Safety and Environmental Protection. 84 (2006) 247–252. https//doi.org/10.1205/psep.05207

[22] V.K. Tyagi, S.-L. Lo, Application of physico-chemical pretreatment methods to enhance the sludge disintegration and subsequent anaerobic digestion: an up to date review, Reviews in Environmental Science and Bio/Technology. 10 (2011) 215–242. https//doi.org/10.1007/s11157-011-9244-9

[23] G. Montes-Hernandez, N. Concha-Lozano, F. Renard, E. Quirico, Removal of oxyanions from synthetic wastewater via carbonation process of calcium hydroxide: Applied and fundamental aspects, Journal of Hazardous Materials. 166 (2009) 788–795. https//doi.org/10.1016/j.jhazmat.2008.11.120

[24] I. Ali, New generation adsorbents for water treatment, Chemical Reviews. (2012). https//doi.org/10.1021/cr300133d

[25] K. Ariga, S. Ishihara, H. Abe, M. Li, J.P. Hill, Materials nanoarchitectonics for environmental remediation and sensing, J. Mater. Chem. 22 (2012) 2369–2377. https//doi.org/10.1039/C1JM14101E

[26] Z. Zhou, M. Hartmann, Progress in enzyme immobilization in ordered mesoporous materials and related applications, Chemical Society Reviews. 42 (2013) 3894. https//doi.org/10.1039/c3cs60059a

[27] Y. Tao, H. Kanoh, L. Abrams, K. Kaneko, Mesopore-Modified Zeolites: Preparation, Characterization, and Applications, Chemical Reviews. 106 (2006) 896–910. https//doi.org/10.1021/cr040204o

[28] R.T. Yang, Desulfurization of Transportation Fuels with Zeolites Under Ambient Conditions, Science. 301 (2003) 79–81. https//doi.org/10.1126/science.1085088

[29]   B. Lee, Y. Kim, H. Lee, J. Yi, Synthesis of functionalized porous silicas via templating method as heavy metal ion adsorbents: the introduction of surface hydrophilicity onto the surface of adsorbents, Microporous and Mesoporous Materials. 50 (2001) 77–90. https//doi.org/10.1016/S1387-1811(01)00437-1

[30]   H.-C. Zhou, J.R. Long, O.M. Yaghi, Introduction to Metal–Organic Frameworks, Chemical Reviews. 112 (2012) 673–674. https//doi.org/10.1021/cr300014x

[31]   I. Ahmed, S.H. Jhung, Composites of metal–organic frameworks: Preparation and application in adsorption, Materials Today. 17 (2014) 136–146. https//doi.org/10.1016/j.mattod.2014.03.002

[32]   N.A. Khan, Z. Hasan, S.H. Jhung, Adsorptive removal of hazardous materials using metal-organic frameworks (MOFs): A review, Journal of Hazardous Materials. 244–245 (2013) 444–456. https//doi.org/10.1016/j.jhazmat.2012.11.011

[33]   H. Furukawa, K.E. Cordova, M. O'Keeffe, O.M. Yaghi, The Chemistry and Applications of Metal-Organic Frameworks, Science. 341 (2013) 1230444–1230444. https//doi.org/10.1126/science.1230444

[34]   H.-C.J. Zhou, S. Kitagawa, Metal–Organic Frameworks (MOFs), Chem. Soc. Rev. 43 (2014) 5415–5418. https//doi.org/10.1039/C4CS90059F

[35]   O.M. Yaghi, G. Li, H. Li, Selective binding and removal of guests in a microporous metal–organic framework, Nature. 378 (1995) 703–706. https//doi.org/10.1038/378703a0

[36]   J.-R. Li, R.J. Kuppler, H.-C. Zhou, Selective gas adsorption and separation in metal–organic frameworks, Chemical Society Reviews. 38 (2009) 1477. https//doi.org/10.1039/b802426j

[37]   R. Seetharaj, P.V. Vandana, P. Arya, S. Mathew, Dependence of solvents, pH, molar ratio and temperature in tuning metal organic framework architecture, Arabian Journal of Chemistry. (2016). https//doi.org/10.1016/j.arabjc.2016.01.003

[38]   J. Duan, Y. Pan, G. Liu, W. Jin, Metal-organic framework adsorbents and membranes for separation applications, Current Opinion in Chemical Engineering. 20 (2018) 122–131. https//doi.org/10.1016/j.coche.2018.04.005

[39]   C. Pettinari, F. Marchetti, N. Mosca, G. Tosi, A. Drozdov, Application of metal – organic frameworks, Polymer International. 66 (2017) 731–744. https//doi.org/10.1002/pi.5315

[40]   N.C. Burtch, H. Jasuja, K.S. Walton, Water Stability and Adsorption in Metal–Organic Frameworks, Chemical Reviews. 114 (2014) 10575–10612. https//doi.org/10.1021/cr5002589

[41]   M. Eddaoudi, H. Li, O.M. Yaghi, Highly Porous and Stable Metal−Organic Frameworks: Structure Design and Sorption Properties, Journal of the American Chemical Society. 122 (2000) 1391–1397. https//doi.org/10.1021/ja9933386

[42]   S. Noro, Metal–Organic Frameworks, in: Comprehensive Inorganic Chemistry II, Elsevier, 2013: pp. 45–71. https//doi.org/10.1016/B978-0-08-097774-4.00503-9

[43]   H. Cai, Y.-L. Huang, D. Li, Biological metal–organic frameworks: Structures, host–guest chemistry and bio-applications, Coordination Chemistry Reviews. 378 (2019) 207–221. https//doi.org/10.1016/j.ccr.2017.12.003

[44]   W.-X. Zhang, P.-Q. Liao, R.-B. Lin, Y.-S. Wei, M.-H. Zeng, X.-M. Chen, Metal cluster-based functional porous coordination polymers, Coordination Chemistry Reviews. 293–294 (2015) 263–278. https//doi.org/10.1016/j.ccr.2014.12.009

[45]   M. Eddaoudi, Systematic Design of Pore Size and Functionality in Isoreticular MOFs and Their Application in Methane Storage, Science. 295 (2002) 469–472. https//doi.org/10.1126/science.1067208

[46]   V. V Butova, M.A. Soldatov, A.A. Guda, K.A. Lomachenko, C. Lamberti, Metal-organic frameworks: structure, properties, methods of synthesis and characterization, Russian Chemical Reviews. 85 (2016) 280–307. https//doi.org/10.1070/RCR4554

[47]   H. Sato, W. Kosaka, R. Matsuda, A. Hori, Y. Hijikata, R. V. Belosludov, S. Sakaki, M. Takata, S. Kitagawa, Self-Accelerating CO Sorption in a Soft Nanoporous Crystal, Science. 343 (2014) 167–170. https//doi.org/10.1126/science.1246423

[48]   J.L.C. Rowsell, O.M. Yaghi, Strategies for Hydrogen Storage in Metal-Organic Frameworks, Angewandte Chemie International Edition. 44 (2005) 4670–4679. https//doi.org/10.1002/anie.200462786

[49]   W.J. Koros, C. Zhang, Materials for next-generation molecularly selective synthetic membranes, Nature Materials. 16 (2017) 289–297. https//doi.org/10.1038/nmat4805

[50]   A.U. Czaja, N. Trukhan, U. Müller, Industrial applications of metal–organic frameworks, Chemical Society Reviews. 38 (2009) 1284. https//doi.org/10.1039/b804680h

[51]   M.H. Yap, K.L. Fow, G.Z. Chen, Synthesis and applications of MOF-derived porous nanostructures, Green Energy & Environment. 2 (2017) 218–245. https//doi.org/10.1016/j.gee.2017.05.003

[52]   S. Bhattacharya, S. Bala, R. Mondal, Design of chiral Co(<scp>ii</scp>)-MOFs and their application in environmental remediation and waste water treatment, RSC Advances. 6 (2016) 25149–25158. https//doi.org/10.1039/C5RA26154F

[53] K.P. Lillerud, U. Olsbye, M. Tilset, Designing Heterogeneous Catalysts by Incorporating Enzyme-Like Functionalities into MOFs, Topics in Catalysis. 53 (2010) 859–868. https//doi.org/10.1007/s11244-010-9518-4

[54] T.M. Al-Jadir, F.R. Siperstein, The influence of the pore size in Metal−Organic Frameworks in adsorption and separation of hydrogen sulphide: A molecular simulation study, Microporous and Mesoporous Materials. 271 (2018) 160–168. https//doi.org/10.1016/j.micromeso.2018.06.002

[55] C. Dey, T. Kundu, B.P. Biswal, A. Mallick, R. Banerjee, Crystalline metal-organic frameworks (MOFs): synthesis, structure and function., Acta Crystallographica Section B, Structural Science, Crystal Engineering and Materials. 70 (2014) 3–10. https//doi.org/10.1107/S2052520613029557

[56] A. Halder, D. Ghoshal, Structure and properties of dynamic metal–organic frameworks: a brief accounts of crystalline-to-crystalline and crystalline-to-amorphous transformations, CrystEngComm. 20 (2018) 1322–1345. https//doi.org/10.1039/C7CE02066J

[57] H. Li, M. Eddaoudi, M. O'Keeffe, O.M. Yaghi, Design and synthesis of an exceptionally stable and highly porous metal-organic framework, Nature. 402 (1999) 276–279. https//doi.org/10.1038/46248

[58] N. Stock, S. Biswas, Synthesis of Metal-Organic Frameworks (MOFs): Routes to Various MOF Topologies, Morphologies, and Composites, Chemical Reviews. 112 (2012) 933–969. https//doi.org/10.1021/cr200304e

[59] K.S. Park, Z. Ni, A.P. Cote, J.Y. Choi, R. Huang, F.J. Uribe-Romo, H.K. Chae, M. O'Keeffe, O.M. Yaghi, Exceptional chemical and thermal stability of zeolitic imidazolate frameworks, Proceedings of the National Academy of Sciences. 103 (2006) 10186–10191. https//doi.org/10.1073/pnas.0602439103

[60] S.H. Jhung, J. Lee, J.-S. Chang, Microwave Synthesis of a Nanoporous Hybrid Material, Chromium Trimesate, Bulletin of the Korean Chemical Society. 26 (2005) 880–881. https//doi.org/10.5012/bkcs.2005.26.6.880

[61] G. Férey, C. Serre, C. Mellot-Draznieks, F. Millange, S. Surblé, J. Dutour, I. Margiolaki, A Hybrid Solid with Giant Pores Prepared by a Combination of Targeted Chemistry, Simulation, and Powder Diffraction, Angewandte Chemie International Edition. 43 (2004) 6296–6301. https//doi.org/10.1002/anie.200460592

[62] K.M.L. Taylor-Pashow, J. Della Rocca, Z. Xie, S. Tran, W. Lin, Postsynthetic Modifications of Iron-Carboxylate Nanoscale Metal−Organic Frameworks for Imaging and Drug Delivery, Journal of the American Chemical Society. 131 (2009) 14261–14263. https//doi.org/10.1021/ja906198y

[63]   N.A. Khan, I.J. Kang, H.Y. Seok, S.H. Jhung, Facile synthesis of nano-sized metal-organic frameworks, chromium-benzenedicarboxylate, MIL-101, Chemical Engineering Journal. 166 (2011) 1152–1157. https//doi.org/10.1016/j.cej.2010.11.098

[64]   U. Mueller, M. Schubert, F. Teich, H. Puetter, K. Schierle-Arndt, J. Pastré, Metal–organic frameworks—prospective industrial applications, J. Mater. Chem. 16 (2006) 626–636. https//doi.org/10.1039/B511962F

[65]   M. Hartmann, S. Kunz, D. Himsl, O. Tangermann, S. Ernst, A. Wagener, Adsorptive Separation of Isobutene and Isobutane on Cu 3 (BTC) 2, Langmuir. 24 (2008) 8634–8642. https//doi.org/10.1021/la8008656

[66]   O. Fleker, A. Borenstein, R. Lavi, L. Benisvy, S. Ruthstein, D. Aurbach, Preparation and Properties of Metal Organic Framework/Activated Carbon Composite Materials, Langmuir. 32 (2016) 4935–4944. https//doi.org/10.1021/acs.langmuir.6b00528

[67]   S.L. James, C.J. Adams, C. Bolm, D. Braga, P. Collier, T. Friščić, F. Grepioni, K.D.M. Harris, G. Hyett, W. Jones, A. Krebs, J. Mack, L. Maini, A.G. Orpen, I.P. Parkin, W.C. Shearouse, J.W. Steed, D.C. Waddell, Mechanochemistry: opportunities for new and cleaner synthesis, Chem. Soc. Rev. 41 (2012) 413–447. https//doi.org/10.1039/C1CS15171A

[68]   A. Pichon, A. Lazuen-Garay, S.L. James, Solvent-free synthesis of a microporous metal–organic framework, CrystEngComm. 8 (2006) 211. https//doi.org/10.1039/b513750k

[69]   W. Yuan, A.L. Garay, A. Pichon, R. Clowes, C.D. Wood, A.I. Cooper, S.L. James, Study of the mechanochemical formation and resulting properties of an archetypal MOF: Cu3(BTC)2 (BTC = 1,3,5-benzenetricarboxylate), CrystEngComm. 12 (2010) 4063. https//doi.org/10.1039/c0ce00486c

[70]   D. Lv, Y. Chen, Y. Li, R. Shi, H. Wu, X. Sun, J. Xiao, H. Xi, Q. Xia, Z. Li, Efficient Mechanochemical Synthesis of MOF-5 for Linear Alkanes Adsorption, Journal of Chemical & Engineering Data. 62 (2017) 2030–2036. https//doi.org/10.1021/acs.jced.7b00049

[71]   F. Bigdeli, H. Ghasempour, A. Azhdari Tehrani, A. Morsali, H. Hosseini-Monfared, Ultrasound-assisted synthesis of nano-structured Zinc(II)-based metal-organic frameworks as precursors for the synthesis of ZnO nano-structures, Ultrasonics Sonochemistry. 37 (2017) 29–36. https//doi.org/10.1016/j.ultsonch.2016.12.031

[72]   O. Abuzalat, D. Wong, M. Elsayed, S. Park, S. Kim, Sonochemical fabrication of Cu(II) and Zn(II) metal-organic framework films on metal substrates, Ultrasonics Sonochemistry. 45 (2018) 180–188. https//doi.org/10.1016/j.ultsonch.2018.03.012

[73]   F. Mojtabazade, B. Mirtamizdoust, A. Morsali, P. Talemi, Ultrasonic-assisted synthesis and the structural characterization of novel the zig-zag Cd(II) metal-organic polymer and their nanostructures, Ultrasonics Sonochemistry. 42 (2018) 134–140. https//doi.org/10.1016/j.ultsonch.2017.11.018

[74]   L.-G. Qiu, Z.-Q. Li, Y. Wu, W. Wang, T. Xu, X. Jiang, Facile synthesis of nanocrystals of a microporous metal–organic framework by an ultrasonic method and selective sensing of organoamines, Chemical Communications. (2008) 3642. https//doi.org/10.1039/b804126a

[75]   M. Schlesinger, S. Schulze, M. Hietschold, M. Mehring, Evaluation of synthetic methods for microporous metal–organic frameworks exemplified by the competitive formation of [Cu2(btc)3(H2O)3] and [Cu2(btc)(OH)(H2O)], Microporous and Mesoporous Materials. 132 (2010) 121–127. https//doi.org/10.1016/j.micromeso.2010.02.008

[76]   R.J. Kuppler, D.J. Timmons, Q.-R. Fang, J.-R. Li, T.A. Makal, M.D. Young, D. Yuan, D. Zhao, W. Zhuang, H.-C. Zhou, Potential applications of metal-organic frameworks, Coordination Chemistry Reviews. 253 (2009) 3042–3066. https//doi.org/10.1016/j.ccr.2009.05.019

[77]   P. Kumar, A. Pournara, K.-H. Kim, V. Bansal, S. Rapti, M.J. Manos, Metal-organic frameworks: Challenges and opportunities for ion-exchange/sorption applications, Progress in Materials Science. 86 (2017) 25–74. https//doi.org/10.1016/j.pmatsci.2017.01.002

[78]   H. Saleem, U. Rafique, R.P. Davies, Investigations on post-synthetically modified UiO-66-NH 2 for the adsorptive removal of heavy metal ions from aqueous solution, Microporous and Mesoporous Materials. 221 (2016) 238–244. https//doi.org/10.1016/j.micromeso.2015.09.043

[79]   Ş. Tokalıoğlu, E. Yavuz, S. Demir, Ş. Patat, Zirconium-based highly porous metal-organic framework (MOF-545) as an efficient adsorbent for vortex assisted-solid phase extraction of lead from cereal, beverage and water samples, Food Chemistry. 237 (2017) 707–715. https//doi.org/10.1016/j.foodchem.2017.06.005

[80]   Y. Pan, J. Wang, X. Guo, X. Liu, X. Tang, H. Zhang, A new three-dimensional zinc-based metal-organic framework as a fluorescent sensor for detection of cadmium ion and nitrobenzene, Journal of Colloid and Interface Science. 513 (2018) 418–426. https//doi.org/10.1016/j.jcis.2017.11.034

[81]   I.M. El-Sewify, M.A. Shenashen, A. Shahat, H. Yamaguchi, M.M. Selim, M.M.H. Khalil, S.A. El-Safty, Dual colorimetric and fluorometric monitoring of Bi3+ ions in water using supermicroporous Zr-MOFs chemosensors, Journal of Luminescence. 198 (2018) 438–448. https//doi.org/10.1016/j.jlumin.2018.02.028

Materials Research Forum LLC
https://doi.org/10.21741/9781644900291-1

[82]   Y. Pi, X. Li, Q. Xia, J. Wu, Y. Li, J. Xiao, Z. Li, Adsorptive and photocatalytic removal of Persistent Organic Pollutants (POPs) in water by metal-organic frameworks (MOFs), Chemical Engineering Journal. 337 (2018) 351–371. https//doi.org/10.1016/j.cej.2017.12.092

[83]   M.R. Azhar, H.R. Abid, H. Sun, V. Periasamy, M.O. Tadé, S. Wang, One-pot synthesis of binary metal organic frameworks (HKUST-1 and UiO-66) for enhanced adsorptive removal of water contaminants, Journal of Colloid and Interface Science. 490 (2017) 685–694. https//doi.org/10.1016/j.jcis.2016.11.100

[84]   H. Ramezanalizadeh, F. Manteghi, Synthesis of a novel MOF/CuWO4 heterostructure for efficient photocatalytic degradation and removal of water pollutants, Journal of Cleaner Production. 172 (2018) 2655–2666. https//doi.org/10.1016/j.jclepro.2017.11.145

[85]   Y. Zhou, Q. Yang, D. Zhang, N. Gan, Q. Li, J. Cuan, Detection and removal of antibiotic tetracycline in water with a highly stable luminescent MOF, Sensors and Actuators B: Chemical. 262 (2018) 137–143. https//doi.org/10.1016/j.snb.2018.01.218

[86]   D. Chen, C. Chen, W. Shen, H. Quan, S. Chen, S. Xie, X. Luo, L. Guo, MOF-derived magnetic porous carbon-based sorbent: Synthesis, characterization, and adsorption behavior of organic micropollutants, Advanced Powder Technology. 28 (2017) 1769–1779. https//doi.org/10.1016/j.apt.2017.04.018

[87]   X. Liu, N.K. Demir, Z. Wu, K. Li, Highly Water-Stable Zirconium Metal–Organic Framework UiO-66 Membranes Supported on Alumina Hollow Fibers for Desalination, Journal of the American Chemical Society. 137 (2015) 6999–7002. https//doi.org/10.1021/jacs.5b02276

[88]   P.G. Ingole, M. Sohail, A.M. Abou-Elanwar, M. Irshad Baig, J.-D. Jeon, W.K. Choi, H. Kim, H.K. Lee, Water vapor separation from flue gas using MOF incorporated thin film nanocomposite hollow fiber membranes, Chemical Engineering Journal. 334 (2018) 2450–2458. https//doi.org/10.1016/j.cej.2017.11.123

[89]   H. Zhan, Y. Bian, Q. Yuan, B. Ren, A. Hursthouse, G. Zhu, Preparation and Potential Applications of Super Paramagnetic Nano-Fe3O4, Processes. 6 (2018) 33. https//doi.org/10.3390/pr6040033

[90]   T.J. Bandosz, C. Petit, MOF/graphite oxide hybrid materials: exploring the new concept of adsorbents and catalysts, Adsorption. 17 (2011) 5–16. https//doi.org/10.1007/s10450-010-9267-5

Metal-Organic Framework Composites - Volume I
Materials Research Foundations **53** (2019) 29-54

Materials Research Forum LLC
https://doi.org/10.21741/9781644900291-2

Chapter 2

Metal-Organic Frameworks and their Composites for the Development of Electrochemical Sensors for Environmental Applications

Ankit Kumar Singh and Ida Tiwari*

Department of Chemistry, (Centre of Advanced Study), Institute of Science, Banaras Hindu University, Varanasi, India

idatiwari@bhu.ac.in

Abstract

The demand for accurate monitoring of environmental pollutants and their control has increased the need to develop some novel sensing techniques with high accuracy and lower limit of detection. Number of electrochemical sensors have been developed which can provide such platform for the determination of variety of chemical as well as biological pollutants. Metal-organic frameworks (MOFs) synthesized by coordinating the metal ions with the organic moieties are extensively used in the electrochemical sensing. Detection and removal of heavy metal ions and several inorganic as well as organic ions can be done by using electrochemical sensors based on MOFs and their composites. MOFs are used for constructing highly sensitive and reliable electrochemical sensor because MOFs are highly porous and have large surface area that helps in concentrating the analyte which results in strong signal intensity and higher sensitivity. Here advantages of MOFs and their composites in the construction of electrochemical sensors as well as their applications in the determination of several environmental contaminants are discussed.

Keywords

Metal-Organic Frameworks, Electrochemical Sensor, Environmental Applications, Sensing

Contents

Metal-Organic Framework Composites - Volume I   Materials Research Forum LLC
Materials Research Foundations **53** (2019) 29-54   https://doi.org/10.21741/9781644900291-2

1.   Introduction

Metal-organic frameworks (MOFs) are porous crystalline material having large internal surface area, uniform but adjustable cavity and reasonable thermal and mechanical stability (Fig. 1) [1-9]. They show various applications in different fields such as catalysis [10-25], gas separation, gas storage [26-38], sensing [39-42], drug delivery [43, 44], luminescence [45, 46], and other which are based on pore size and shape as well as those where host-guest interactions are involved. In addition to these applications, in recent years fascinating applications of MOFs in supercapacitors, batteries, fuel cell, hydrogen evolution reaction (HER), oxygen evolution reaction (OER) and oxygen reduction reaction (ORR) are reported [2].

The concepts related to MOFs was firstly introduced by Yaghi group [2] in 1995 and they provided MOF structure (MOF-5 based on Zinc) in 1999 which have large surface area of about 2900 m^2 g^{-1} and having porosity of 60%. After this large numbers of MOFs were also reported that showed structural, magnetic, optical, catalytic and electrical properties based on the choice of metal ions and organic ligands. .

Recently several investigations have been performed in order to determine the potential of MOF as sensors. However, in order to achieve certain potential as a sensor, MOF may require some functionalization. Three different pathways have been identified in order to create some modifications in the MOFs so that it can be utilized in various fields [47]. The first method involves modification in the specific organic ligand or doping in the metal ions within the framework of the MOFs. Generally lanthanide metal ions are involved in doping so that luminescent MOFs are created for optical sensing [48]. The second approach is post-synthesis modification (PSM), which involves the organic moiety with functional groups that can be used for subsequent chemical grafting. For the preparation of single-crystal MOF, Kurmoo et al. [49] tried to design a trifunctional (carboxylate, imidazolate, and hydroxyethyl groups) tag and for the improvement in the gas uptake of the MOF two consecutive PSMs of elimination and bromination were performed. The third approach involves the synthesis of composites having multifunctionalities for molecular recognition and transduction of signal by entrapping of functional molecules and nanoparticles (NPs) within the framework. These three approaches form the principle for functioning of the MOF i.e. any changes of MOF properties on the basis of incoming guest could be measured as a sensing signal [47].

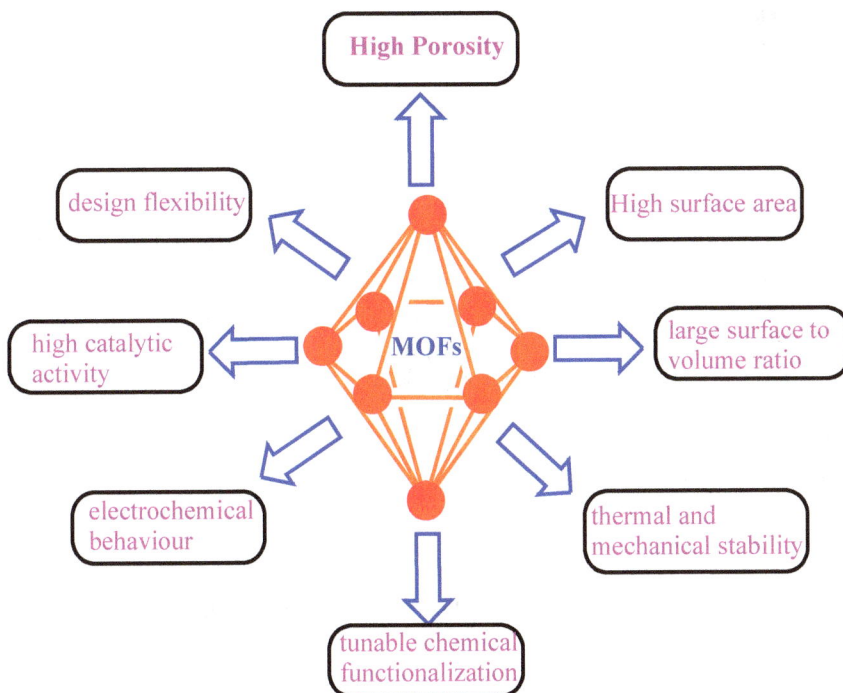

Figure 1. *Useful properties of Metal-organic frameworks (MOFs)*

Materials Research Forum LLC
https://doi.org/10.21741/9781644900291-2

Environmental pollution due to emission and discharge of chemical pollutants have increased the demand for advanced chemical sensor devices however electrochemical sensors show great promise for this task. Metal-organic frameworks (MOFs) generally prepared by coordinating metal ions with organic ligands in suitable solvent are widely used in electrochemical sensing. But initially MOF materials were considered that they were generally not suitable for various electrochemical applications as they have poor electron conductive behavior. However various MOFs and MOF composites are used now-a-days in the manufacture of electrode material for rechargeable batteries and electrochemical sensors by making some modification in MOF so that its electron conductive behavior gets improved [50-52]. This can be achieved moreover by converting MOF to several nano structures, such as metal nanoparticles, porous carbon, metal oxides and their composites by treating with different methods. In these ways nanostructures may have maximum exposure of their active sites i.e. they have large surface area of MOF precursors and may exhibit good response as electrodes. Therefore the most important thing is to design MOFs and MOF composites with desired electrochemical behavior [1]. The efficient electrochemical signal can be achieved by preparing MOF with high redox activity and electrical conductivity. This can be achieved in large number of ways; one useful way is by the incorporation of several functional groups into MOFs so that it can be utilized for various applications in environmental cleaning [53,54].

The working of electrochemical sensor is based on the redox reactions of the analytes taking place in an electrochemical system. An electrochemical sensor consists of three components that are a working (or sensing) electrode, a counter electrode and usually a reference electrode as well. These electrodes are enclosed along with a liquid electrolyte [55-57]. Several modification can be done in working electrode for specific recognition and for determination of concentration of metal ions [58, 59]. The presence of heavy metals as well as several inorganic and organic components as pollutant can cause the change of current, potential, capacitance, electrochemical impedance, electrochemiluminiscence and these changes can be utilized in their detection [60, 61]. On the basis of these detection signals, the electrochemical sensing can be classified as amperometry, potentiometry, capacitance, electrochemical impedance and electrochemiluminiscent methods.

In the upcoming 50 years, the energy and environmental issues will rank top priorities among the various global issues. In order to satisfy the energy needs the fossil fuel can act as dominant source. However depletion of these resources are leading to several environmental hazards such as air pollution and global warming due to emission of several toxic gases such as methane, nitrous oxide, carbon monoxide, carbon dioxide and other gases having volatile organic compounds [62, 63]. Some other environmental issues are related with the availability of fresh water because the industrial waste water causes severe problem and the water become polluted. Industrial effluents mainly contains heavy metals as well as some inorganic and organic pollutants that are responsible for the contamination of the fresh water and in this way living organisms on earth are suffering from the scarcity of fresh water [64].

Heavy metals are highly toxic and are considered as "Environmental health hazards".

These heavy metals are non-biodegradable and causes harm to human health and environment [64, 65]. These heavy metals are easily accumulated in the biosphere and can enter the living organisms through the alimentary chain and in this way they can affect the human health [66]. Heavy metals can exert their toxic effect through bond formation with thiol group of proteins but when they enter the living cell can alter the biological lifecycle. Various heavy metals like mercury (Hg), cadmium (Cd), lead (Pb), chromium (Cr) and arsenic (As) are the most probable for causing most of the heavy metal-related diseases. But some of the heavy metals like iron, cobalt, zinc, copper, manganese, etc. are required by the living organisms in lower quantity however its presence at higher concentration causes various toxic effects [67-69]. For this purpose many nanomaterials such as porous carbon, transition metal oxides (TMOs), mixed TMOs, and transition-metal oxide-carbon (TMO@C) composites are studied so that they can be utilized in environmental cleaning due to their ability of detection of heavy metals [62]. However the researches performed in recent years have shown that a better result can be obtained by using MOF and their composites as an electrochemical sensor for the detection of several heavy metals, some inorganic and organic species as well as in their removal mechanism [70,71].

2. Synthesis of MOFs and MOF composites

Due to various applications of MOFs it becomes necessary to pay much attention on its synthesis. Extensive reviews have been published that cover partially or wholly several approaches for the synthesis of MOFs and their composites. Liquid phase synthesis of MOF is widely used which involve the mixing of separate solution of metal salt and ligand in a glass vial. This can also be done in one other way that is to a solid mixture of metal salt and ligand in glass vial solvent is added. A suitable solvent should be taken based on different aspects such as reactivity, solubility, redox potential, stability constant, etc. The solid phase synthesis has also been worked on as it is quick and easy in comparison to liquid phase synthesis however it shows several difficulties in obtaining single crystal and thereby in determining product structure which is quite easy in the case of liquid phase synthesis. The end result is formation of metal-organic frameworks by the

Materials Research Forum LLC
https://doi.org/10.21741/9781644900291-2

interaction of metal ions or clusters with the organic moieties. (Fig. 2) These MOF formed may be one dimensional, two dimensional or three dimensional [72, 73].

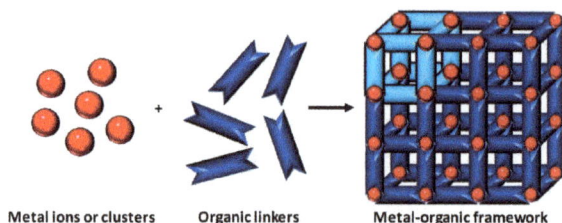

Metal ions or clusters   Organic linkers   Metal-organic framework

Figure 2. *Scheme showing the formation of MOF by mixing metal ions or clusters with organic linkers in the presence of some suitable solvent [74].*

Table 1. Different conditions required for various synthesis method.

S.No.	Synthesis Method	Different conditions for synthesis			References
		Type of Energy	Time	Temperature (K)	
1.	Solvothermal	Thermal Energy	48-96 hours	353-453	62, 72, 73
2.	Microwave	Microwave Radiation	4 minutes to 4 hours	303-373	62, 72, 73
3.	Sonochemical	Ultrasonic Radiation	30-180 minutes	273-313	62, 72, 73
4.	Slow Evaporation	No External Energy	7 days to 7 months	298	62, 72, 73
5.	Mechanochemical	Mechanical Energy	30 minutes to 2 hours	298	62, 72, 73
6.	Electrochemical	Electrical Energy	10 to 30 minutes	273-303	62, 72, 73

Several methods such as slow evaporation method, solvothermal method, microwave-assisted synthesis, electrochemical synthesis, mechanochemical synthesis, and sonochemical synthesis have been applied for the synthesis of MOF [62]. All these synthesis methods require different conditions that are represented in Table 1. However these techniques have several limitations as they require instrumentations that are highly expensive and they are energy as well as time consuming and also require high temperatures. Therefore it becomes necessary to develop a simple, ecologically-friendly preparation method in order to meet the industrial production needs. And hence the reflux method act as one of the promising method for the large-scale synthesis as it requires simple instrumentation and the reaction occurs at atmospheric pressure. Some method involve mixing of MOF precursors with composing material that is presynthesized and then further procedure of synthesis can be carried out. One of the widely reported such type of material is graphite oxide/MOF (GO/MOF) due to its various applications [75-78].

3.   Sensors based on MOF for environmental applications

The increase in pollution due to variety of pollutants introduced into the environment lead to the development of sensors that can detect these pollutants. Heavy metals, anions, organics, inorganics, bacteria and antibiotics in water are highly harmful to both the human health and the ecological environment. Therefore the demand for development of sensitive and reliable sensors in order to determine the water contamination has increased. The chemical detections in an aqueous solution are widely utilizing MOFs due to their unique characters for selective detection and determination of analytes. They can show reversible adsorption and release of target molecules due to their large surface area as well as porosity. These unique properties of MOF and their composites help to develop sensors based on MOF that can utilize luminescent, colorimetric or electrochemical signals. The sensing performances of these sensor have shown promising results in the determination of several water pollutants, although these sensors are based on different sensing mechanisms.

4.   Electrochemical sensor based on MOF for environmental applications

MOF based electrochemical sensors are widely involved in most areas of technological society such as medicine, public health, energy control and environmental quality monitoring. In environmental pollution monitoring MOF based electrochemical sensors are used in three main areas such as toxics in water, organics and gases including volatile organics (VOCs). One of the major reasons for environmental pollution is the chemical route of pollution. Chemicals are released into the environment and they disturb the balance of our ecosystem, affecting our health, polluting the air we breathe and also contaminating our food. Therefore, the electrochemical sensors based on MOF are developed that are based on the measurement of electric current, electric potential, or any other electrical signal [79, 80]. MOFs are showing wide potential to be used as the surface for sensing based on electrochemical signals because they show high surface area as well as volume of the pore. Catalytic activity is also high as they are showing good absorbability [81, 82]. Several MOF based electrochemical sensors are developed that are used in the determination of several contaminants present in the environment (Table 2).

Metal-Organic Framework Composites - Volume I   Materials Research Forum LLC
Materials Research Foundations **53** (2019) 29-54   https://doi.org/10.21741/9781644900291-2

4.1   Heavy metal ion detection

Due to the rapid industrialization, heavy metal ion pollution is becoming a challenging issue as it effects the survival of living organism on our earth. Heavy metals are defined as metals with atomic weight between 63.5 and 200.6 g mol^{-1} and a specific gravity greater than 5 g cm^{-3} [83-86]. The major sources through which heavy metal ion pollution created are industrial wastewater, lead-acid batteries, fertilizers, smelting, tanneries, metal plating, etc. [87]. They can easily accumulate in foods such as aquatic products, land animals and vegetables as they are non biodegradable. These heavy metals can enter plant, animal and human tissues through diet, air inhalation, and also can enter through manual handling. Through the uptake of water heavy metals enter the plants; through plants they can enter the animal bodies that are feeding on plants [88-92]. In this way plant and animal based foods are among the largest sources of heavy metal in humans [84, 85,]. These heavy metals can bind our cellular components such as protein, nucleic acid, enzymes, etc. and can disturb their functioning.

For metal ion analysis several analytical techniques including inductively coupled plasma mass spectrometry, atomic absorption spectrometry, atomic emission spectrometry and anodic stripping voltammetry have been used [93-96]. One of the highly sensitive and selective techniques for trace heavy metal detection is electrochemical stripping voltammetric technique that involves advanced electrochemical measurement of analytes accumulated in an effective preconcentration step [97-99]. Most of the electrode materials show a number of limitations. Among these several limitations the major ones are low surface areas, limited possibility for surface modification at which the process of electron transfer takes place and lack in structural design. Behind these limitations, the electrode materials behave as the heart of electrochemical sensors. Therefore, the researchers are fully devoted for designing advanced electrode materials having well defined physical and chemical properties that will improve the electrochemical performances. In the last few decades one of the important advances in materials science has occurred due to the development in the design of metal-organic frameworks (MOFs) with high surface areas, chemical tunability and uniform pores [100, 101]. MOFs with these characteristics act as a unique materials with potential for diverse applications [102-105]. MOFs show multiple issues of poor conductivity and instability in aqueous solution and due to this the use of MOFs as novel electrode materials have been under explored. Coupling of MOFs with carbon materials such as carbon spheres [106], metal nanoparticles (Au [107], Pt [108]) and macroporous carbon [109] showed an effective approach in order to improve the mechanical strength and conductive performance of the MOFs and have shown some promise in area of electrochemical biosensor. But as the

Metal-Organic Framework Composites - Volume I   Materials Research Forum LLC
Materials Research Foundations **53** (2019) 29-54   https://doi.org/10.21741/9781644900291-2

synthesis processes of these materials are not very easy their wider practical applications were greatly inhibited in the electrochemical field.

4.1.1 Detection of lead ion (Pb^{2+})

One of the most commonly found heavy metal ions in aquatic systems is lead ion (Pb^{2+}) and it is also considered as one of the most toxic heavy metal also. High intake of lead ions and its exposure for longer duration can cause memory loss, irritability, anemia, reproductive, cardio-vascular, neurological, developmental and several other disorders. Thus it becomes necessary to monitor even trace amount of Pb^{2+} in the water environment as well as in food industry regarding public health [110]. Several conventional techniques such as atomic absorption spectroscopy (AAS), atomic fluorescence spectrometry (AFS) and inductively coupled plasma mass spectrometry (ICP-MS) are widely used for determination of Pb^{2+}. Although these methods are sufficiently sensitive but they are also highly complicated. Thus it becomes an important task to develop a rapid, highly sensitive and convenient method for Pb^{2+} ion analysis and for this MOFs based electrochemical sensors show many advantages over other electrochemical sensors.

For the detection of Pb^{2+}, the He group [111] developed a strategy using Pd-Pt alloy-modified Fe-MOFs (Fe-MOFs/PdPt NPs) along with hairpin DNA that was immobilized on the surface as a signal tag. For the immobilization of more DNAzyme they used streptavidin modified reduced graphene oxide-tetraethylene pentamine-gold nanoparticles (rGO-TEPA-Au) as a sensor platform. The new single-DNA can be produced by cleaving the ribonucleotide at its specific site by using DNAzyme in the presence of Pb^{2+}. Then the hybrid strand of hairpin DNA is matched with its complement to the single- DNA and was employed for modifying the Fe-MOFs/PdPt NPs bioconjugates for amplifying the signal. The Fe-MOFs/PdPt NPs can catalyze the production of electrochemical signal from the hydrogen peroxide (H$_2$O$_2$) which can be recorded by chronoamperometry. In this way by using this method Pb^{2+} can be determined selectively in the presence of other metal ions and this selectivity is due to Pb^{2+} dependent DNAzyme. The developed biosensor corresponds to a low detection limit of 2 pM (S/N =3) and also exhibited a good linear relationship ranging from 0.005 to 1000 nmol L^{-1} for the sensing of Pb^{2+} ion. High sensitivity and selectivity is shown by this newly designed ultrasensitive biosensor based on Pb^{2+} dependent DNAzyme.

Based on MOFs another electrochemical sensor was also developed for the sensing of Pb^{2+} ion and it was designed by Guo et al. [112] They developed a novel metal-organic framework (MOF) material that was flake-like NH$_2$-MIL-53(Cr), by using a simple reflux method. This prepared MOF can behave as an effective material for electrochemical determination of Pb^{2+} ion in aqueous sample even at the microgram levels. Square wave anodic stripping voltammetry is used as the detection technique. The surface of glassy carbon electrode (GCE) was modified by NH_2-MIL-53(Cr) that showed an excellent stability and selectivity for the determination of Pb^{2+} ion. This modified GCE is used for the study of electrochemical properties and applications. Some operational parameters such as deposition time, deposition potential and pH were optimized to detect even trace amount of metal ions. In this way this method developed a new way for heavy metal detection using MOF materials. Excellent electronic response is obtained for Pb^{2+} ion by using this sensor. As the concentration of Pb^{2+} is increased under optimal conditions in the range of 0.4-80 µM the oxidation current is increased linearly. It also showed a very good detection limit of 30.5 nM.

4.1.2 Detection of copper ion (Cu^{2+})

The biological availability of copper ions in water is more than in food. Copper ions are also highly toxic and can cause several disorders in living organisms. Several methods have been designed for the determination of copper ions however the MOF based electrochemical sensor play a very crucial role in the determination of lead ion. These MOF based electrochemical sensor show high accuracy, sensitivity and precision. Therefore the way used in determination of lead ion (Pb^{2+}) is also used for the electrochemical measurements of copper (II) in the presence of MOF-based electrochemical sensor. During the measurement of copper (II), an anionic MOF i.e. AuNPs/Me_2NH_2@MOF-1/GCE was dipped in solution containing copper (II) so that preconcentration of copper can takes place. The square wave anodic stripping voltammetry (SW ASV) is the technique used for the electrochemical measurement of copper ion [113,114].

4.2 Sensing of inorganic ions

Among several inorganic pollutants present in water, most of them are not very toxic but due to their extensive use they are still dangerous for our environment. These inorganic pollutants may include nitrates and phosphates that are extensively present in fertilizers and hence they can reach to the water body very easily. Besides these inorganic pollutants several other pollutants are also present in water and they are dangerous both for human being and their environment. Thus it becomes necessary to monitor these inorganic pollutants. Several techniques have been employed for this purpose but most of them show several limitations of poor selectivity, sensitivity and detection limit. Therefore the development of MOF based electrochemical sensor have tendency to remove these limitations. Several inorganic pollutants can be detected by using MOF based electrochemical sensor.

Metal-Organic Framework Composites - Volume I
Materials Research Foundations **53** (2019) 29-54

Materials Research Forum LLC
https://doi.org/10.21741/9781644900291-2

4.2.1 Detection of Nitrite ion (NO_2^-)

One of the important contaminant in water is nitrite ion. Nitrite ions have been extensively used by human beings for several purposes such as in drinking water, vegetables, as fertilizing agents and as food preservatives. Nitrites are highly toxic to both humans and animals, especially infants. Nitrite is a nutrient which is essential for plant growth and it can enter the body of living organisms as nitrate which can be easily converted to nitrite through various mechanisms. It disrupts the various biochemical processes of living organisms by affecting the ability of hemoglobin in oxygen delivering process in the bloodstream. If an infant is exposed with nitrite in water or formula mixed with water that is contaminated with nitrate then the disorder known as blue baby syndrome (methemoglobinemia) may be develop in the body of infant that behave as a life threatening disorder. Several other disorders such as hypertension, stomach cancer can also occur due to the adverse effect of nitrite ion. Therefore it became an important requirement to develop nitrite sensors for regular monitoring of the harmful nitrite ions with accuracy. Although number of analytical techniques has been already developed for the determination of toxic nitrites but most of these techniques have no reliability because these techniques are very tedious and time consuming. Electrochemical sensors however provide several advantages such as simple handling, safety, reliability, stability, rapidity, selectivity and sensitivity and therefore due to these features they have gained significant research interest. But the traditional electrodes cannot be used for accurate and rapid sensing of nitrite ion due to very high oxidation potential of nitrite. Recently MOF based electrochemical sensors have been developed that serve as a better tool for nitrite ion determination than the other developed nitrite sensors.

A MOF/rGO hybrid electrode was developed by Mobin group [115] for the electrocatalytic oxidative determination of nitrite. In this study, by using simple ultrasonication method Cu-MOFs were stacked with rGO. In order to achieve better electrocatalytic performance for nitrite ion determination the GCE is modified with Cu-MOF/rGO composites because GCE modified with Cu-MOF/rGO composites exhibited better LOD of 0.033mM than those of an MOF electrode or bare electrode. The sensing performance of the modified GCE is better due to the increase in conductivity of MOF with rGO. In addition to this improvement in performance it also showed good selectivity for the nitrite ions in the presence of some common salts such as $NaClO_4$, $CaCl_2$, $MgSO_4$, KCl, CH_3COONa and KNO_3. The activity of this sensor was also checked for pond water spiked with nitrite ion and the recoveries were observed up to 100-120%.

Materials Research Forum LLC
https://doi.org/10.21741/9781644900291-2

4.3   Sensing of organic compounds

Like inorganic ions and heavy metals, there are also large numbers of organic molecules that can act as pollutants in water environment. Generally these organic moieties are highly toxic and show lower degradability and hence they can cause very bad effects to our environment. Therefore it becomes imperative to perform some studies so that it becomes easy to monitor their effect on the environment and living organisms. Several researchers have tried to study their origin, effect on the environment and their control. One such study show that some dihydroxybenzene isomers (DBIs) of phenolic compounds such as catechol (CT), resorcinol (RS), and hydroquinone (HQ) can act as water pollutants [116]. They are toxic and showing lower degradability and hence they are hazardous both for living organisms and their environment. So it becomes demanding to develop some rapid and simple analytical technology for determination of these organic components present as pollutants in nature. Today several analytical methods such as fluorescence, chemiluminiscence, high performance liquid chromatography, spectrophotometry, mass spectrometry, capillary electrochromatography and electrochemical methods have been employed in quantitative determination of DBIs. Most of these techniques are not efficient in sensitive and selective determination. However, due to several advantages such as fast response, low cost, high sensitivity, excellent selectivity and lower detection limit, the electrochemical methods are used widely in this field. But the simultaneous determination of these three isomers that are CT, RS and HQ are not so easy because these isomers have similar stereochemical structure and having close redox potentials on common electrode. In order to overcome these limitations in simultaneous determination of these DBIs, some functional materials such as carbon nanotubes, metal sulfides, quantum dots, graphene have been utilized as electrochemical sensing materials. But we do not use these materials in order to solve the problems because these materials involve very complicated synthesis process, high toxicity and/or poor analytical response. Therefore to develop a rapid and convenient sensing platform for simultaneous determination of DBIs is still a challenge. However the development of MOF-based electrochemical sensor has shown very promising result in the determination of these organic moieties [117].

One chitosan (CS) coated electrode along with the doping of GO was developed for the detection of these DBIs [117]. This method involves the electrochemical reduction of GO to rGO in the CS/GO composite by using some electrorduction method. This rGO was used as supporting carrier for the grafting of electroactive MOF i.e. $Cu_3(BTC)_2$ and it also showed high electrical conductivity. Electrochemical measurement shows good selectivity because the reduction peaks of RS, CT, and HQ are well separated from each other. The current response of the CS/rGO matrix is improved because it shows high conductivity and hence good LODs of 0.44, 0.41, and 0.33 µM can be achieved for HQ, CT, and RS respectively [110, 117]. This MOF based electrochemical sensor has been successfully applied for the determination of DBIs in real water sample with high accuracy. That's why the MOFs based electrochemical sensors are widely used in the organic compound sensing.

Another electrochemical sensor based on MOFs was designed by the Wang group [118] for the detection of DBIs. Here they synthesized magnetic Ni@graphene composites having core-shell structure, C-SNi@G through thermal annealing of Ni-BTC MOF. By using bare electrode HQ and CT cannot be differentiated because the oxidation peaks of HQ and CT overlap each other. Hence during the cyclic voltammetric studies only one oxidation peak was observed by using bare electrode. However by using the C-SNi@G/MGCE as an electrode, two well separated peaks were observed at 0.085V for HQ and 0.190 V for CT vs Ag/AgCl [110,118].

A highly sensitive electrochemical sensor was developed by the Liu group [119] for the simultaneous determination of HQ and CT in water sample and it was based on copper-centered MOF-graphene composites [Cu-MOF-GN, Cu-MOF: $Cu_3(BTC)_2$]. However excellent elecrocatalytic activity and high selectivity can be observed for HQ and CT when Cu-MOF-GN electrode was used under optimized conditions. This MOF based electrode show very low detection limit 0.59M for HQ and 0.331M for CT, respectively [110,119].

Besides these DBIs, some other phenolic compounds are also harmful to humans and our environment. These include chlorinated phenol contaminants such as 2,4-dichlorophenol (2,4-DCP) that are present in water and they may also enter the human body through the food chain. This 2,4-DCP can cause harm for human health even when present in very low concentration [120]. 2,4-DCP can be selectively determined by using MOF based electrochemical sensor i.e. 1,3,5-benzenetricarboxylic acid copper. This developed sensor for 2, 4-DCP is not only showing high selectivity but it also show broad range for sensing i.e. from 0.04 to 1.01M and very low LOD of 9nM. This copper based MOF sensor show several advantages: (i) It has large surface area, (ii) It shows high adsorption capacity and (iii) It has also good electron transfer efficiency. These advantages help in enhancing the performance of electrochemical sensor. This developed sensor is also applied for the determination of 2, 4-DCP present in raw water samples [121].

Materials Research Forum LLC
https://doi.org/10.21741/9781644900291-2

Table 2 MOF-based electrochemical sensors for determination of contaminants present in water.

S.No.	MOF-based sensing material	Analytes to be detected	LOD	Sample used for testing	References
1.	$Zn_4O(BDC)_3$	Pb^{2+}	4.9 nM	Real water	122
2.	Fe-MOFs/PdPt NPs	Pb^{2+}	2 pM	Reservoir water, well water, tap water	111
3.	NH_2-MIL-53(Cr)	Pb^{2+}	30.5 nM		112
4.	$Me_2NH_2@MOF-1$	Cu^{2+}	1pM	river water	113,114
5.	Cu-MOF/rGO	NO^{2-}	33 nM	pond water	117
6.	UiO-66-NH_2	NO^{2-}	0.01 μM		123
7.	$Cu_3(BTC)_2$	Hydroquinone	0.44 μM		110
		Catechol	0.41 μM		110
		Resorcinol	0.33 μM		110
		2,4-dichlorophenol	9 nM	reservoir raw water	121
8.	Cu-MOF-199/SWCTs	Hydroquinone	0.08 μM	river water	124
		Catechol	1 μM	river water	124

Conclusion

Environmental pollutions are increasing day-by-day, due to several inorganic and organic pollutant thus it becomes demanding for the development of sensors that can detect these pollutants and helps in their removal. Several chemical, fluorescence, UV, electrochemical sensors are developed for the determination of these pollutants even when present at low concentration. Among these sensors, the electrochemical sensors developed show very promising result in the determination of these inorganic or organic pollutants. But the conventional electrochemical sensors developed do not show very low detection limit. They also do not have high selectivity, specificity and accuracy. Therefore Metal-organic frameworks (MOFs) are used in designing electrochemical sensors that are showing high selectivity, specificity and accuracy and also they have a lower limit of detection. These MOFs used in electrochemical sensing are prepared by chelating metal ions with organic moieties. MOFs have large surface area; they are rich in porosity and active sites; also the pore sizes are highly flexible. Hence due to these behaviors they are used in the construction of electrochemical sensor. Several MOFs

Metal-Organic Framework Composites - Volume I   Materials Research Forum LLC
Materials Research Foundations **53** (2019) 29-54   https://doi.org/10.21741/9781644900291-2

based electrochemical sensors are developed such as NH_2-MIL-53(Cr) for the determination of Pb^{2+} ion, MOF/rGO hybrid electrode for the determination of nitrite ions, Cu-MOF-GN, Cu-MOF: $Cu_3(BTC)_2$ are used for sensing catechol and hydroquinone. Generally MOFs are widely used for removal of pollutants through adsorption and only few MOF based electrochemical sensors are available for the determination of inorganic as well as organic pollutants. However, due to its wide potential in the field of electrochemistry, materials based on MOF will show lot of promise for the detection of different pollutants. Therefore, research is fast in progress to develop MOF based electrochemical sensor in order to overcome the limitations of other electrochemical sensors.

References

[1] F.Y. Yi, R. Zhang, H. Wang, L.F. Chen, L. Han, H. L. Jiang, and Q. X, Metal-Organic Frameworks and Their Composites: Synthesis and Electrochemical Applications, Small Methods 1 (2017) 1700187.
https://doi.org/10.1002/smtd.201700187

[2] O. M. Yaghi, G. M. Li, H. L. Li, Selective binding and removal of guests in a microporous metal–organic framework, Nature 378 (1995) 703.
https://doi.org/10.1038/378703a0

[3] B. F. Hoskins, R. Robson, Infinite polymeric frameworks consisting of three dimensionally linked rod-like segments, J. Am. Chem. Soc. 111(1989) 5962.
https://doi.org/10.1021/ja00197a079

[4] B. F. Hoskins, R. Robson, Design and construction of a new class of scaffolding-like materials comprising infinite polymeric frameworks of 3D-linked molecular rods. A reappraisal of the zinc cyanide and cadmium cyanide structures and the synthesis and structure of the diamond-related frameworks $[N(CH_3)_4][Cu(I)Zn(II)(CN)_4]$ and $Cu(I)[4,4',4'',4'''$-tetracyanotetraphenylmethane]$BF_4.xC_6H_5NO_2$, J. Am. Chem. Soc. 112 (1990) 1546. https://doi.org/10.1021/ja00160a038

[5] D. Venkataraman, G. B. Gardner, S. Lee, J. S. Moore, Zeolite-like Behavior of a Coordination Network, J. Am. Chem. Soc. 117 (1995) 11600.
https://doi.org/10.1021/ja00151a034

[6] G. B. Gardner, D. Venkataraman, J. S. Moore, S. Lee, Spontaneous assembly of a hinged coordination network, Nature 374 (1995) 792.
https://doi.org/10.1038/374792a0

[7] G. Férey, Hybrid porous solids: past, present, future, Chem. Soc. Rev. 37 (2008) 191.
https://doi.org/10.1039/b618320b

[8] S. Horike, S. Shimomura, S. Kitagawa, Soft porous crystals, Nat. Chem. 1(2009) 695.

[9] H. Furukawa, K. E. Cordova, M. O'Keeffe, O. M. Yaghi, The Chemistry and Applications of Metal-Organic Frameworks, Science341 (2013) 974.

[10] J. M. Yoon, R. Srirambalaji, K. Kim, Homochiral, Metal–Organic Frameworks for Asymmetric Heterogeneous Catalysis, Chem. Rev. 112 (2012) 1196. https://doi.org/10.1021/cr2003147

[11] J. Liu, L. Chen, H. Cui, J. Zhang, L. Zhang, C.-Y.Su, Applications of metal–organic frameworks in heterogeneous supramolecular catalysis, Chem. Soc.Rev. 43 (2014) 6011. https://doi.org/10.1039/c4cs00094c

[12] G. Huang, Y.-Z.Chen, H.-L.Jiang, Metal-organic frameworks for Catalysis, Acta. Chim. Sin. 74 (2016) 113.

[13] Y.-Z. Chen, Z. U. Wang, H. Wang, J. Lu, S.-H.Yu, H.-L. Jiang, Singlet Oxygen-Engaged Selective Photo-Oxidation over Pt Nanocrystals/Porphyrinic MOF: The Roles of Photothermal Effect and Pt Electronic State, J. Am. Chem. Soc. 139(2017) 2035. https://doi.org/10.1021/jacs.6b12074

[14] S. Ou, C.-D.Wu, Rational construction of metal–organic frameworks for heterogeneous catalysis, Inorg. Chem. Front. 1 (2014) 721. https://doi.org/10.1039/c4qi00111g

[15] L. Zeng, X. Guo, C. He, C. Duan, Metal–Organic Frameworks: Versatile Materials for Heterogeneous Photocatalysis, ACS Catal. 6 (2016) 7935. https://doi.org/10.1021/acscatal.6b02228

[16] A. Aijaz, A. Karkamkar, Y. J. Choi, N. Tsumori, E. Rönnebro,T. Autrey, H. Shioyama, Q. Xu, Immobilizing Highly Catalytically Active Pt Nanoparticles inside the Pores of Metal–Organic Framework: A Double Solvents Approach, J. Am. Chem. Soc. 134 (2012) 13926. https://doi.org/10.1021/ja3043905

[17] R. Q. Zou, H. Sakurai, S. Han, R. Q. Zhong, Q. Xu, Probing the Lewis Acid Sites and CO Catalytic Oxidation Activity of the Porous Metal−Organic Polymer [Cu(5 methylisophthalate)], J. Am. Chem. Soc. 129 (2007) 8402. https://doi.org/10.1021/ja071662s

[18] J. Y. Lee, O. K. Farha, J. Roberts, K. A. Scheidt, S. T. Nguyen, J. T. Hupp, Metal–organic framework materials as catalysts, Chem. Soc. Rev 38(2009) 1450. https://doi.org/10.1039/b807080f

[19] A. Corma, H. García, F. X. Llabrés-Xamena, Engineering Metal Organic Frameworks for Heterogeneous Catalysis, Chem. Rev. 110 (2010) 4606. https://doi.org/10.1002/chin.201046237

[20] J.-L. Wang, C. Wang, W. Lin, Metal–Organic Frameworks for Light Harvesting and Photocatalysis, ACS Catal. 2 (2012) 2630. https://doi.org/10.1021/cs3005874

[21] Q.-L. Zhu, Q. Xu, Immobilization of Ultrafine Metal Nanoparticles to High-Surface-Area Materials and Their Catalytic Applications, Chem 1 (2016) 220. https://doi.org/10.1016/j.chempr.2016.07.005

[22] X. Gu, Z.-H. Lu, H.-L. Jiang, T. Akita, Q. Xu, Synergistic Catalysis of Metal–Organic Framework-Immobilized Au–Pd Nanoparticles in Dehydrogenation of Formic Acid for Chemical Hydrogen Storage, J. Am. Chem. Soc. 133 (2011) 11822. https://doi.org/10.1021/ja200122f

[23] P.-Z. Li, K. Aranishi, Q. Xu, ZIF-8 immobilized nickel nanoparticles: highly effective catalysts for hydrogen generation from hydrolysis of ammonia borane, Chem. Commun. 48 (2012) 3173. https://doi.org/10.1039/c2cc17302f

[24] Q.-L. Zhu, J. Li, Q. Xu, Immobilizing Metal Nanoparticles to Metal–Organic Frameworks with Size and Location Control for Optimizing Catalytic Performance, J. Am. Chem. Soc. 135 (2013) 10210. https://doi.org/10.1021/ja403330m

[25] P. Pachfule, X. Yang, Q.-L. Zhu, N. Tsumori, T. Uchidaa, Q. Xu,From Ru nanoparticle-encapsulated metal–organic frameworks to highly catalytically active Cu/Ru nanoparticle-embedded porous carbon, J. Mater. Chem. A. 5 (2017) 4835. https://doi.org/10.1039/c6ta10748f

[26] J.-P. Zhang, X.-M. Chen, Exceptional Framework Flexibility and Sorption Behavior of a Multifunctional Porous Cuprous Triazolate Framework, J. Am. Chem. Soc. 130 (2008) 6010. https://doi.org/10.1021/ja800550a

[27] J. R. Li, R. J. Kuppler, H. C. Zhou, Selective gas adsorption and separation in metal–organic fameworks Chem. Soc. Rev. 38 (2009) 1477. https://doi.org/10.1039/b802426j

[28] M. P. Suh, H. J. Park, T. K. Prasad, D. W. Lim, Hydrogen Storage in Metal–Organic Frameworks, Chem. Rev. 112(2012) 782.

[29] H. H. Wu, Q. H. Gong, D. H. Olson, J. Li, Commensurate Adsorption of Hydrocarbons and Alcohols in Microporous Metal Organic Frameworks, Chem. Rev. 112 (2012) 836. https://doi.org/10.1021/cr200216x

[30] S. H. Yang, X. Lin, W. Lewis, M. Suyetin, E. Bichoutskaia,J. E. Parker, C. C. Tang, D. R. Allan, P. J. Rizkallah, P. Hubberstey, N. R. Champness, K. M. Thomas, A. J. Blake, M. Schröder, A partially interpenetrated metal-organic framework for selective hysteretic sorption of carbon dioxide, Nat. Mater. 11 (2012) 710. https://doi.org/10.1038/nmat3343

[31] P. Nugent, Y. Belmabkhout, S. D. Burd, A. J. Cairns, R. Luebke,K. Forrest, T. Pham, S. Ma, B. Space, L. Wojtas, M. Eddaoudi,M. J. Zaworotko, Porous materials

with optimal adsorption thermodynamics and kinetics for CO_2 separation, Nature 495 (2013) 80. https://doi.org/10.1038/nature11893

[32] Y. Peng, V. Krungleviciute, I. Eryazici, J. T. Hupp, O. K. Farha,T. Yildirim, Methane Storage in Metal–Organic Frameworks: Current Records, Surprise Findings, and Challenges, J. Am. Chem. Soc. 135 (2013) 11887. https://doi.org/10.1021/ja4045289

[33] K. Sumida, D. L. Rogow, J. A. Mason, T. M. McDonald, E. D. Bloch, Z. R. Herm, T. H. Bae, J. R. Long, Carbon Dioxide Capture in Metal–Organic Frameworks, Chem. Rev. 112 (2012) 724. https://doi.org/10.1021/cr2003272

[34] Y. He, W. Zhou, G. Qian, B. Chen, Methane storage in metal–organic frameworks, Chem. Soc. Rev. 43(2014) 5657. https://doi.org/10.1039/c4cs00032c

[35] J. R. Li, J. Sculley, H. C. Zhou, Metal–Organic Frameworks for Separations, Chem. Rev. 112(2012)869. https://doi.org/10.1021/cr200190s

[36] B. V. de Voorde, B. Bueken, J. Denayer, D. De Vos, Adsorptive separation on metal organic frameworks in the liquid phase, Chem. Soc.Rev.43 (2014) 5766. https://doi.org/10.1039/c4cs00006d

[37] S. Qiu, M. Xue, G. Zhu, Metal–organic framework membranes: from synthesis to separation application, Chem. Soc. Rev.43(2014) 6116. https://doi.org/10.1039/c4cs00159a

[38] Q.-L. Zhu, Q. Xu, Liquid organic and inorganic chemical hydrides for high-capacity hydrogen storage, Energy Environ. Sci. 8 (2015) 478. https://doi.org/10.1039/c4ee03690e

[39] F.-Y. Yi, D. Chen, M.-K.Wu, L. Han, H.-L. Jiang, Chemical Sensors Based on Metal–Organic Frameworks, Chem Plus Chem 81 (2016) 675.

[40] B. L. Chen, S. C. Xiang, G. D. Qian, Metal-organic frameworks with functional pores for recognition of small molecules, Acc. Chem. Res. 43 (2010) 1115. https://doi.org/10.1021/ar100023y

[41] L. E. Kreno, K. Leong, O. K. Farha, M. Allendorf, R. P. V. Duyne,J. T. Hupp, Metal–Organic Framework Materials as Chemical Sensors, Chem. Rev. 112 (2012) 1105. https://doi.org/10.1021/cr200324t

[42] Z. Hu, B. J. Deibert, J. Li, Luminescent metal–organic frameworks for chemical sensing and explosive detection, Chem. Soc. Rev. 43 (2014) 5815. https://doi.org/10.1039/c4cs00010b

[43] P. Horcajada, R. Gref, T. Baati, P. K. Allan, G. Maurin, P. Couvreur,G. Férey, R. E. Morris, C. Serre, Metal–Organic Frameworks in Biomedicine, Chem. Rev. 112 (2012) 1232. https://doi.org/10.1021/cr200256v

[44] C. He, D. Liu, W. Lin, Nanomedicine Applications of Hybrid Nanomaterials Built from Metal–Ligand Coordination Bonds: Nanoscale Metal–Organic Frameworks and Nanoscale Coordination Polymers, Chem. Rev. 115 (2015) 11079. https://doi.org/10.1021/acs.chemrev.5b00125

[45] Y. Cui, Y. Yue, G. Qian, B. Chen, Luminescent Functional Metal–Organic Frameworks, Chem. Rev. 112 (2012) 1126. https://doi.org/10.1021/cr200101d

[46] P. Cheng, Lanthanide Metal–Organic Frameworks, Structure and Bonding Series, Springer, New York, USA (2015)

[47] J Lei, R. Qian, P. Ling, L. Cui, H.Ju, Design and sensing applications of metal–organic framework composites, Trends in Analytical Chemistry, 58 (2014) 71–78. https://doi.org/10.1016/j.trac.2014.02.012

[48] L.E. Kreno, K. Leong, O.K. Farha, M. Allendorf, R.P. Van Duyne, J.T. Hupp, Metal organic framework materials as chemical sensors, Chem. Rev. 112 (2012)1105–1125. https://doi.org/10.1021/cr200324t

[49] F. Sun, Z. Yin, Q.Q. Wang, D. Sun, M.H. Zeng, M. Kurmoo, Tandem postsynthetic modification of a metal–organic framework by thermal elimination and subsequent bromination: effects on absorption properties and photoluminescence, Angew. Chem. Int. Ed. 52 (2013) 4538–4543. https://doi.org/10.1002/anie.201300821

[50] X.Q. Wu, J.G. Ma, H. Li, D.M. Chen, W. Gu, G.M. Yang, P. Cheng, Metal–organic framework biosensor with high stability and selectivity in a bio-mimic environment., Chem. Commun. 51(44) (2015)9161–9164. https://doi.org/10.1039/c5cc02113h

[51] D. Sheberla, L. Sun, M.A. Blood-Forsythe, S. Er, C.R. Wade,C.K. Brozek, A. Aspuru-Guzik, M. Dinca, High electrical conductivity in Ni(3)(2,3,6,7,10,11-hexaiminotriphenylene)(2), a semiconducting metal–organic graphene analogue, J. Am. Chem. Soc. 136(25),(2014) 8859–8862. https://doi.org/10.1021/ja502765n

[52] M.G. Campbell, D. Sheberla, S.F. Liu, T.M. Swager, M. Dinca, Cu(3)(hexaiminotriphenylene)(2): an electrically conductive 2Dmetal–organic framework for chemiresistive sensing, Angew. Chem. Int. Ed. 54(14) (2015) 4349–4352. https://doi.org/10.1002/anie.201411854

[53] X. Wang, Q.X. Wang, Q.H. Wang, F. Gao, Y.Z. Yang,H.X. Guo, Highly dispersible and stable copper terephthalate metal–organic framework-graphene oxide nanocomposite for an electrochemical sensing application, ACS Appl. Mater.123Nano-Micro Lett. (2018) 10:64 Page 17 of 19 64Interfaces 6(14), 11573–11580 (2014). https://doi.org/10.1021/am5019918

[54] Z.D. Xu, L.Z. Yang, C.L. Xu, Pt@UiO-66 heterostructures for highly selective detection of hydrogen peroxide with an extended linear range, Anal. Chem. 87(6) (2015) 3438–3444. https://doi.org/10.1021/ac5047278

[55] P. Falcaro, R. Ricco, A. Yazdi, I. Imaz, S. Furukawa, D. Maspoch,R. Ameloot, J.D. Evans, C.J. Doonan, Application of metal and metal oxide nanoparticles@MOFs, Coord. Chem. Rev. 307 (2016) 237–254. https://doi.org/10.1002/chin.201609231

[56] J. Wang, J.T. Jiu, T. Araki, M. Nogi, T. Sugahara, S. Nagao, H.Koga, P. He, K. Suganuma, Silver nanowire electrodes: conductivity improvement without post-treatment and application in capacitive pressure sensors, Nano-Micro Lett. 7(1) (2015) 51–58. https://doi.org/10.1007/s40820-014-0018-0

[57] Z. Yang, Z.H. Li, M.H. Xu, Y.J. Ma, J. Zhang, Y.J. Su, F. Gao,H. Wei, L.Y. Zhang, Controllable synthesis of fluorescent carbon dots and their detection application as nanoprobes, Nano-Micro Lett. 5(4) (2013) 247–259. https://doi.org/10.1007/bf03353756

[58] I.Bontidean, C. Berggren, G. Johansson, E. Csoregi, B. Mattiasson, J.R. Lloyd, K.J Jakeman, N.L Brown, Detection of heavy metal ions at femtomolar levels using protein-based biosensors, Anal.Chem.70 (1998) 4162–4169. https://doi.org/10.1021/ac9803636

[59] D. Pan, Y. Wang, Z. Chen, T. Lou, W. Qin, Nanomaterial/Ionophore-Based Electrode for Anodic Stripping Voltammetric Determination of Lead: An Electrochemical Sensing Platform toward Heavy Metals, Anal. Chem. 81 (2009) 5088–5094. https://doi.org/10.1021/ac900417e

[60] C. Combellas, F. Kanoufi, J. Pinson, F.I. Podvorica, Sterically Hindered Diazonium Salts for the Grafting of a Monolayer on Metals, J.Am.Chem.Soc.130 (27) (2008) 8576–8577. https://doi.org/10.1021/ja8018912

[61] L. Fan, J. Chen, S. Zhu, M. Wang , G. Xu, Determination of Cd2+ and Pb2+ on glassy carbon electrode modified by electrochemical reduction of aromatic diazonium salts, Electrochem. Commun. 11 (2009) 1823– 1825. https://doi.org/10.1016/j.elecom.2009.07.026

[62] Z. Xie, W. Xu, X. Cui, and Y. Wang, Recent Progress in Metal–Organic Frameworks and Their Derived Nanostructures for Energy and Environmental Applications, Chem. Sus. Chem 10 (2017), 1645 – 1663. https://doi.org/10.1002/cssc.201601855

[63] M. Z. Jacobson, Review of solutions to global warming, air pollution, and energy security, Energy Environ. Sci. 2 (2009) 148–173. https://doi.org/10.1039/b809990c

[64] V. Thavasi, G. Singh, S. Ramakrishna, Electrospun nanofibers in energy and environmental applications, Energy Environ. Sci. 1 (2008) 205 –221. https://doi.org/10.1039/b809074m

[65] Z. Q. Xie, X. D. Cui, W. W. Xu, Y. Wang, Metal-Organic framework Derived CoNi@CNTs embedded Carbon Nanocages for efficient dye-sensitized solar cells, Electrochim. Acta 229 (2017) 361–370. https://doi.org/10.1016/j.electacta.2017.01.145

[66] Z. L. Li, J. Chen, H. Y. Guo, X. Fan, Z. Wen, M. H. Yeh, C. W. Yu, X. Cao, Z. L. Wang, Triboelectrification-enabled self-powered detection and removal of heavy metal ions in wastewater, Adv. Mater. 28 (2016) 2983–2991. https://doi.org/10.1002/adma.201504356

[67] R. K. Sharma, M. Agrawal, Biological effects of heavy metals: an overview, J. Environ. Biol. 26 (2005) 301–313.

[68] S. H. Hsu, C. T. Li, H. T. Chien, R. R. Salunkhe, N. Suzuki, Y. Yamauchi,K. C. Ho, K. C. W. Wu, Platinum- free counter electode comprised of metal- organic-framework(MOF)- derived cobalt sulfide nanoparticles for efficient dye-senitized solar cells (DSSCs) Sci. Rep.4 (2014) 6983. https://doi.org/10.1038/srep06983

[69] G. Aragay, J. Pons, A. Merkoci, Recent trends in macro-, micro-, and nanomaterial-based tools and strategies for heavy-metal detection, Chem. Rev. 111 (2011) 3433–3458. https://doi.org/10.1021/cr100383r

[70] E. Tahmasebi, M. Y. Masoomi, Y. Yamini, A. Morsali, Application of Mechano synthesized Azine-Decorated Zinc(II)Metal- Organic frameworks for highly efficient removal and extraction of some heavy-metal ions from aqueous samples: A comparative study, Inorg. Chem. 54 (2015) 425–433. https://doi.org/10.1021/ic5015384

[71] J.-N. Hao, B. Yan, A water-stable lanthanide- functionalized MOF as a highly selective and sensitive fluorescent probe for Cd^{2+}, Chem. Commun. 51 (2015) 7737–7740. https://doi.org/10.1039/c5cc01430a

[72] N. Stock, S. Biswas, Synthesis of Metal-Organic Frameworks (MOFs): routes to various MOF Topologies, Morphologies, and composites, Chem. Rev. 112 (2012) 933–969. https://doi.org/10.1021/cr200304e

[73] C. Dey, T. Kundu, B. P. Biswal, A. Mallick and R. Banerjee,Crystalline metal-organic frameworks (MOFs): synthesis, structure and function, Acta. Cryst. B 70 (2014) 3–10. https://doi.org/10.1002/chin.201426232

[74] Sergio Carrasco , Metal-Organic Frameworks for the Development of Biosensors: A Current Overview, Biosensors 8 (2018), 92. https://doi.org/10.3390/bios8040092

[75] C. Petit, et al. Langmuir, Toward Understanding reactive adsorption of ammonia on Cu-MOF/Graphite Oxide Nanocomposites 27 (2011) 13043–13051. https://doi.org/10.1021/la202924y

[76] C. Petit, B. Mendoza, T.J. Bandosz, Hydrogen sulfide adsorption on MOFs and MOF/ graphite oxide composites, ChemPhysChem 11 (2010) 3678–3684. https://doi.org/10.1002/cphc.201000689

[77] C. Petit, T.J. Bandosz, Synthesis, Characterization, and Ammonia Adsorption Properties of mesoporous metal-organic framework (MIL(Fe))-Graphite Oxide Composites: Exploring the limits of Materials Fabrication, Adv. Funct. Mater. 21 (2011) 2108–2117. https://doi.org/10.1002/adfm.201002517

[78] C. Petit, T.J. Bandosz, Exploring the coordination chemistry of MOF- graphite oxide composites and their applications as adsorbents, Dalton Trans. 41 (2012) 4027–4035. https://doi.org/10.1039/c2dt12017h

[79] X. Chen, Y. Wang, Y. Zhang, Z. Chen, Y. Liu, Z. Li, J. Li,Sensitive electrochemical aptamer biosensor for dynamic cell surface N-glycan evaluation featuring multivalent recognition and signal amplification on a dendrimer-graphene electrode interface, Anal. Chem. 86(9) (2014)4278–4286. https://doi.org/10.1021/ac404070m

[80] X. Fang, J.F. Liu, J. Wang, H. Zhao, H.X. Ren, Z.X. Li, Dual signal amplification strategy of Au nanopaticles/ZnO nanorods hybridized reduced grapheme nanosheet and multi enzyme functionalized Au@ZnO composites for ultrasensitive electrochemical detection of tumor biomarker, Biosens. Bioelectron. 97 (2017)218–225. https://doi.org/10.1016/j.bios.2017.05.055

[81] Y. Wang, C. Hou, Y. Zhang, F. He, M.Z. Liu, X.L. Li, Preparation of grapheme nano-sheet bonded PDA/MOF micro capsules with immobilized glucose oxidase as a mimetic multi-enzyme system for electrochemical sensing of glucose, J. Mater. Chem. B 4(21)(2016) 3695–3702. https://doi.org/10.1039/c6tb00276e

[82] C. Zhang, X.R. Wang, M. Hou, X.Y. Li, X.L. Wu, J. Ge, Immobilization on metal–organic framework engenders high sensitivity for enzymatic electrochemical detection, ACS Appl. Mater. Interfaces 9(16) (2017) 13831–13836. https://doi.org/10.1021/acsami.7b02803

[83]F. Fu, Q. Wang, Removal of heavy metal ions from wastewaters: A review, Journal of Environmental Management 92 (2011) 407-418. https://doi.org/10.1016/j.jenvman.2010.11.011

[84]N.K. Srivastava, C.B. Majumder, Novel biofiltration methods for the treatment of heavy metals from industrial wastewater, J. Hazard. Mater. 151 (2008) 1-8.

[85] N. Tekaya, O. Saiapina, H. Ben Ouada, F. Lagarde, H. Ben Ouada, N.Jaffrezic-Renault, Ultra-sensitive conductometric detection of heavy metalsbased on inhibition of alkaline phosphatase activity from Arthrospiraplatensis, Bioelectrochemistry 90 (2013) 24–29. https://doi.org/10.1016/j.bioelechem.2012.10.001

[86] G.L. Turdean, Design and development of biosensors for the detection of heavy metal toxicity, Int. J. Electrochem. (2011) 1–15.

[87] W. S. Wan Ngah, M. A. K. M. Hanafiah, Removal of heavy metal ions from wastewater by chemically modified plant wastes as adsorbents: A review, Bioresour. Technol. 99 (2008) 3935–3948. https://doi.org/10.1016/j.biortech.2007.06.011

[88] A. Singh, R.K. Sharma, M. Agrawal, F.M. Marshall, Health risk assessment of heavy metals via dietary intake of food stuffs from the wastewater irrigated site of a dry tropical area of India, Food Chem. Toxicol. 48 (2010) 611–619. https://doi.org/10.1016/j.fct.2009.11.041

[89] C. Gao, X.Y. Yu, S.Q. Xiong, J.-H. Liu, X.J. Huang, Electrochemical detection of arsenic (III) completely free from noble metal: Fe3O4microspheres-roomtemperature ionic liquid composite showing better performance than gold, Anal. Chem. 85 (2013) 2673–2680. https://doi.org/10.1021/ac303143x

[90] K. Tag, K. Riedel, H.-J. Bauer, G. Hanke, K.H.R. Baronian, G. Kunze, Amperometric detection of Cu2+by yeast biosensors using flow injection analysis (FIA),Sens. Actuators B: Chem. 122 (2007) 403–409. https://doi.org/10.1016/j.snb.2006.06.007

[91] X. Rajaganapathy, M.P. Sreekumar, Heavy metal contamination in soil,water and fodder and their presence in livestock and products: a review, J. Environ. Sci. Technol. 4 (2011) 234–249.

[92] M.R. Guascito, C. Malitesta, E. Mazzotta, A. Turco, Inhibitive determination of metal ions by an amperometric glucose oxidase biosensor, Sens. Actuators B:Chem. 131 (2008) 394–402. https://doi.org/10.1016/j.snb.2007.11.049

[93] M. Li, H. Gou, I. Al-Ogaidi, and N. Wu, Nanostructured Sensors for Detection of Heavy Metals: A Review, ACS Sustainable Chem. Eng. 2013, 1, 713−723. https://doi.org/10.1021/sc400019a

[94] K.E.Lorber, Monitoring of heavy metals by energy dispersive X-ray fluorescence spectrometry, Waste Manage. Res. 4 (1986) 3−13. https://doi.org/10.1177/0734242x8600400102

[95] R. Kunkel, S.E.Manahan, Atomic absorption analysis of strong heavy metal chelating agents in water and waste water, Anal. Chem 45(1973) 1465−1468. https://doi.org/10.1021/ac60330a024

[96] M. Lopez-Artiguez, A. Cameán, M. Repetto, Preconcentration of heavy metals in urine and quantification by inductively coupled plasma atomic emission spectrometry, J. Anal. Toxicol. 17 (1993) 18−22. https://doi.org/10.1093/jat/17.1.18

[97] J. Wang, Stripping Analysis, VCH Publishers, New York, 1985.

[98] J. Buffle and M.L. Tercier-Waeber, Trends Anal. Chem., 24 (2005) 172.

[99] J. Wang, Analytical Electrochemistry, 3rd ed, Wiley, New York, 2006.

[100] O. K. Farha and J. T. Hupp, Rational design, synthesis, purification, and activation of metal-organic framework materials, Acc. Chem. Res. 43 (2010) 1166-1175. https://doi.org/10.1021/ar1000617

[101]. D. Zhao, D. J. Timmons, D. Q. Yuan and H. C. Zhou, Tuning the topology and functionality of metal-organic frameworks by ligand design, Acc.Chem. Res., 44 (2010) 123-133. https://doi.org/10.1021/ar100112y

[102] B. Seoane, J. Coronas, I. Gascon, M. Etxeberria Benavides, O. Karvan, J. Caro, F. Kapteijnand J. Gascon, Metal-organic framework based mixed matrix membranes: a solution for highly efficient CO_2 capture? Chem. Soc. Rev. 44 (2015) 2421-2454. https://doi.org/10.1039/c4cs00437j

[103]. J. L. C. Rowsell and O. M. Yaghi, Effects of functionalization, catenation, and variation of the metal oxide and organic linking units on the low-pressure hydrogen adsorption properties of metal-organic frameworks, J. Am. Chem. Soc. 128 (2006) 1304-1315. https://doi.org/10.1021/ja056639q

[104] K. Schlichte, T. Kratzke and S. Kaskel, Improved synthesis, thermal stability and catalytic properties of the metal-organic framework compound $Cu_3(BTC)_2$, MicroporousMesoporous Mater.73 (2004) 81-88. https://doi.org/10.1016/j.micromeso.2003.12.027

[105] B. Liu and B. Smit, Comparative Molecular simulation study of CO_2/N_2 and CH_4/N_2 Separation in zeolites and metal-organic frameworks, J. Am. Chem. Soc. 25 (2009) 5918-5926. https://doi.org/10.1021/la900823d

[106] Y. Wang, H. Ge, G. Ye, H. Chen and X. Hu, Carbon functionalized metal organic framework/ Nafion composites as novel electrode materials for ultrasensitive determination of dopamine, J. Mater. Chem. B 3 (2015) 3747-3753. https://doi.org/10.1039/c4tb01869a

[107] Z. Xu, L. Yang and C. Xu, Pt@UiO-66 heterostructures for highly selective detection of hydrogen peroxide with an extended linear range, Anal. Chem. 87 (2015) 3438-3444. https://doi.org/10.1021/ac5047278

[108] Y. Wang, L. Wang, H. Chen, X. Hu and S. Ma, Fabrication of highly sensitive and stable hydroxylamine electrochemical sensor based on gold nanoparticles and metal-

metalloporphyrin framework modified electrode, ACS Appl. Mater. Interfaces 8 (2016) 18173-18181. https://doi.org/10.1021/acsami.6b04819

[109] Y. Zhang, X. Bo, C. Luhana, H. Wang, M. Li and L. Guo, Facile synthesis of a Cu-based MOF confined in macroporous carbon hybrid material with enhanced electrocatalytic ability, Chem.Commun.49 (2013) 6885-6887. https://doi.org/10.1039/c3cc43292k

[110] X. Fang, B. Zong and S. Mao, Metal–Organic Framework-Based Sensors for Environmental Contaminant Sensing, Nano-Micro Lett. 10 (2018) 64. https://doi.org/10.1007/s40820-018-0218-0

[111] Y.J. Yu, C. Yu, Y.Z. Niu, J. Chen, Y.L. Zhao, Y.C. Zhang, R.F.Gao, J.L. He, Target triggered cleavage effect of DNAzyme: relying on Pd–Pt alloys functionalized Fe-MOFs for amplified detection of Pb^{2+}, Biosens. Bioelectron. 101 (2018) 297–303. https://doi.org/10.1016/j.bios.2017.10.006

[112] H.X. Guo, D.F. Wang, J.H. Chen, W. Weng, M.Q. Huang, Z. S. Zheng, Simple fabrication of flake-like NH_2-MIL-53(Cr) and its application as an electrochemical sensor for the detection of Pb^{2+}, Chem. Eng. J. 289 (2016) 479–485. https://doi.org/10.1016/j.cej.2015.12.099

[113] J.C. Jin, J. Wu, G.P. Yang, Y.L. Wu, Y.Y. Wang, A microporous anionic metal–organic framework for a highly selective and sensitive electrochemical sensor of Cu^{2+} ions. Chem. Commun. 52 (2016) 8475–8478. https://doi.org/10.1039/c6cc03063g

[114] J. C. Jin, J. Wu, G. P. Yang, Y. L. Wua and Y. Y. Wang, A microporous anionic metal–organic framework for highly selective and sensitive electrochemical sensor of Cu^{2+} ion, Electronic Supplementary Material (ESI) for Chem. Comm. J.

[115] M. Saraf, R. Rajak, S.M. Mobin, A fascinating multitasking Cu-MOF/rGO hybrid for high performance supercapacitors and highly sensitive and selective electrochemical nitrite sensors, J. Mater. Chem. A 4, 42 (2016)16432–16445. https://doi.org/10.1039/c6ta06470a

[116] D. A. Perry, T. M. Razer, K. M. Primm, T. Chen, J. B. Shamburger et al., Surface-enhanced infrared absorption and density functional theory study of dihydroxybenzene isomer adsorption on silver nanostructures, J. Phys. Chem. C 117(16) (2013) 8170–8179. https://doi.org/10.1021/jp3121462

[117] Y. Yang, Q. Wang, W. Qiu, H. Guo, F. Gao, Covalent immobilization of $Cu_3(btc)_2$ at chitosan–electro reduced grapheme oxide hybrid film and its application for simultaneous detectionof dihydroxybenzene isomers, J. Phys. Chem. C 120(18)(2016)9794–9803. https://doi.org/10.1021/acs.jpcc.6b01574

[118] X. Zhou, X. Yan, Z. Hong, X. Zheng, F. Wang, Design of magnetic core–shell Ni@graphene composites as a novel electrochemical sensing platform, Sens. Actuators B-Chem. 255(2018)2959–2962. https://doi.org/10.1016/j.snb.2017.09.117

[119] J. Li, J. Xia, F. Zhang, Z. Wang, Q. Liu, An electrochemical sensor based on copper-based metal–organic frameworks-graphene composites for determination of dihydroxybenzene isomers in water, Talanta 181 (2018) 80–86. https://doi.org/10.1016/j.talanta.2018.01.002

[120] S.S. Huang, Y.X. Qu, R.N. Li, J. Shen, L.W. Zhu, Biosensor based on horseradish peroxidase modified carbon nanotubes for determination of 2,4-dichlorophenol,Microchim. Acta., 162 (1–2) (2008) 261–268. https://doi.org/10.1007/s00604-007-0872-2

[121] S. Dong, G. Suo, N. Li, Z. Chen, L. Peng, Y. Fu, Q. Yang, T. Huang, A simple strategy to fabricate high sensitive 2,4-dichlorophenol electrochemical sensor based on metal organic framework $Cu_3(BTC)_2$, Sens. Actuators B-Chem. 222 (2016) 972–979. https://doi.org/10.1016/j.snb.2015.09.035

[122] Y. Wang, Y.C. Wu, J. Xie, X.Y. Hu, Metal–organic framework modified carbon paste electrode for lead sensor. Sens. Actuators B-Chem. 177 (2013) 1161–1166. https://doi.org/10.1016/j.snb.2012.12.048

[123] J. Yang, L.T. Yang, H.L. Ye, F.Q. Zhao, B.Z. Zeng, Highly dispersed Au-Pd alloy nanoparticles immobilized on UiO-66-NH_2 metal–organic framework for the detection of nitrite. Electrochim. Acta 219 (2016) 647–654. https://doi.org/10.1016/j.electacta.2016.10.071

[124]. J. Zhou, X. Li, L.L. Yang, S.L. Yan, M.M. Wang et al., The Cu-MOF-199/single-walled carbon nanotubes modified electrode for simultaneous determination of hydroquinone and catechol with extended linear ranges and lower detection limits. Anal. Chim. Acta 899 (2015) 57–65. https://doi.org/10.1016/j.aca.2015.09.054

Metal-Organic Framework Composites - Volume I
Materials Research Foundations **53** (2019) 55-72

Materials Research Forum LLC
https://doi.org/10.21741/9781644900291-3

Chapter 3

Metal-Organic Frameworks for Wastewater Treatment

Luyu Wang[1], Junkuo Gao[1]*

[1]Institute of Fiber based New Energy Materials, The Key Laboratory of Advanced Textile Materials and Manufacturing Technology of Ministry of Education, College of Materials and Textiles, Zhejiang Sci-Tech University, Hangzhou 310018, P. R. China

jkgao@zstu.edu.cn

Abstract

With the increasing demand of human beings for clean water, wastewater treatment technology has received extensive attention and achieved rapid development. Metal-organic frameworks (MOFs) have been introduced into the research of water pollution control, and gradually become one of the effective strategies in the field of wastewater treatment. In this chapter, we summarize the results of a large number of literature in recent years on the adsorption of pollutants in wastewater by MOFs, including decontamination strategies and practical applications.

Keywords

Metal-Organic Framework, Wastewater Treatment, Cations, Anions, Organic Dyes, Drugs, Harmful Organisms

Contents

1.   Introduction

Water is an indispensable substance for human beings and all living things. It is an invaluable natural resource that is irreplaceable for industrial production, agricultural planting, economic development and environmental improvement [1,2]. Freshwater resources that are currently easier to use by humans include river water, fresh lake water and shallow groundwater. The above fresh water storage accounts for only a very small fraction of all fresh water [3]. Fresh water in nature is not static, but is constantly circulating [4]. In the water circulation system, some non-aqueous pollutants are incorporated into the water, called pollutants [5]. When the amount of pollutants entering the water exceeds its self-cleaning ability, the water body will lose its practical value and become wastewater [6].

Wastewater contains inorganic and organic pollutants, such as heavy metals, organic dyes, etc. These pollutants are always found in industrial streams, dyeing process, textiles and many other industries. Owing to the environmental impact and health threats of these pollutants, the methods to remove them from wastewater have intrigued many scientists [7]. Many methods for the removal of pollutants in wastewater have been reported in the existing literature, including adsorption, oxidation, electro-dialysis, ion exchange, electrolysis etc. [8-12]. These methods other than adsorption are also good technologies, but require a lot of money. Considering the removal efficiency and cost comprehensively, adsorption has the unique advantage. Adsorption is an ideal method for wastewater treatment because of its inexpensiveness and easy operation. Besides, adsorption can synchronously remove insoluble and soluble pollutants. The optimal removal capacity by adsorption can be up to 99.9%. Thanks to these advantages, the adsorption method has been used to remove a variety of pollutants from various wastewater sources. Theoretically, adsorption is the step-by-step accumulation of a substance on the interface or surface of the solid material. This process occurs at the interface between the adsorbent and the wastewater during the process of wastewater treatment [13]. The adsorbed pollutant is called as adsorbate, and the solid material is called as adsorbent.

Metal-organic frameworks (MOFs) are a class of three-dimensional functional materials with a porous structure [15]. In recent years, the research on these materials has achieved molecular level assembly, and gradually moved to the height of controllable molecular

Metal-Organic Framework Composites - Volume I   Materials Research Forum LLC
Materials Research Foundations **53** (2019) 55-72   https://doi.org/10.21741/9781644900291-3

design and synthesis. MOFs are attractive because of huge porosity, as well as the facile tunability of pore structure, including pore size and pore shape, by altering the connectivity of the organic linkers and inorganic cations. In addition, the diversity and modifiability of organic linkers offer countless possibilities for MOFs [16]. Both the pore structure and the type of organic linkers have great influence on the adsorption ability of porous material [17]. Based on these characteristics of MOFs, a variety of gases, liquids and solids can be adsorbed. In particular, using the similar molecular sieve type structure, MOFs are always applied to the field of wastewater treatment by adsorption method, and have been taken seriously by many scientific researchers around the world [18].

2.   Treatment of inorganic pollutants in wastewater by MOFs

Harmful inorganic pollutants in various industrial wastewater mainly include heavy metals, fluorides, acid-base salts, etc., and most of them exist in the form of ions in water, such as $Cr_2O_7^{2-}$, Hg^{2+}, Pb^{2+}, etc. Some of these inorganic pollutants are extremely toxic and pose a serious threat to the environment and human health. At present, researchers have carried out some strategies to remove the above inorganic pollutants by using MOFs.

2.1   Cations

Wang and coworkers functionalized $Cu_3(BTC)_2$, a common MOF, by using sulfonic acid groups ($-SO_3H$), and utilized it to adsorb Cd^{2+} ion in wastewater [19]. The functionalized MOF, named as $Cu_3(BTC)_2-SO_3H$, demonstrated a satisfactory Cd^{2+} uptake capacity (88.7 mg/g). As shown in Fig. 1, Cd^{2+} ion was adsorbed by chelation between $-SO_3H$ groups and itself. Besides, the adsorption rate of $Cu_3(BTC)_2-SO_3H$ was extremely fast, which was more than ten times faster than most of the existing Cd^{2+} adsorbents. In addition, in the presence of extra background metal ions, The Cd^{2+} ion was selectively adsorbed. This material can be regenerated and recycled without losing overmuch Cd^{2+} absorption capacity.

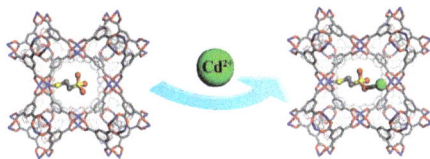

Figure 1. Mechanism for Cd^{2+} adsorption in $-SO_3H$ functionalized $Cu_3(BTC)_2$. Reproduced with permission [19]. Copyright 2015, Royal Society of Chemistry.

Chen et al. calcined the Fe^{3+} modified MOF-5 to obtain a composite, $ZnO/ZnFe_2O_4/C$, as the adsorbent for removing Pb^{2+} ion in wastewater [20]. When the PH value of the solution was 5, the Pb^{2+} adsorbent reached the best adsorption capacity. The maximum adsorption capacity of the MOF-derived ZnO/ZnFe2O4/C composite was 344.83 mg/g. The adsorption of Pb^{2+} on the composite was based on the ion exchange of Pb^{2+} and Zn^{2+} on the surface of ZnO nanocrystals. The composite was derived from a porous MOF-5, so its specific surface was relatively large, which was beneficial to the adsorption of ions. Conveniently, the composite filled with Pb^{2+} can be easily collected from the treated solution by utilizing a magnet.

Queen and coworkers reported a water stable MOF/polydopamine composite, called Fe-BTC/PDA, to remove Hg^{2+} ion in wastewater rapidly [21]. PDA was attached to the surface of porous MOF, and obtained porous structure. The MOF/PDA composite can remove 99.8% of Hg^{2+} from the 1 ppm solution. Besides, when a small amount of Hg^{2+} ions were added to the river water, the MOF/PDA composite can also remove them. Besides, even if the concentrations of Na^+ , K^+ and Ca^{2+} ions in the water were much higher than that of Hg^{2+} ion, the MOF/PDA composite would hardly adsorb these interfering ions, but only adsorb Hg^{2+}. It is indicating that the selectivity of the Hg^{2+} adsorbent is favorable. The MOF/PDA composite was further tested for resistance to fouling after testing in high concentration organic interference and was fully regenerated in multiple cycles.

In addition, some MOF materials and their derivatives can remove many kinds of metal cations in wastewater together. For example, Demir and coworkers modified zirconium metal-organic framework UiO-66 membranes on the surface of alumina hollow fibers to remove Ca^{2+}, Mg^{2+}, and Al^{3+} ions in wastewater [22]. The MOF based membrane can remove these ions efficiently (86.3% for Ca^{2+} ion, 98.0% for Mg^{2+} ion, and 99.3% for Al^{3+} ion). Peng et al used EDTA to substitute the ordered HCHOOH in MOF-88. Then, the substituted material, named as BS-HMT, was used to remove a wide variety of cations, including La^{3+}, Nd^{3+}, Pr^{3+}, Ce^{3+},Eu^{3+}, Hg^{2+}, Pd^{2+}, Pt^{2+}, Pb^{2+}, Sb^{3+}, Cu^{2+}, etc. [23]. These cations can be removed by BS-HMT, and the removal efficiency was all close to 100%. This excellent property of BS-HMT is undoubtedly able to treat the multi-polluted water resources and make them available for secondary use.

2.2   Anions

In industrial treatment process of drinking water, it is difficult to avoid the low concentration of ClO_4^- ion residue. Oliver and coworkers reported a cationic MOF, [Ag-bipy$^+$][NO_3^-](SBN), to trap ClO_4^- in water [24]. The NO_3^- released by SBN can substitute ClO_4^- in water. After this substitution treatment, the concentration of ClO_4^- ion in water decreased from 10 ppb to 6 ppb. Reducing the concentration of ClO_4^- would result in an increase in the concentration of NO_3^-, but it was still below the regulated level and could be used in the irrigation system.

In order to remove TcO_4^-, Xiao et al. synthesized SCU-100, a cationic MOF with open Ag^+ sites [25]. The material can remove TcO_4^- in wastewater with fast uptake kinetics, high capacity, and excellent selectivity (Fig. 2). SCU-100 can rapidly and quantitatively remove TcO_4^- from wastewater within 0.5 h. The removal efficiency can reach 87%. Besides, SCU-100 can capture TcO_4^- in the presence of other competitive anions, such as NO_3^-, SO_4^{2-}, CO_3^{2-}, and PO_4^{3-}.

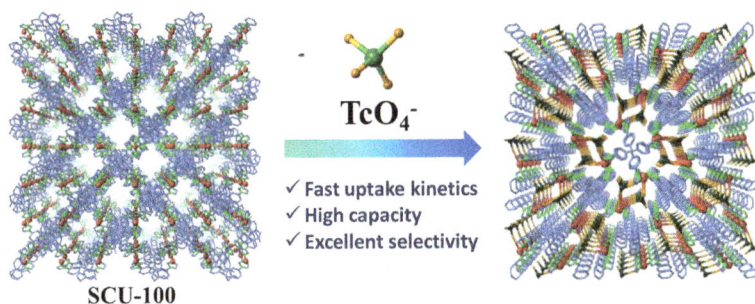

Figure. 2 Uptake of TcO_4^- in SCU-100. Reproduced with permission [25]. Copyright 2017, American Chemical Society.

One layered MOF (FIR-53) had been synthesized for $Cr_2O_7^{2-}$ adsorption. FIR-53 showed largest capacity of 74 mg/g (Fig. 3a), and retained its good adsorption capacity after five cycles (Fig. 3b). As shown in Fig. 3c, the $Cr_2O_7^{2-}$ anions adsorbed by FIR-53 were fixed into the double layers of MOF through weak C-H·O interaction [26].

In order to remove $Cr_2O_7^{2-}$, Hayat and coworkers have also finished some exploration work. They synthesized well-defined core-double-shell structured magnetic polydopamine (MP)@ZIF-8, named as MP@ZIF-8, by using a designed method (Fig. 4a). The material displayed a favorable $Cr_2O_7^{2-}$ uptake capacity of 136.56 mg/g, as shown in Fig. 4b. Fig. 4c shows that Cr (VI) of $Cr_2O_7^{2-}$ can be converted into harmfulless Cr (III), and then immobilized on MP@ZIF-8 [27].

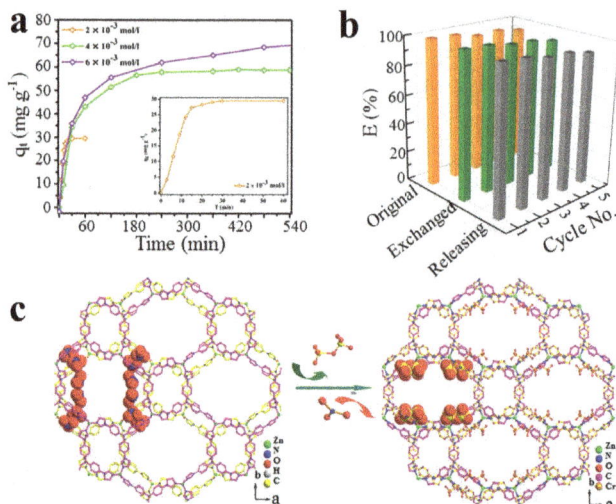

Figure 3. (a) Adsorption kinetics of $Cr_2O_7^{2-}$ at different concentrations for FIR-53. (b) The exchanged ratio (green pillars) and released ratio (gray pillars) of $Cr_2O_7^{2-}$ ions during five reuse cycles. (c) Structures of FIR-53 before and after $Cr_2O_7^{2-}$ adsorption. Reproduced with permission [26]. Copyright 2014, American Chemical Society.

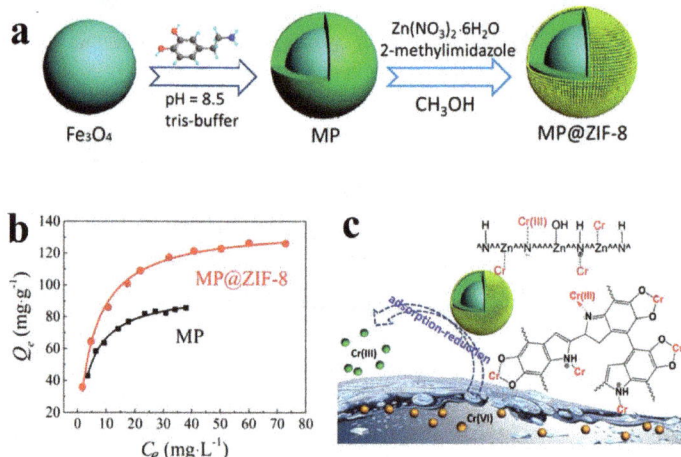

Figure 4. (a) The schematic illustration of the formation procedure of MP@ZIF-8. (b) The adsorption isotherms of $Cr_2O_7^{2-}$ on MP and MP@ZIF-8. (c) $Cr_2O_7^{2-}$ reduction and immobilization by MP@ZIF-8. Reproduced with permission [27]. Copyright 2017, American Chemical Society.

Metal-Organic Framework Composites - Volume I   Materials Research Forum LLC
Materials Research Foundations **53** (2019) 55-72   https://doi.org/10.21741/9781644900291-3

3. Treatment of organic pollutants in wastewater by MOFs

There are many types of organic pollutants in wastewater, including organic dyes, drugs, etc. They may have various types of alcohols, acids, or aromatic compounds, etc. in their molecular structures. Most of these organisms are toxic, difficult to biodegrade in the environment, then migrate with environmental media, ultimately endangering human health. MOF materials and their derivatives can remove organic pollutants from various types of wastewater mainly by utilizing their porous properties, surface charge and open metal sites. These characteristics enhance the interaction between MOFs and organic pollutants for adsorption processing.

3.1 Organic dyes

Zhang and coworkers synthesized MIL-53(Fe)-polyvinylidene fluoride blend ultrafiltration membranes to adsorb methylene blue (MB) in wastewater [28]. The membrane showed an effective treatment for more than 75% MB removal. Guo et al. Synthesized a amine-functionalized MOF, MIL-125 (Ti), to adsorb crystal violet (CV) [29]. The MOF was superhydrophobic, and retained high porosity after soaked in water for one week. CV adsorption capacity of functionalized MOF (129.87 mg/g) was higher than that of initial MOF. The significance of this work is providing the possibility of adsorbing dyes in water for water-sensitive MOF. In order to remove the Congo red (CR), Luo and coworkers synthesized Ni-MOF/graphene oxide (GO) composite by the method of ultrasonic wave-assisted ball milling [30]. The CR adsorption process was an endothermic and spontaneous process. The adsorption capacity reached 2489 mg/g, higher than other reports. Chen et al. studied the adsorptive removal of Rhodamine B (RhB) by utilized a magnetic MOF composite: Fe_3O_4/MIL-100(Fe), which can fast separate RhB from wastewater with stability and repeatability [31]. This magnetic MOF composite can be used more than five times by washing with methanol solution including HCl (0.001 mol/L).

Lin's group prepared ZIF-67 for the removal of Malachite Green (MG) in wastewater and studied the effect of temperature on adsorption capacity [32]. In the temperature range of 20 to 60 °C, the adsorption capacity of ZIF-67 increased with the increase of temperature. Besides, the recyclability of ZIF-67 was also excellent. After treating the ZIF-7 with ethanol to desorb the adsorbed MG, the material maintained good adsorption performance during four adsorption cycles. The possible mechanism of MG adsorption was that the ligand of ZIF-67 and dye all contained benzene ring structure, so π-π stacking interaction was easy to occur.

In order to facilitate the application of synthesizing MOF to treat wastewater, and meet the convenient recovery demand of wastewater treatment chemicals, Park and Oh had proposed a new strategy [33]. As shown in Fig. 5a, they purchased a commercially available carboxymethylated filter paper (CMFP), then synthesized ZIF-8 in situ on the filter paper to remove the methyl orange (MO⁻) in wastewater. Scanning electron microscope (SEM) images (Fig. 5b and Fig. 5c) show that ZIF-8 grows evenly on the surface of the CMFP and has good crystallinity. As shown in Fig. 5d and Fig. 5e, ZIF-8 modified CMFP (CMFP/ZIF-8) could be recycled by a methanol washing process. After recycling it four times, its preeminent MO⁻ filtration rate was well preserved.

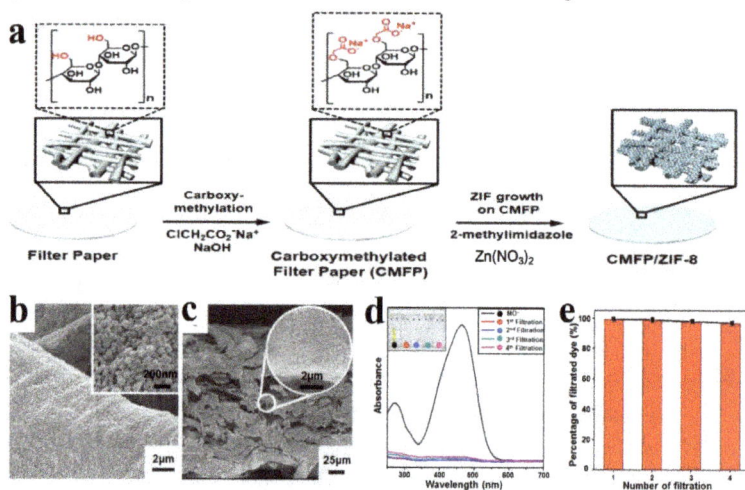

Figure 5. (a) The schematic representation of CMFP/ZIF-8. (b-c) SEM images of CMFP/ZIF-8. (d) UV-Vis spectra showing the effective capture of MO⁻ when the regenerated CMFP/ZIF-8 was used. (e) The percentage of MO⁻ captured in each subsequent filtration (four times) using the regenerated CMFP/ZIF-8. Reproduced with permission [33]. Copyright 2017, Royal Society of Chemistry.

3.2 Drugs

Thermodynamic stability of MOF materials used as wastewater tools is a prerequisite for their recycling. Thermodynamically stable materials do not collapse or decompose. Azhar and coworkers synthesized a thermodynamically stable copper-based MOF, HKUST-1, and studied its adsorptive removal of sulfachloropyridazine (SCP) [34]. SCP adsorption capacity of the MOF was 384 mg/g at 25 °C, which was higher than most of the previously reported materials for adsorbing SCP. Azhar and coworkers believed that π-π stacking interaction, hydrogen bonding interaction, and electrostatic interaction all contributed greatly to excellent adsorption performance.

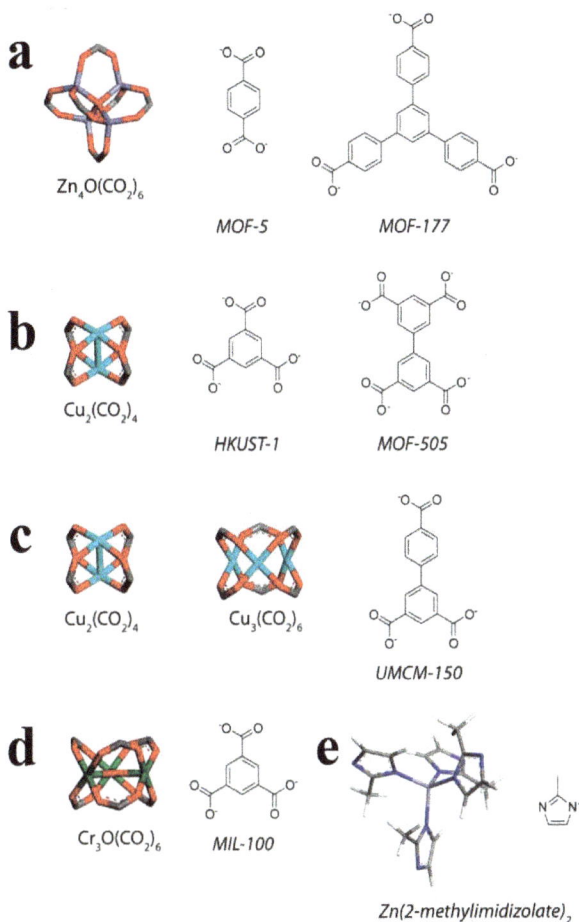

Figure 6. (a) $Zn_4O(CO_2R)_6$ based MOFs and corresponding organic linker for MOF-5 and MOF-177. (b) $Cu_2(CO_2R)_4$ paddlewheel based MOFs and corresponding organic linker for HKUST-1 and MOF-505. (c) UMCM-150 comprised of $Cu_2(CO_2R)_4$ paddlewheel and trinuclear copper metal clusters with corresponding organic linker. (d) MIL-100 comprised of trinuclear chromium metal clusters and corresponding organic linker. (e) Zn(2-methylimidizolate)$_2$ comprised of Zn^{2+} coordinated to four imidizolate nitrogens. Reproduced with permission [35]. Copyright 2010, American Chemical Society.

Metal-Organic Framework Composites - Volume I   Materials Research Forum LLC
Materials Research Foundations **53** (2019) 55-72   https://doi.org/10.21741/9781644900291-3

Although many water-stable MOFs contain heavy metal, the heavy metal is always coordinated with the ligands in the framework material and do not dissolve or decompose into the water. After the adsorption process, the MOF can be separated from the treated wastewater by filtration or other methods, and is harmless to the water environment. As shown in Fig. 6, Matzger et al. evaluated the water stability of seven kinds of MOF materials and found that water stability was related to metal clusters [35]. The order was: $Cr_3O(CO_2)_6$>$Cu_2O(CO_2)_4$>$Zn_4O(CO_2)_6$. Notably, MIL-100 containing $Cr_3O(CO_2)_6$ could be stably stored in pure water for nearly two years after being activated. Therefore, they used MIL-100, which had the best water stability, to adsorb drugs in wastewater. The two drugs were furosemide (7.5 µg/ml) and sulfasalazine (1.4 µg/ml). The adsorption capacity of MIL-100 for furosemide and sulfasalazine was 11.8 mg/g and 6.2 mg/g, respectively. Under the same conditions, neither of these two drugs can be adsorbed by Na (Y) zeolite.

Setoodeh et al. utilized MOF-5 to remove tetracycline antibiotic in water [36]. The maximum adsorption capacity was about 233 mg/g. In addition, when the pH of the solution was between 3 and 11, the removal efficiency can reach 80% or more. Especially, when the value of pH was 6, the removal efficiency can reach a maximum value of 91%. It is indicating that MOF-5 has the strongest adsorption capacity for tetracycline antibiotic in a weakly acidic environment.

To remove nitroimidazole antibiotics in wastewater efficiently, Jhung and coworkers synthesized urea grafted MIL-101 (urea-MIL-101) [37]. The N atoms of urea groups on the surface of urea-MIL-101 can adsorb nitroimidazole antibiotics by weak hydrogen bond adsorption. Dimetridazole was one of typical nitroimidazole antibiotics. The dimetridazole adsorption ability of urea-MIL-101 was better than MIL-101, indicating the significant effect of grafted urea. Moreover, the adsorption capacity of these two MOF materials all decreased with the increase of PH.

3.3 Other harmful organisms

In addition to organic dyes and drugs mentioned above, there are additional kinds of harmful organisms present in water. Bisphenol A (BPA), an endocrine-disrupting compound, is toxic to aquatic organisms and humans even at 1 mg/m^3 concentration. Bhadra and coworkers synthesized a novel bio-MOF-1 derived porous carbon to remove BPA in wastewater. This MOF was $Zn_8(adenine)_4(biphenyldicarboxylate)_6O$ [38]. The obtained carbon was a highly porous material. Experimental results indicated that the carbonization time had a profound influence on the BPA adsorption performance of the material. When the carbonization time was 12 h, the specific surface area of the obtained material was the largest, and the adsorption capacity of the material to BPA can reach more than 500 mg/g.

Kim et al. designed a microporous organic network (MON) to remove toluene in water by using UIO-66-NH$_2$ [39]. As shown in Fig. 7a, UIO-66-NH$_2$ was coated by the MON through Sonogashira coupling of tetra (4-ethynylphenyl) methane with 4,4'-diiodobiphenyl or 1,4-diiodobenzene. Then, the MOF@MON composite was immersed in HF solution, and the interior UIO-66-NH$_2$ can be disassembled. Finally, hollow MON can be obtained (Fig. 7b). Compared with UIO-66-NH$_2$ and UIO-66-NH$_2$ coated with MON, the adsorption ability of hollow MON for toluene was improved significantly.

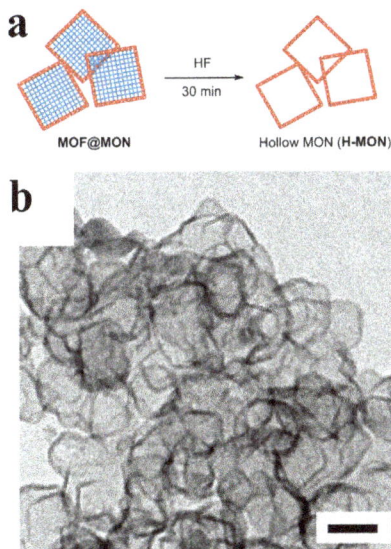

Figure 7. (a) The Synthetic scheme for hollow MON via MOF etching from MOF@MON. (b) The transmission electron microscope (TEM) image of hollow MON. Scale bars in panel = 200 nm. Reproduced with permission [39]. Copyright 2014, American Chemical Society.

P-nitrophenol (PNP) is flammable, toxic and allergic to skin. In order to degrade PNP in water, Gao and coworkers loaded ZIF-9 and ZIF-12 on cellulose aerogels to activate peroxymonosulfate (PMS) to produce sulfate radicals (SO$_4$·$^-$), which played the key role in PNP degradation [40]. Two aerogels combined with activated PMS could remove more than 80% of PNP in 60 min. The green aerogel composites can be easily separated from the treated solution and possessed good recyclability.

Yan et al. reported the adsorptive removal of 1-naphthol from water with ZIF-67 [41]. The ZIF-67 is hexagonal. ZIF-67 has the strongest adsorption capacity for 1-naphthol when the temperature is 313 K. Besides, when the pH was between 9 and 11, the adsorption capacity of ZIF-67 to 1-naphthol was advantageous. After three cycles, ZIF-67 adsorbed 1-naphthol well (more than 200mg/g), indicating its good recyclability.

Illegal cultivation of crops is inseparable from artificial sweeteners (ASWs), and some ASWs can pollute fresh water. By using urea as the modifier, Seo and coworkers synthesized modified MIL-101 to remove ASWs in wastewater [42]. The modifying process is shown in Fig. 8a. As shown in Fig. 8b, the adsorptive removal abilities of three ASWs, including saccharin, cyclamate, and acesulfame, were studied using the above MOFs. The adsorbed quantities under various conditions decreased in the order urea-MIL-101>MIL-101. Especially for saccharin, urea-MIL-101 had the highest adsorption capacity (about 80 mg/g). The adsorption mechanism of saccharin, cyclamate, and acesulfame over urea-MIL-101 should be attributed to the H-bonding interaction between -NH$_2$ groups on the superficial MOF and ASWs.

a   **b**

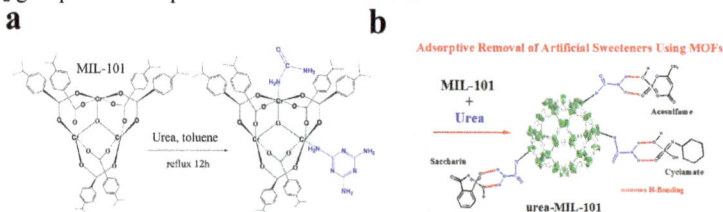

Figure 8. (a) The grafting methods to introduce urea on MIL-101. (b) Adsorptive removal of artificial sweeteners by using urea-MIL-101. Reproduced with permission [42]. Copyright 2016, American Chemical Society.

Zhang and coworkers reported an ice-templating process to fabricate chitosan/UIO-66 monolith [43]. As shown in Fig. 9a, three freeze-dried chitosan/UIO-66 monoliths can be obtained after washing with NaOH (chitosan-W), solvent exchange (chitosan/UIO-66-1-W), and drying (chitosan/UIO-66-2-W). The purpose of designing this material was to easily remove the methylchlorophenoxypropionic acid (MCPP) from the wastewater. As shown in Fig. 9b, among all these four materials used for comparison, the adsorbed quantities of chitosan/UIO-66-2-W and UIO-66 were the largest. However, UIO-66 is powdered, which cannot be easily removed from the treated wastewater. Therefore, chitosan/UIO-66-2-W is the most ideal MCPP adsorption material.

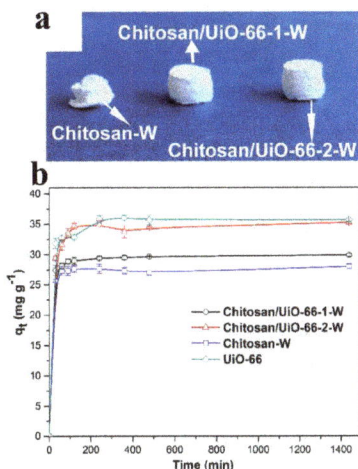

Figure 9. (a) The photo shows the three freeze-dried monoliths after washing with NaOH (chitosan-W), solvent exchange (chitosan/UIO-66-1-W), and drying (chitosan/UIO-66-2-W). (b) Profiles of the adsorbed quantity of MCPP versus soaking time by immersing the composites (10 mg) or dispersing UiO-66 nanoparticles in 10 ml of aqueous solution of MCPP (60 ppm). Reproduced with permission [43]. Copyright 2017, American Chemical Society.

Conclusions and perspectives

Scientists have made a lot of important progresses and breakthrough achievements in the research of MOFs in wastewater treatment. In this chapter, the authors briefly introduce the research progresses of the removal of harmful cations, anions, organic dyes, drugs and harmful organisms in wastewater by using MOFs. In addition, the authors also briefly summarize the development of some synthesis design research of water-stable MOFs, and the preliminary exploratory work of some practical wastewater treatment applications of MOFs, such as MOFs based aerogels. These works can accelerate the speed of MOFs going out of the laboratory and entering the industrial field.

However, it should also be recognized that many challenges remain in using MOFs for wastewater treatment. One challenge is the water stability of MOFs. Besides, the PH value of wastewater is also very important. It is significant to adjust appropriate PH to optimize the wastewater treatment performance of MOFs. The adsorption capability of MOFs also needs to be improved to completely remove specific harmful pollutants in wastewater. In theory, these challenges can be overcome by modifying the organic linkers and enlarging the specific surface area of MOFs. In addition, the effects of morphology, microstructure and defects on the adsorption properties of MOFs still need to be explored. Overall, MOFs are promising to overcome these challenges in future because of their unlimited possibilities for structural control.

References

[1]   T. Chidambaram, Y. Oren, M. Noel, Fouling of nanofiltration membranes by dyes during brine recovery from textile dye bath wastewater, CHEM. ENG. J. 262 (2015) 156-168. https://doi.org/10.1016/j.cej.2014.09.062

[2]   T. Oki, S. Kanae, Global hydrological cycles and world water resources, Science 313 (2006) 1068-1072. https://doi.org/10.1126/science.1128845

[3]   J. S. Baron, N. L. Poff, P. L. Angermeier, Meeting ecological and societal needs for freshwater, ECOL. APPL. 12 (2002) 1247-1260.

[4]   T. Chidambaram, Y. Oren, M. Noel, Fouling of nanofiltration membranes by dyes during brine recovery from textile dye bath wastewater, CHEM. ENG. J. 262 (2015) 156-168. https://doi.org/10.1016/j.cej.2014.09.062

[5]   B. Vrana, I.J. Allan, R. Greenwood, G.A. Mills, E. Dominiak, K. Svensson, et al., Passive sampling techniques for monitoring pollutants in water, TRAC-TREND. ANAL. CHEM. 24 (2005) 845-868. https://doi.org/10.1016/j.trac.2005.06.006

[6]   A. Sonune, R. Ghate, Developments in wastewater treatment methods, Desalination 167 (2004) 55-63. https://doi.org/10.1016/j.desal.2004.06.113

[7]   V. K. Gupta, I. Ali, T. A. Saleh, A. Nayak, S. Agarwal, Chemical Treatment Technologies for Waste-Water Recycling-An Overview, RSC ADV. 2 (2012) 6380-6388. https://doi.org/10.1039/c2ra20340e

[8]   S. Gisi, G. Lofrano, M. Grassi, M. Notarnicola, Characteristics and adsorption capacities of low-cost sorbents for wastewater treatment: A review, Sustain. Energy Technol. 9 (2016) 10-40. https://doi.org/10.1016/j.susmat.2016.06.002

[9]   M. A. Oturan, J. J. Aaron, Advanced Oxidation Processes in Water/Wastewater Treatment: Principles and Applications. A Review, CRIT. REV. ENV. SCI. TEC. 44 (2014) 2577-2641.

[10]   M. Sadrzadeh, A. Ghadimi, T. Mohammadi, Coupling a mathematical and a fuzzy logic-based model for prediction of zinc ions separation from wastewater using electrodialysis, CHEM. ENG. J. 151 (2009) 262-274. https://doi.org/10.1016/j.cej.2009.03.003

[11] S. Rengaraj, K. H. Yeon, S. H. Moon, Removal of chromium from water and wastewater by ion exchange resins, J. HAZARD. MATER. 87 (2015) 273-287. https://doi.org/10.1016/s0304-3894(01)00291-6

[12] H. Cheng, W. Xu, J. Liu, H. Wang, Y. He, C. Gang, Pretreatment of wastewater from triazine manufacturing by coagulation, electrolysis, and internal microelectrolysis, J. HAZARD. MATER. 146 (2007) 385-392. https://doi.org/10.1016/j.jhazmat.2006.12.038

[13] I. Ali, M. Asim, T.A. Khan, Low cost adsorbents for the removal of organic pollutants from wastewater, J. ENVIRON. MANAGE. 113 (2012) 170-183. https://doi.org/10.1016/j.jenvman.2012.08.028

[14] I. Ali, V. K. Gupta, Advances in water treatment by adsorption technology, NAT. PROTOC. 1 (2006) 2661-2667. https://doi.org/10.1038/nprot.2006.370

[15] Y. Cui, B. Li, H. He, W. Zhou, B. Chen, G. Qian, Metal-Organic Frameworks as Platforms for Functional Materials, Acc. Chem. Res. 49 (2016) 483-493. https://doi.org/10.1021/acs.accounts.5b00530

[16] H. Huang, J.R. Li, K. Wang, T. Han, M. Tong, L. Li, et al., Anin situself-assembly template strategy for the preparation of hierarchical-pore metal-organic frameworks, NAT. COMMUN. 6 (2015) 8847. https://doi.org/10.1038/ncomms9847

[17] H. Li, Q. Zheng, J. Luo, Z. Cheng, J. Xu, Impacts of meso-structure and organic loadings of fluoroalcohol derivatives/SBA-15 hybrids on nerve agent simulant sensing, SENSOR. ACTUAT. B-CHEM. 187 (2013) 604-610. https://doi.org/10.1016/j.snb.2013.04.122

[18] P. A. Kobielska, A. J. Howarth, O. K. Farha, S. Nayak, Metal-organic frameworks for heavy metal removal from water, COORDIN. CHEM. REV. 358 (2018) 92-107. https://doi.org/10.1016/j.ccr.2017.12.010

[19] Y. Wang, G. Ye, H. Chen, X. Hu, Z. Niu, S. Ma, Functionalized Metal-Organic Framework as a New Platform for Efficient and Selective Removal of Cadmium (II) from Aqueous Solution, J. MATER. CHEM. A 3 (2015) 15292-15298. https://doi.org/10.1039/c5ta03201f

[20] D. Chen, W. Shen, S. Wu, C. Chen, X. Luo, L. Guo, Ion exchange induced removal of Pb(II) by MOF-derived magnetic inorganic sorbents, Nanoscale 8 (2016) 7172-7179. https://doi.org/10.1039/c6nr00695g

[21] D. T. Sun, L. Peng, W. S. Reeder, S. M. Moosavi, D. Tiana, D. K. Britt, et al., Rapid, Selective Heavy Metal Removal from Water by a Metal-Organic

Framework/Polydopamine Composite, ACS CENTRAL. SCI. 4 (2018) 349-356. https://doi.org/10.1021/acscentsci.7b00605

[22] X. L. Liu, N. K. Demir, et al., Highly Water-Stable Zirconium Metal-Organic Framework UiO-66 Membranes Supported on Alumina Hollow Fibers for Desalination, J. AM. CHEM. SOC. 137 (2015) 6999-7002. https://doi.org/10.1021/jacs.5b02276

[23] Y. Peng, H. Huang, Y. Zhang, C. Kang, S. Chen, L. Song, et al., A versatile MOF-based trap for heavy metal ion capture and dispersion, NAT. COMMUN. 9 (2018) 187. https://doi.org/10.1038/s41467-017-02600-2

[24] I. R. Colinas, R. C. Silva, S. R. Oliver, Reversible, selective trapping of perchlorate from water in record capacity by a cationic metal-organic framework, ENVIRON. SCI. TECHNOL. 50 (2016) 1949-1954. https://doi.org/10.1021/acs.est.5b03455

[25] D. Sheng, L. Zhu, C. Xu, C. Xiao, Y. Wang, Y. Wang, et al., Efficient and Selective Uptake of TcO4(-) by a Cationic Metal-Organic Framework Material with Open Ag(+) Sites, ENVIRON. SCI. TECHNOL. 51 (2017) 3471-3479. https://doi.org/10.1021/acs.est.7b00339

[26] H. R. Fu, Z. X. Xu, J. Zhang, Water-Stable Metal–Organic Frameworks for Fast and High Dichromate Trapping via Single-Crystal-to-Single-Crystal Ion Exchange, CHEM. MATER. 27 (2014) 205-210. https://doi.org/10.1021/cm503767r

[27] K. Zhu, C. Chen, H. Xu, Y. Gao, X. Tan, A. Alsaedi, et al., Cr (VI) Reduction and Immobilization by Core-Double-Shell Structured Magnetic Polydopamine@Zeolitic Idazolate Frameworks-8 Microspheres, ACS SUSTAIN. CHEM. ENG. 5 (2017) 6795-6802. https://doi.org/10.1021/acssuschemeng.7b01036

[28] R. Yi, T. Li, W. M. Zhang, et al., MIL-PVDF blend ultrafiltration membranes with ultrahigh MOF loading for simultaneous adsorption and catalytic oxidation of methylene blue, J. HAZARD. MATER. 365 (2019) 312-321. https://doi.org/10.1016/j.jhazmat.2018.11.013

[29] G. Wen, Facile modification of NH2-MIL-125(Ti) to enhance water stability for efficient adsorptive removal of crystal violet from aqueous solution, COLLOID. SURFACE. A 541 (2018) 58-67. https://doi.org/10.1016/j.colsurfa.2018.01.011

[30] S. Zhao, D. Chen, F. Wei, N. Chen, Z. Liang, Y. Luo, Removal of Congo red dye from aqueous solution with nickel-based metal-organic framework/graphene oxide

composites prepared by ultrasonic wave-assisted ball milling, ULTRASON. SONOCHEM. 39 (2017) 845-852. https://doi.org/10.1016/j.ultsonch.2017.06.013

[31] H. Liu, X. Ren, L. Chen, Synthesis and characterization of magnetic metal–organic framework for the adsorptive removal of Rhodamine B from aqueous solution, J. IND. ENG. CHEM. 34 (2016) 278-85.

[32] K. Y. A. Lin, H. A. Chang, Ultra-high adsorption capacity of zeolitic imidazole framework-67 (ZIF-67) for removal of malachite green from water, Chemosphere 139 (2015) 624-631. https://doi.org/10.1016/j.chemosphere.2015.01.041

[33] J. Park, M. Oh, Construction of flexible metal-organic framework (MOF) papers through MOF growth on filter paper and their selective dye capture, Nanoscale 9 (2017) 12850-12854. https://doi.org/10.1039/c7nr04113f

[34] M. Azhar, H. Abid, H. Sun, V. Periasamy, M. Tadé, S. Wang, Excellent performance of copper based metal organic framework in adsorptive removal of toxic sulfonamide antibiotics from wastewater, J. COLLOID INTERF. SCI. 478 (2016) 344-352. https://doi.org/10.1016/j.jcis.2016.06.032

[35] K. Cychosz, A. J. Matzger, Water Stability of Microporous Coordination Polymers and the Adsorption of Pharmaceuticals from Water, Langmuir 26 (2010) 17198-17202. https://doi.org/10.1021/la103234u

[36] M. Malakootian, S. Bahraini, et al., Removal of Tetracycline Antibiotic From Aqueous Solutions Using Modified Pumice With Magnesium Chloride, Adv. Environ. Biol. 10 (2016) 46-56. https://doi.org/10.17795/jjhr-37583

[37] P. W. Seo, N. A. Khan, S. H. Jhung, Removal of nitroimidazole antibiotics from water by adsorption over metal-organic frameworks modified with urea or melamine, CHEM. ENG. J. 315 (2017) 92-100. https://doi.org/10.1016/j.cej.2017.01.021

[38] B. N. Bhadra, C. W. Cho, et al., Remarkably efficient adsorbent for the removal of bisphenol A from water: Bio-MOF-1-derived porous carbon, CHEM. ENG. J. 343 (2018) 225-234. https://doi.org/10.1016/j.cej.2018.03.004

[39] C. Jiseul, K. Sungah, P. Nojin, et al., Metal-organic framework@microporous organic network: hydrophobic adsorbents with a crystalline inner porosity, J. AM. CHEM. SOC. 136 (2014) 6786-6789. https://doi.org/10.1021/ja500362w

[40] W. Ren, J. Gao, C. Lei. Y. Xie, Y. Cai, Q. Ni, J. Yao, Recyclable metal-organic framework/cellulose aerogels for activating peroxymonosulfate to degrade organic pollutants, CHEM. ENG. J. 349 (2018) 766-774. https://doi.org/10.1016/j.cej.2018.05.143

[41]   X. Yan, X. Hu, T. Chen, S. Zhang, M. Zhou, Adsorptive removal of 1-naphthol from water with Zeolitic imidazolate framework-67, J. PHYS. CHEM. SOLIDS 107 (2017) 50-54. https://doi.org/10.1016/j.jpcs.2017.03.024

[42]   P. W. Seo, N. A. Khan, Z. Hasan, S. H. Jhung, Adsorptive Removal of Artificial Sweeteners from Water Using Metal-Organic Frameworks Functionalized with Urea or Melamine, ACS APPL. MATER. INTER. 8 (2016) 29799-29807. https://doi.org/10.1021/acsami.6b11115

[43]   Q. Fu, L. Wen, L. Zhang, X. Chen, D. Pun, A. Ahmed, et al., Preparation of Ice-templated MOF-polymer composite monoliths and their application for wastewater treatment with high capacity and easy recycling, ACS APPL. MATER. INTER. 9 (2017) 33979-33988. https://doi.org/10.1021/acsami.7b10872

Metal-Organic Framework Composites - Volume I
Materials Research Foundations 53 (2019) 73-104

Materials Research Forum LLC
https://doi.org/10.21741/9781644900291-4

Chapter 4

Metal-Organic-Framework (MOFs) and Environmental Application

Hasan Ay[1], Sümeyye Karakuş[1], Hakan Burhan[1], Anish Khan[2,3], Fatih Sen[1]*

[1]Sen Research Group, Biochemistry Department, Faculty of Arts and Science, Dumlupınar University, Evliya Çelebi Campus, 43100 Kütahya, Turkey.

[2]Chemistry Department, Faculty of Science, King Abdulaziz University, Jeddah-21589, P.O. Box 80203, Saudi Arabia

[3]Center of Excellence for Advanced Materials Research, King Abdulaziz University, Jeddah 21589, P.O. Box 80203, Saudi Arabia

* fatih.sen@dpu.edu.tr

Abstract

Toxic anions and heavy metals are one of the most important issues of concern in the world because of their exposure to ionic contamination, which is very common, as well as their potential for the environment and human health. Until now, the applications of metal organic frameworks (MOFs) have been studied extensively in many different fields such as filtering, purification, detection and storage. However, for some contaminations (eg organic pollutants, heavy metals and hazardous chemicals), their use in the field of wastewater treatment (WWT) has not yet been fully assessed. In this chapter, we discussed the development and comparison of traditional materials based on metal organic frameworks based wastewater treatment techniques. The process of pollutants in water and the performance and details of MOFs in different treatment areas have been emphasized.

Keywords

Activated Carbon, Contaminant, Environment, Ion Exchange, MOF, Wastewater

Contents

Materials Research Forum LLC
https://doi.org/10.21741/9781644900291-4

1.   Introduction

1.1   Outline

Hazardous organic molecules pose enormous risks to nature and living organisms, and are one of the key problems of the world [1–9].

The rapid increase in the world population and therefore the development and growth of residential areas has exceeded the critical level of environmental pollution and deteriorates every day [10]. As the population increase in many areas as well as the increase in the number of such populations, the establishment of new facilities and factories has resulted in even more dangerous outputs of toxic products [11]. A number of studies have been carried out to reduce these outputs from industry, thus, removing the existing contamination from the environment [12].

Very high amounts of water contaminated by industrial process water and releasing of these contaminated water, environmental cleaning and mixing into the domestic water; large amounts of water will cause contamination. [10–11–13].

In nature, compounds that are generally harmful to health; (H2S, NOy, COy, SOy and derivatives), organic chemicals in suspension, chemicals containing N and S, paints, oil spills and PPCPs pollutes water sources [14–19]. For this reason, it is necessary to make assesments for the existing and still rapidly continuing pollution in our planet. According to these problems, the ideal cleaning processes should be developed.

Hazardous compounds which are present in our nature come from two different sources according to their formation. The first one is the nature's own processes, and the other one is the human induced pollutants. Large amounts of pollutants in the air, soil and water are already present in nature. Besides, contamination from human being such as incineration, chemical reactions causes big enviromental problems. The world's energy needs are mostly supplied from crude oil, natural gas and coal. However, these substances leave many toxic molecules because of the burning process. In general terms, those are the most risky components for air pollution (H_2S, NO_Y, COy, SOy, NH_3 and its derivatives) and other hydrocarbons [20–21].

Until now, various methods have been used to eliminate organic and toxic contamination such as oxidation, chemical treatment, adsorption, combustion and biological oxidation [22–23]. In this context, one of the most effective and simple methods is adsorption [19–24].

When evaluated by other methods, adsorption; it is more attractive than other methods because of its low amount of secondary hazardous materials, economic, practical use and design [25]. Due to the discovery of variable porous materials and accelerated development of adsorption materials, adsorption has been the method of treatment in the treatment sector. Porous products such as zeolite [25] derivatives, MOFs (Metal Organics Frameworks) [26–28] and activated carbon [29] were analyzed for adsorption or filtration of different liquids and gas molecules. For ideal adsorption, it is necessary to select the appropriate material to hold porosity, porous structure and distribution and pollutants [25].

MOFs are known to consist mainly of two components. One of them is a group of metal ions or ions and another is an organic component that acts as a binder (Fig. 1).

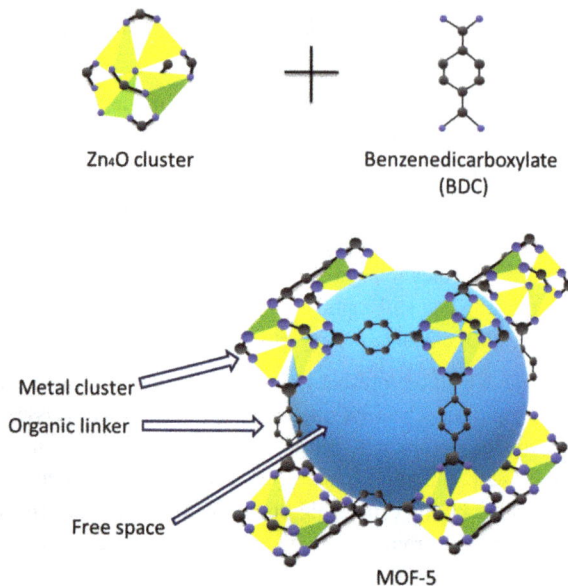

Zn₄O cluster   Benzenedicarboxylate
(BDC)

Metal cluster
Organic linker
Free space
MOF-5

Fig 1. A Classic Metal-Organic Framework (MOF / 5)

MOFs attract attention because of their different application areas. Metal-organic frameworks have, better porosity, porousity size and distribution, and their composition is better than different porous materials. In this context, a wide range of studies have been done to filter and absorb the components of different gases and liquids with Metal-organic frameworks.

Many different porous adsorbents such as meso-porous materials [30–33], activated carbons [34–38], MOFs [26–39–43], zeolites [44–46] have been studied for the treatment of hazardous components (Fig. 2).

Effectively, the process of adsorption is important for the areas of purification, porosity, pore geometry [41–47] and specific adsorption [40–48–49]. Naturally, porous structures are known as absorbers which play a crucial role in the filtration and separation processes. Therefore, a great interest has been given to the research and development of modified porous structures in recent years [7–50–52]. The emergence of MOFs is based on centuries ago and is of remarkable crystal structure with high porosity [12–53–55].

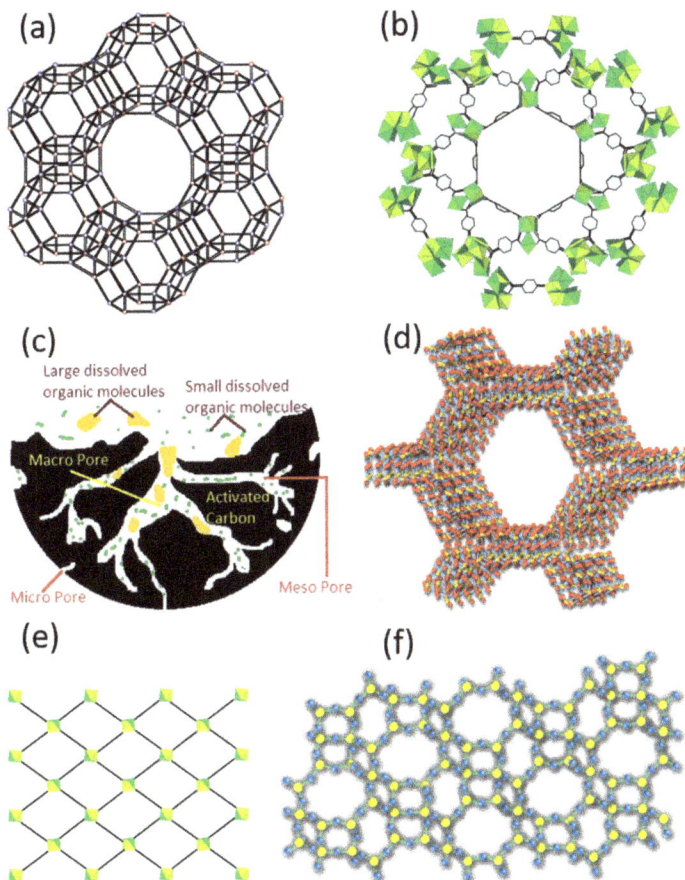

Fig. 2. *Generally Used Porous Absorbers: (a) HY Zeolite, (b) MIL/101 (Cr), (c) Activated Carbon, (d) MCM/41, (e) MIL/47-53 and (f) Silicalite (MFI Framework)*

1.2 Background on MOFs

The formation of MOFs is composed of metallic cations and groups with coordination bonds and these metallic cations are bound by polyotopic organics. The fact that organic and metal ions are bound and have very different structures may cause the derivatives of

MOFs to be unlimited [45]. Particularly, in the production of MOFs without any limitation, Ca^{+2}, Zn^{+2}, Co^{+2}, Zr^{+4}, Ln^{+3}, Cu^{+2}, Al^{+3}, Cd^{+2} metals are found, and these are crystal structure can host all variants [46]. While a number of metal groups useful in the synthesis of MOFs have toxicity (Co^{+2}, Cd^{+2}), they are not economical in a few groups (Eu^{+3}, Ag^{+1}), but are useful for the development of the main structures of MOFs [56–57]. Due to the natural nature of the metal complexes, the formation of bonds between metal ions and organic binders may be reversed, thereby allowing rearrangement of the bond structures in the polymerization process and forming a number of regular MOFs [58–59]. Some organic ligands (Amines, Sulfonates, Carboxylates and Phosphates) play a critical role in the construction of MOFs [60–64]. There are two main methods of synthesizing MOFs. We can directly determine the use of (1) an aqueous or non-aqueous method and the use of (2) these two methods in a certain proportion. Solvothermal, microwave, ultraconic, hydroelectric, layered growth, electrochemical processes are common and efficient methods of MOF synthesis [58–65–69]. MOFs are also known to be involved in catalysis, luminescence detection, gas storage and drug delivery [58–65–66–70–74]. There are also superior aspects that make MOFs attractive for water treatment processes, such as porosity, ion exchange capacity and adsorption. For this reason, in this chapter, the causes of waste water pollution, wastewater treatment methods and the use of MOFs in wastewater will be explained.

2.   Problems associated with wastewater treatment

Water contamination is a social problem as it poses a serious threat to human health. Water contamination has a factor related to the supply-demand relation to the source of water [75–76]. Many environmental pollutants such as pharmaceutical outputs, solid household waste, heavy metals and so on contribute to water contamination [75–77–78]. These contaminations in the water should be treated in an ideal way so as not to cause health problems of living beings and the increasing and permanent problems for the ecosystem [79]. Under this subject, we will examine the contamination of the water, its danger and factors contributing to flour under short headings.

2.1   Diverse categories of contaminants

2.1.1   Microbe-derived contaminants

The lack of proper cleaning and pathogenic organisms caused by dirty pipes are important factors in the pollution of waste water. The most effective pathogens in water pollution are *Clostridium, Rotaviruses, Cryptosporidium parvum, Bacteroides, Entamoeba histolytica, Vibrio cholera* and *C. Jejuni* [80–81]

2.1.2 Common water pollution index

There are a lot of sources that pollute water, such as particulate debris, sewage ponds, household waste refuse, paper waste and ashes [82–83]. The sources of environmental water pollution include aspirin-derived drugs, drug clofibrate, chlorophenoxyisobutyrate and drug metabolites. [84].

In the areas near agricultural industrial establishments, organic contaminants, various inorganic, various organic and their mixtures, polyaromatic structures and pesticides are contaminated with very intense (metallic ions (Cr, Pb, Ni and Cd)) [85–86]. The plants grown with water contaminated with heavy metals and their consumption have a high risk for human health in the long term and cause chronic problems to the eco system.

2.1.3 Emerging pollutants

Wastewater contaminations are classified in the literature as hormones, organic pollutants, hazardous chemicals, surfactants, drugs and residues, steroids, electronic wastes [69–82–87–88]. From these potential waste groups, electronic contamination is an example of the emerging wastes group [69–82–87–89]. The technological advancement of MOFs has created a new area not only for today but also for the cleaning and / or filtering of different contamination in the future. New components, like MOFs, are expected to allow wastewater treatment applications to be more advantageous in the case of contamination in water and contamination in different environments.

2.1.4 Toxicity and symptoms of WWT (Waste Water Treatment) poisoning

The presence of heavy metal ions and contamination in water increases the risk of intoxication that provides continuity within the human body. The source of this risk was caused by heavy metal ions (Fe, Pb, Cu, Cd, Cr, Zn, and Ag) that passed through water and foodstuffs. This heavy metal ion contamination in water causes biochemical problems in living things and causes difficulties in their development and growth. Recent studies have shown that there are deficiencies in brain function, synapses, and some metabolic systems.

These contaminations show a continuous rate of increase in wastewater, and they are (1) diclofenac, hydroxy- and carboxy-ibuprofen, (2) 2- (4-chloropeoxy) clofibric acid) -2-methyl propionic acid, (3) carbamazepine, (4) many components such as carboxy-hydrotrophic acid (ibuprofen) (Fig. 3) [90].

Materials Research Forum LLC

https://doi.org/10.21741/9781644900291-4

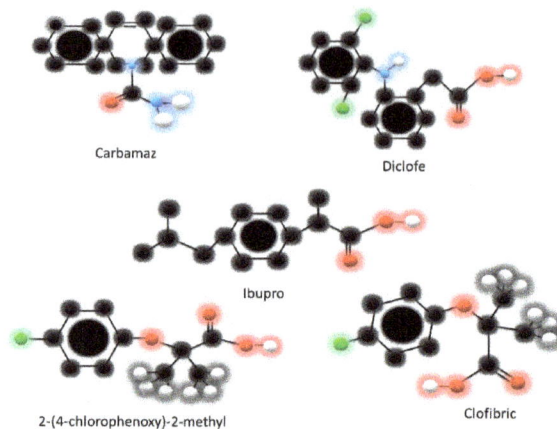

Carbamaz

Diclofe

Ibupro

2-(4-chlorophenoxy)-2-methyl

Clofibric

Fig. 3. Generally, the drug and chemical components encountered in wastewater [90].

Most studies, pathological conditions in the cortex and phospholipid metabolism were found in cases and disruptions. For example, in alzhemier disease and similar cases plasminethanolamine and phosphatidylserine limit values were measured [75–89–92]. These heavy metal ions are transmitted through water, air and food to living things and to the human body, and may increase in bio-systems and in the eco-system for a long time. In the mining sector, these heavy metal ions, with leachate from ground water; water from lakes to rivers and from there to the sea soluble in systems. In addition, non-degrading drugs are another contamination with biologically toxic risks and have the potential to accumulate in the water system continuously [90]. This type of chemical and/or drugs consumes contaminated water; can cause bone injury and liver diseases and similar diseases [93]. For all this, it is very urgent to improve the treatment processes in order to eliminate risky contaminants in wastewater and to reduce the risk of healthy living in the world.

3.   Treatment methods in WWT

As a standardized process in the world, waste water treatment processes are formed from four stages. Figure 3 shows the treatment process stages commonly used in the world.

Generally, wastewater treatment processes are composed of four main consecutive stages namely the preliminary, primary, secondary, and tertiary stages. Fig. 4 is a typical/overall layout for describing WWT that has been commonly adopted by the EPA and UNEP.

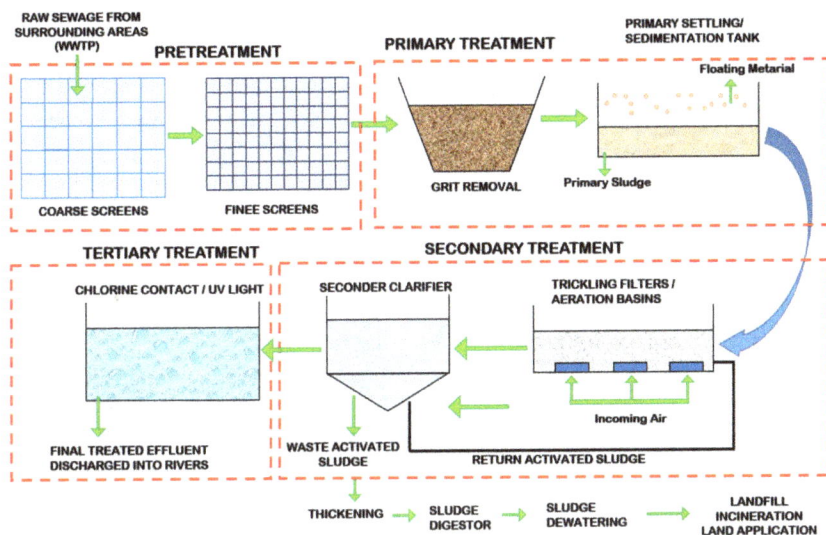

Fig. 4. *Overview of treatment stages in a wastewater treatment plant (EPA 1997 and UNEP 2012) [92].*

The main procedures for the treatment of contaminants in water are summarized below [92]. In this section, according to the latest studies and reports, Waste Water Treatment methods will be used in which traditional materials are used with their boundaries. The following techniques significantly increase the capacity of the treatment processes and the advantageous properties of the materials.

3.1 Activated sludge treatment (AST)

The first method of treatment of biological waste water is activated sludge treatment [94]. According to AST standards, treatment of drugs and chemical materials similar to acetaminophen, bezafibrate, ibuprofen and naproxen worldwide [95–96].

Oxidation mechanism and the formation of flocculation, activated sludge treatment of active materials is often provided with adsorption and bio-chemical cycles [97]. In addition, the mechanisms that operate the raw materials are one of the leading methods for removing metal ions, although different methods are also available.

Ibuprofen, aspirin, crotamitone, endocrine disrupters, triclosan and nonylphenol have been used to remove and purify the pharmaceutical components contained in water in the process. With waste water treatment, some pharmaceutical compounds have not been fully effective in refining. To improve this, nitrate removal and/or nitrification processes can be an effective way. Due to the very high porosity of the materials used, they are more efficient than known methods and cause very good absorption of organic components such as carbohydrates [98]. The modified cation exchange amount thus makes this type of harmful waste a very important method for absorbent treatment.

3.2 Advanced wastewater treatment

First times, advanced techniques such as biological and non-biological (membrane bio-reactors/reverse osmosis and ultrafilitations) have been tried to remove small amounts of contaminated components. MBRs (Membrane Bioreactors) are one of the most effective methods for water treatment in microbiological terms. During waste water treatment, MBRs are used in chemicals such as TiO_2, UV radiation and hydrogen peroxide [99]. Compared with reverse osmosis nanfiltration, the nano-filtration steps provided good performance in the purification of hormones and pharmaceuticals and the like. Active carbon is a suitable vehicle for filtering the drugs and similar wastes on the surface of water. In addition to that, active carbon is not used alone but if it is applied together with the ozonation method, another treatment technique which is very effective is obtained [100]. In light of all this, in the future it is expected to reach a wide range of technologies combining MBR, MOF and traditional WWTs.

3.3 Miscellaneous treatments

3.3.1 Activated carbon (AC)

As absorbant; one of the materials whose surface area varies between 600-1600 m^2 and one of the most studied materials is Active Carbon [101]. Activated carbon is generally an adsorbent used to separate contaminated components with carboxylic groups. Hexavalent chromium from the most important pollutants is purified using activated carbon. There are four different types of activated carbon. These; "(AAC) activated cloth type, (PAC) activated powder type, (GAC) activated granular type (AFC) activated fibrous"[101–102].

3.3.2 Ion-exchange

One of the most frequently used methods for the treatment of heavy metal ions from industrial waters and use water is the ion exchange process [103]. Lead with mercury is one of the most toxic metals, although it has the least amount of industrial contamination.

Materials Research Forum LLC
https://doi.org/10.21741/9781644900291-4

If the material to be used for the treatment using the ion exchange process is mesoporus, the choice and displacement will be highly efficient. Modified Metal Organic Frameworks, with its adsorption property; It is thought to be due to the fact that it is in order and high porosity. In addition, studies for ion exchange processes with MOF-based materials are still in their early stages. It is envisaged that the ion exchange studies related to MOF will have a more functional design which can be adjusted for the material to be used for treatment [104–105].

4.   MOF performance in diverse WWT applications

The contiguous physicochemical behaviors of metal organic frameworks (adjustable band porosities, crystal structure, thermal stability, easily changeable structure) have been used as contaminated structures compared to traditional materials (sawdust, activated carbon, etc.) [104–106]. Essentially, MOF is a crystal structure that is self-assembling with organic organics with single-solid crystals or metal cations, binding ends [90–107–108]. In general, organic ligands contain conjugated or aromatic networks, which form 1-2 or 3-dimensional elongated coordination bonds [105–109].

According to limited studies on MOF for wastewater treatment, it will help to increase and improve the application areas of MOFs. For WWT, there is a lot of refinement with MOFs, and they have provided new ways of research (Figure 5). In the literature, the three basic methods (photocatalytic hydrorene production, adsorption, photocatalytic degradation), as determined by MOFs for waste water treatment, are generally shown in Figure 6.

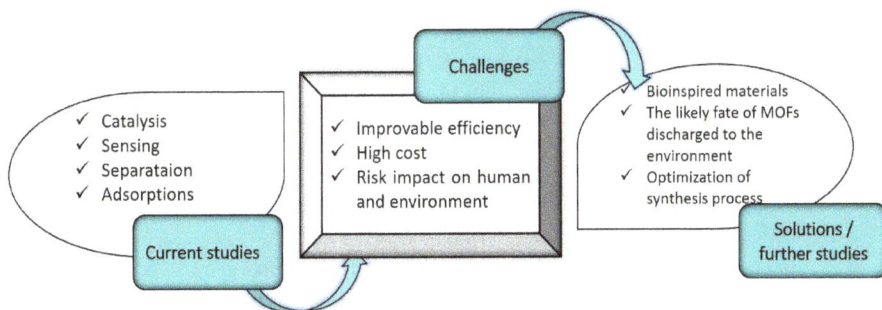

Fig. 5. An outline for diverse management approaches for WWT using MOFs.

Materials Research Forum LLC

https://doi.org/10.21741/9781644900291-4

Accordingly, there is a direct correlation between adsorption and catalytic performance. For this reason, the great adsorption properties of MOFs, especially for photocatalytic performance, should be studied in order to improve their catalytic properties.

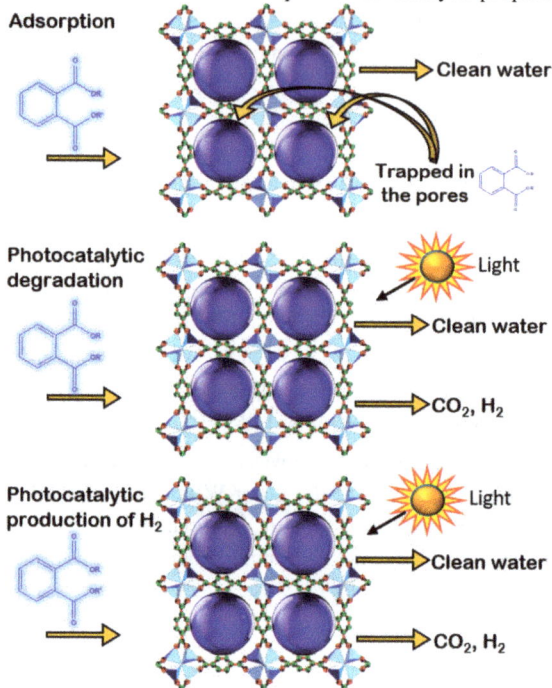

Fig. 6. *Three basic methods for water treatment with Metal Organic Frameworks (a) treatment of organics using MOFs (b) Catalytically water degradation of organics using MOFs (c) Using organic and water to produce photocatalic H2 using MOFs [101].*

In the current literature, it is quite complex to compare the relative wastewater treatment between conventional porous materials and MOFs (eg, polymers, zeolite, charcoal, nanocomposites). In this chapter, we shared up-to-date information on WWT applications using MOFs: (a) Current perspectives for wastewater treatment (b) Decontamination processes for wastewater treatment and (c) the performance of MOFs for different wastewater treatment.

Metal-Organic Framework Composites - Volume I   Materials Research Forum LLC
Materials Research Foundations **53** (2019) 73-104   https://doi.org/10.21741/9781644900291-4

4.1 MOFs in adsorption applications for WWT

The basic structures of MOFs used to adsorb risky contaminants (including crystalline powder and gas forms) were analyzed and the reports were evaluated [12–104–105–110–112]. In light of these, there are very few studies using the liquid phase of MOFs in wastewaetr treatment systems. Metal Organic Frameworks, which are used in WWT based on adsorbents, can be applied in wide area, low toxic intermediates, simple design, reuse, and easy to use; WWT is also superior to nano and traditional materials. The adsorption is compared to the porous structure, the availability of adsorbent materials, including accessibility and specific selectivity (Figure 7).

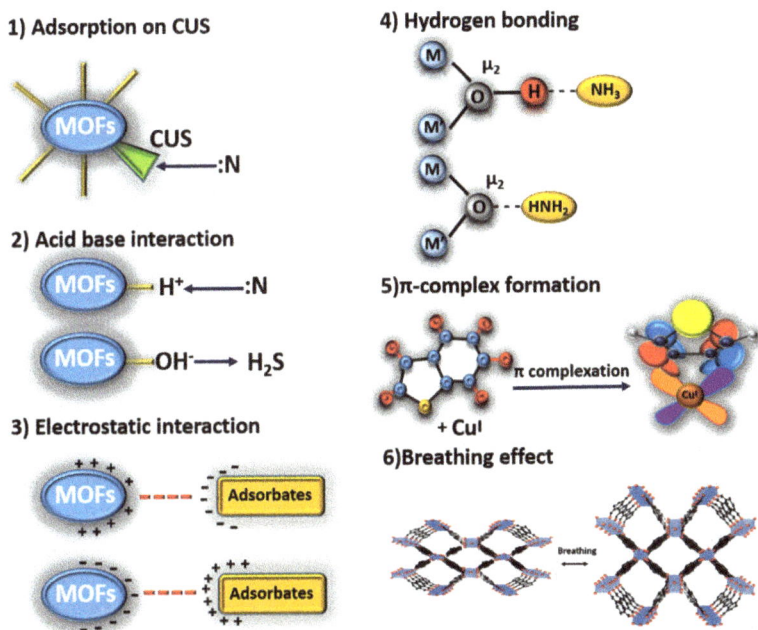

Fig. 7. *Different adsorption mechanisms using MOFs for a diverse range of hazardous compounds and gases in real wastewater samples [111].*

Physically adsorbed contaminations adhere to the cavities of MOFs by Van der Waals [111–115]. For this reason, the MOFs used as adsorbents can be separated from these contamination by easy removal methods or sonic waves. However, since the bond structures of chemically bound contaminants are chemical, their reusability requires very difficult processes [113–116–119]. In this way, the repeated use of adsorbent MOFs is a difficult subject as it requires an additional treatment/cleaning. Accordingly, the efficacy, selectivity, adsorbent capacity, chemical-mechanical strength and re-usability of the adsorbent materials should be evaluated. Meso-porous structures, zeolites, activated carbons and composite materials have also been frequently studied for wastewater treatmeant as adsorbent [112–120–123]. However, metal organic frameworks have become very important as the absorbent reactant in wastewater treatment. For example, the performance of copper-MOF adsorbants has been reported to be favorable for the removal of excess toxic sulfonamide antimicotics from water [124].

4.2 MOFs in separation applications for WWT

If the efficiency of the treatment with MOFs is regulated, it can be ensured that a highly efficient process can be created for the environment and energy. MOF-based membranes can be given as a very practical example of the separation of organic components due to their heavy metal requirements and low maintenance requirements [125–126]. In the meantime, waste water treatment can provide many different areas and innovations as MOFs. Similarly, Wang et al. have studied UiO/66 for the purification of arsenic in water. These crystals (UiO / 66) performed very well between pH 1-10, and at pH 2 yielded the treatment capacity of arsenate with the highest peak value. In the last equation, with the occurrence of Zr-O-As bonds, a seven equivalent arsenic variant was obtained which was found to exhibit a broad contact area active per unit area. The binding sites at the ends of the adsorbent structure (benzenedicarboxylate ligands and hydroxyl groups) gave a very good absorbance value of as (arsenate) 280 mg g_1. Similarly, Lee et al. have studied porous matrix membranes that can be used with pressure-controlled MOF components for water purification. Copper-based templates of MOFs were used to improve porosity and interconnectivity in pressurized system membranes. In very recent times, water has been studied in 5 different main areas (catalysis, adsorption, proton conduction, membrane separation, sensation) of MOFs [127–128].

It is used for the separation of biological based pollutants in water/contamine water treatment by using MOFs together with membranes. Prince et al. modified the membrane surfaces and found a new process that could kill or clean biological contamination. These structures are generally produced using antibacterial silver nanoparticles [130]. The membrane-based MOFs developed with the new understanding of this type of surface structure have been found to have an approximate 40% increase in water flow compared to the membrane structure, whereas the contact angle has been reduced by about 50% (Fig. 8) [130]. Agln (ina), which is similar to the zeolite structure and which has two tatrahedral crystal structures, was used for the treatment of organic dyes [131]. Ag-Metal organic frameworks were found to release it very quickly after absorption of methyl orange (MO). Looking at all this, the use of metal organic frameworks in wastewater treatment gives a lot of promise for the future.

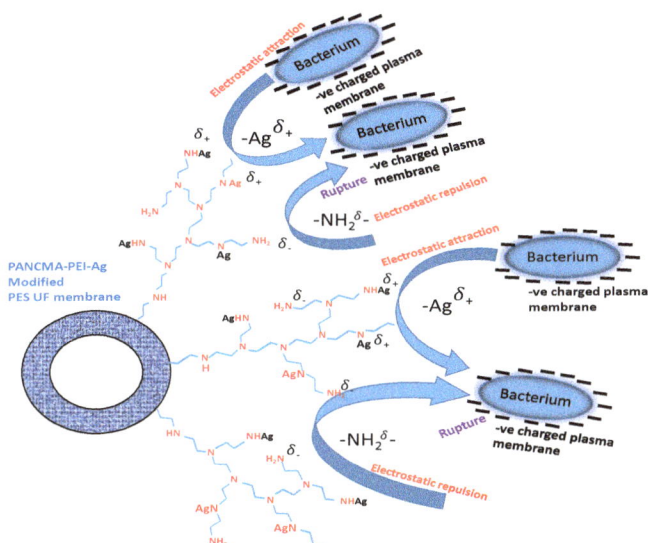

Fig. 8. *Schematic presentation of the self-cleaning property of an MOF membrane [129].*

4.3 MOFs in sensing applications for WWT

Selective applications of metal organic frameworks are well known in the recent literature. Recently, it has been reported by many research groups that it has been used to detect different target compounds, including pesticides, MOFs, heavy metals, VOCs and many other contaminants. In this section, some new research will be discussed for a different perspective on the selectivity performance of MOFs and their benefits.

Metal organic frameworks are promising structures for different selectivity processes, which can be managed by modifying crystal units. Too much organic contamination; TNF, 4-NP, heavy metals (Cd^{+2}, arsenic, PB^{+2}) and organic dyes were studied to selectively increase MOFs. For example, wang et al., Two stable co-ordinated Zr (IV) based MOFs, have been studied for the treatment of organic explosive materials and antibiotics in water [132]. Zr-based metal organic frameworks are well-known for their very good fluorescent specifications and are sensitive enough to respond to ppm levels of NZF and TNPs [132]. Similarly, Liu and colleagues have reported potential fluorescent sensors for ketones. The double-stranded metal organic frameworks showing fluorescence exhibit high stability at pH 3-11 as opposed to the general organic solvent and their solutions. Cd-based MOFs also showed a fluorescent quenching reaction in response to ketones, which could be suggested as a possible fluorescence sensor for acetone in aqueous solutions [133].

Recently, double ligand MOFs or biological MOF/1 have been investigated for the optimal detection and elimination of polycyclic aromatic hydrocarbons (PAHs) in environmental waters [134]. Biological (MOF/1), one of the important roles of refining PAHs is the complex interactions between the aromatics ring in the porosities and bilifyldicarboxylate and pyrimidines. In the water samples taken, it is seen that PAHs are obtained with gains ranging from 80-115% [134].

Extremely durable MOFs are also very important for use in the process of separating and retaining contaminants in water by applying water to thin films and/or membranes. For example, a porous Cu-MOF made from benzene-based copper/MOFs was used as a colorometric sensor for sequential detection of moist formaldehyde by the naked eye [135]. Similarly, the mechanisms of operation of different MOFs (Cu-MOF, Mg-MOF, etc.) have also been confirmed in the current literature for the determination of very little organic solvents and water by a rapid process [135]. As a result, it is certain that MOFs incorporate water molecules in a sensitive and sensitive manner; thus, it allowed the modification of high local densities by using an opening-detection process [74].

4.4   MOFs in miscellaneous WWT applications

As mentioned so far, MOFs can be applied in a variety of ways for wastewater treatment, in addition to the many different options available in the literature. The small size of measurable performance data is a general challenge for the processes described [37–48–136–138] In order to solve this problem in waste water treatment processes, different approaches are used for different contamination (VOCs, insect chemicals and heavy metals) in a wide range of density, and R & D studies are continuing.

5.   Perspectives on the use of MOFs in WWT

Due to the intensive work of MOFs, there are more than ten thousand types developed for different applications [27–128–139–145] Recently, the importance of using MOFs in practice has been studied extensively by many researchers in order to help overcome and limit the waste water treatment limits. In order to manage the industrial waste problem, detailed studies are still needed to modify and / or engineer the structures of the MOF in order to meet the requirements.

So far, much effort has been devoted to the modification and development of many different groups and classes of porous materials. Accordingly, different applications and water-stable and/or soluble, metal-organic-frameworks have been introduced into the field to diversify them. In this way, these various MOFs have been effectively implemented in biological/non-biological treatment areas. In order to increase the stability of MOFs, MOFs with different properties have been tested according to their different methods and their adsorption capabilities in water [27–107–128–139–145]. These initiatives are successful in changing the adsorption (chemical) criteria of MOFs.

The following must be observed in order to apply and extend Metal Organic Frameworks in processes.

1- Selective and fast adsorbing ability of ions,

2- MOF based membrane, thin film or column to ensure the continuous movement of waste,

3- To be able to show mechanical resistance to very high fluid pressures [27–107–128–139–145].

In more detail, in order to ensure that WWT is implemented in real life, it is necessary to dimension these materials and carry out R & D studies of such qualitative studies. Future research should be on the development of complex and complex MOFs that can be inert to the water and can be much more effective with many features.

Conclusion

In general, our chapter covered, the purification and separation processes of chemical components in water which is the importance of the use of MOFs. According to traditional materials (nanometers, silica, zeolites and active carbon), we hope that the use of MOFs can be more stable and performance. In this chapter, MOFs in the field of application are discussed for the economical and dynamic wastewater treatment according to their advanced functions. In order to make metal organic frameworks technology more practical for wastewater treatment, much work has been done to understand the purification capacity and accuracy for many different organic contaminations in the water. Of course, the use of recyclable and water-resistant MOFs or biological MOFs will undoubtedly help in the use and expansion of wastewater treatment and other work areas. Lastly, we hope that the progress of MOF technology will lead to commercial demands in the near future and thus will provide further research.

References

[1] S. D. Richardson & T. A. Ternes, Water Analysis: Emerging Contaminants and Current Issues. *Analytical Chemistry*, **90** (2018) 398–428. https://doi.org/10.1021/acs.analchem.7b04577

[2] S. D. Richardson, Environmental Mass Spectrometry: Emerging Contaminants and Current Issues. *Analytical Chemistry*, **84** (2012) 747–778. https://doi.org/10.1021/ac202903d

[3] Y. Huang & A. A. Keller, Magnetic Nanoparticle Adsorbents for Emerging Organic Contaminants. *ACS Sustainable Chemistry & Engineering*, **1** (2013) 731–736. https://doi.org/10.1021/sc400047q

[4] M. Kuster, M. J. López de Alda, M. D. Hernando, M. Petrovic, J. Martín-Alonso, & D. Barceló, Analysis and occurrence of pharmaceuticals, estrogens, progestogens and polar pesticides in sewage treatment plant effluents, river water and drinking water in the Llobregat river basin (Barcelona, Spain). *Journal of Hydrology*, **358** (2008) 112–123. https://doi.org/10.1016/J.JHYDROL.2008.05.030

[5] N. Bolong, A. F. Ismail, M. R. Salim, & T. Matsuura, A review of the effects of emerging contaminants in wastewater and options for their removal. *Desalination*, **239** (2009) 229–246. https://doi.org/10.1016/J.DESAL.2008.03.020

[6] K. E. Murray, S. M. Thomas, & A. A. Bodour, Prioritizing research for trace pollutants and emerging contaminants in the freshwater environment. *Environmental Pollution*, **158** (2010) 3462–3471. https://doi.org/10.1016/J.ENVPOL.2010.08.009

[7] K. Ariga, S. Ishihara, H. Abe, M. Li, & J. P. Hill, Materials nanoarchitectonics for environmental remediation and sensing. *J. Mater. Chem.*, **22** (2012) 2369–2377. https://doi.org/10.1039/C1JM14101E

[8] Y. Kosaki, H. Izawa, S. Ishihara, K. Kawakami, M. Sumita, Y. Tateyama, Q. Ji, V. Krishnan, S. Hishita, Y. Yamauchi, J. P. Hill, A. Vinu, S. Shiratori, & K. Ariga, Nanoporous Carbon Sensor with Cage-in-Fiber Structure: Highly Selective Aniline Adsorbent toward Cancer Risk Management. *ACS Applied Materials & Interfaces*, **5** (2013) 2930–2934. https://doi.org/10.1021/am400940q

[9] T. Mori, M. Akamatsu, K. Okamoto, M. Sumita, Y. Tateyama, H. Sakai, J. P. Hill, M. Abe, & K. Ariga, Micrometer-level naked-eye detection of caesium particulates in the solid state. *Science and Technology of Advanced Materials*, **14** (2013) 015002. https://doi.org/10.1088/1468-6996/14/1/015002

[10]   C. Bao & C. Fang, Water Resources Flows Related to Urbanization in China: Challenges and Perspectives for Water Management and Urban Development. *Water Resources Management*, **26** (2012) 531–552. https://doi.org/10.1007/s11269-011-9930-y

[11]   X. Zhang, C. Chen, P. Lin, A. Hou, Z. Niu, & J. Wang, Emergency Drinking Water Treatment during Source Water Pollution Accidents in China: Origin Analysis, Framework and Technologies [†]. *Environmental Science & Technology*, **45** (2011) 161–167. https://doi.org/10.1021/es101987e

[12]   N. A. Khan, Z. Hasan, & S. H. Jhung, Adsorptive removal of hazardous materials using metal-organic frameworks (MOFs): A review. *Journal of Hazardous Materials*, **244–245** (2013) 444–456. https://doi.org/10.1016/J.JHAZMAT.2012.11.011

[13]   I. Michael, L. Rizzo, C. S. McArdell, C. M. Manaia, C. Merlin, T. Schwartz, C. Dagot, & D. Fatta-Kassinos, Urban wastewater treatment plants as hotspots for the release of antibiotics in the environment: A review. *Water Research*, **47** (2013) 957–995. https://doi.org/10.1016/J.WATRES.2012.11.027

[14]   A. Finizio, G. Azimonti, & S. Villa, Occurrence of pesticides in surface water bodies: a critical analysis of the Italian national pesticide survey programs. *J. Environ. Monit.*, **13** (2011) 49–57. https://doi.org/10.1039/C0EM00192A

[15]   V. K. Gupta & Suhas, Application of low-cost adsorbents for dye removal – A review. *Journal of Environmental Management*, **90** (2009) 2313–2342. https://doi.org/10.1016/J.JENVMAN.2008.11.017

[16]   Y. Gong, X. Zhao, Z. Cai, S. E. O'Reilly, X. Hao, & D. Zhao, A review of oil, dispersed oil and sediment interactions in the aquatic environment: influence on the fate, transport and remediation of oil spills. *Marine pollution bulletin*, **79** (2014) 16–33. https://doi.org/10.1016/j.marpolbul.2013.12.024

[17]   R. F. Johnson, T. G. Manjreker, & J. E. Halligan, Removal of oil from water surfaces by sorption on unstructured fibers. *Environmental Science & Technology*, **7** (1973) 439–443. https://doi.org/10.1021/es60077a003

[18]   M. He, Y. Sun, X. Li, & Z. Yang, Distribution patterns of nitrobenzenes and polychlorinated biphenyls in water, suspended particulate matter and sediment from mid- and down-stream of the Yellow River (China). *Chemosphere*, **65** (2006) 365–74. https://doi.org/10.1016/j.chemosphere.2006.02.033

[19]   M. Ahmaruzzaman, Adsorption of phenolic compounds on low-cost adsorbents: A review. *Advances in Colloid and Interface Science*, **143** (2008) 48–67. https://doi.org/10.1016/j.cis.2008.07.002

[20]   B. Pawelec, R. M. Navarro, J. M. Campos-Martin, & J. L. G. Fierro, Retracted article: Towards near zero-sulfur liquid fuels: a perspective review. *Catalysis Science & Technology*, **1** (2011) 23. https://doi.org/10.1039/c0cy00049c

[21]   R. . Colvile, E. . Hutchinson, J. . Mindell, & R. . Warren, The transport sector as a source of air pollution. *Atmospheric Environment*, **35** (2001) 1537–1565. https://doi.org/10.1016/S1352-2310(00)00551-3

[22]   M. Hartmann, S. Kullmann, & H. Keller, Wastewater treatment with heterogeneous Fenton-type catalysts based on porous materials. *Journal of Materials Chemistry*, **20** (2010) 9002. https://doi.org/10.1039/c0jm00577k

[23]   M. M. Khin, A. S. Nair, V. J. Babu, R. Murugan, & S. Ramakrishna, A review on nanomaterials for environmental remediation. *Energy & Environmental Science*, **5** (2012) 8075. https://doi.org/10.1039/c2ee21818f

[24]   A. Walcarius & L. Mercier, Mesoporous organosilica adsorbents: nanoengineered materials for removal of organic and inorganic pollutants. *Journal of Materials Chemistry*, **20** (2010) 4478. https://doi.org/10.1039/b924316j

[25]   R. T. Yang, *Adsorbents : fundamentals and applications* (Wiley-Interscience, 2003)

[26]   J.-R. Li, R. J. Kuppler, & H.-C. Zhou, Selective gas adsorption and separation in metal–organic frameworks. *Chemical Society Reviews*, **38** (2009) 1477. https://doi.org/10.1039/b802426j

[27]   B. Van de Voorde, B. Bueken, J. Denayer, & D. De Vos, Adsorptive separation on metal–organic frameworks in the liquid phase. *Chem. Soc. Rev.*, **43** (2014) 5766–5788. https://doi.org/10.1039/C4CS00006D

[28]   M. P. Suh, Y. E. Cheon, & E. Y. Lee, Syntheses and functions of porous metallosupramolecular networks. *Coordination Chemistry Reviews*, **252** (2008) 1007–1026. https://doi.org/10.1016/J.CCR.2008.01.032

[29]   C. Moreno-Castilla, Adsorption of organic molecules from aqueous solutions on carbon materials. *Carbon*, **42** (2004) 83–94. https://doi.org/10.1016/J.CARBON.2003.09.022

[30]   E. Haque, J. W. Jun, S. N. Talapaneni, A. Vinu, & S. H. Jhung, Superior adsorption capacity of mesoporous carbon nitride with basic CN framework for phenol. *Journal of Materials Chemistry*, **20** (2010) 10801. https://doi.org/10.1039/c0jm02974b

[31]   Y. Wang, R. T. Yang, & J. M. Heinzel, Desulfurization of jet fuel by π-complexation adsorption with metal halides supported on MCM-41 and SBA-15

mesoporous materials. *Chemical Engineering Science*, **63** (2008) 356–365.
https://doi.org/10.1016/J.CES.2007.09.002

[32]   L. Zhang, W. Zhang, J. Shi, Z. Hua, Y. Li, & J. Yan, A new thioether
functionalized organic–inorganic mesoporous composite as a highly selective and
capacious Hg2+ adsorbentElectronic supplementary information (ESI) available: Figs.
S1–S3. See http://www.rsc.org/suppdata/cc/b2/b210457a/. *Chemical Communications*,
0 (2003) 210–211. https://doi.org/10.1039/b210457a

[33]   X. Wang, T. Sun, J. Yang, L. Zhao, & J. Jia, Low-temperature H2S removal from
gas streams with SBA-15 supported ZnO nanoparticles. *Chemical Engineering
Journal*, **142** (2008) 48–55. https://doi.org/10.1016/J.CEJ.2007.11.013

[34]   A. Dąbrowski, P. Podkościelny, Z. Hubicki, & M. Barczak, Adsorption of
phenolic compounds by activated carbon—a critical review. *Chemosphere*, **58** (2005)
1049–1070. https://doi.org/10.1016/J.CHEMOSPHERE.2004.09.067

[35]   K. Kadirvelu, K. Thamaraiselvi, & C. Namasivayam, Removal of heavy metals
from industrial wastewaters by adsorption onto activated carbon prepared from an
agricultural solid waste. *Bioresource Technology*, **76** (2001) 63–65.
https://doi.org/10.1016/S0960-8524(00)00072-9

[36]   C. Y. Yin, M. K. Aroua, & W. M. A. W. Daud, Review of modifications of
activated carbon for enhancing contaminant uptakes from aqueous solutions.
Separation and Purification Technology, **52** (2007) 403–415.
https://doi.org/10.1016/J.SEPPUR.2006.06.009

[37]   †,‡ Anning Zhou, † and Xiaoliang Ma, & † Chunshan Song*, Liquid-Phase
Adsorption of Multi-Ring Thiophenic Sulfur Compounds on Carbon Materials with
Different Surface Properties. (2006). https://doi.org/10.1021/JP0550210

[38]   E. Deliyanni, M. Seredych, & T. J. Bandosz, Interactions of 4,6-
Dimethyldibenzothiophene with the Surface of Activated Carbons. *Langmuir*, **25**
(2009) 9302–9312. https://doi.org/10.1021/la900854x

[39]   J.-R. Li, Y. Ma, M. C. McCarthy, J. Sculley, J. Yu, H.-K. Jeong, P. B. Balbuena,
& H.-C. Zhou, Carbon dioxide capture-related gas adsorption and separation in metal-
organic frameworks. *Coordination Chemistry Reviews*, **255** (2011) 1791–1823.
https://doi.org/10.1016/J.CCR.2011.02.012

[40]   A. Samokhvalov & B. J. Tatarchuk, Review of Experimental Characterization of
Active Sites and Determination of Molecular Mechanisms of Adsorption, Desorption
and Regeneration of the Deep and Ultradeep Desulfurization Sorbents for Liquid
Fuels. *Catalysis Reviews*, **52** (2010) 381–410.
https://doi.org/10.1080/01614940.2010.498749

[41] H.-L. Jiang & Q. Xu, Porous metal–organic frameworks as platforms for functional applications. *Chemical Communications*, **47** (2011) 3351. https://doi.org/10.1039/c0cc05419d

[42] C.-Y. Huang, M. Song, Z.-Y. Gu, H.-F. Wang, & X.-P. Yan, Probing the Adsorption Characteristic of Metal–Organic Framework MIL-101 for Volatile Organic Compounds by Quartz Crystal Microbalance. *Environmental Science & Technology*, **45** (2011) 4490–4496. https://doi.org/10.1021/es200256q

[43] J. Liu, P. K. Thallapally, B. P. McGrail, D. R. Brown, & J. Liu, Progress in adsorption-based CO $_2$ capture by metal–organic frameworks. *Chem. Soc. Rev.*, **41** (2012) 2308–2322. https://doi.org/10.1039/C1CS15221A

[44] G. Crini, Non-conventional low-cost adsorbents for dye removal: A review. *Bioresource Technology*, **97** (2006) 1061–1085. https://doi.org/10.1016/J.BIORTECH.2005.05.001

[45] S. Velu, and Xiaoliang Ma, & C. Song*, Selective Adsorption for Removing Sulfur from Jet Fuel over Zeolite-Based Adsorbents. (2003). https://doi.org/10.1021/IE020995P

[46] * Jarkko Helminen, and Joni Helenius, E. Paatero, & I. Turunen, Adsorption Equilibria of Ammonia Gas on Inorganic and Organic Sorbents at 298.15 K. (2001). https://doi.org/10.1021/JE000273+

[47] M. Seredych, E. Deliyanni, & T. J. Bandosz, Role of microporosity and surface chemistry in adsorption of 4,6-dimethyldibenzothiophene on polymer-derived activated carbons. *Fuel*, **89** (2010) 1499–1507. https://doi.org/10.1016/J.FUEL.2009.09.032

[48] C. O. A. and & †,‡ Teresa J. Bandosz*, Importance of Structural and Chemical Heterogeneity of Activated Carbon Surfaces for Adsorption of Dibenzothiophene. (2005). https://doi.org/10.1021/LA050772E

[49] M. Seredych & T. J. Bandosz, Removal of dibenzothiophenes from model diesel fuel on sulfur rich activated carbons. *Applied Catalysis B: Environmental*, **106** (2011) 133–141. https://doi.org/10.1016/J.APCATB.2011.05.016

[50] J. D. Evans, B. Garai, H. Reinsch, W. Li, S. Dissegna, V. Bon, I. Senkovska, R. A. Fischer, S. Kaskel, C. Janiak, N. Stock, & D. Volkmer, Metal–organic frameworks in Germany: From synthesis to function. *Coordination Chemistry Reviews*, **380** (2019) 378–418. https://doi.org/10.1016/j.ccr.2018.10.002

[51] Z. Zhou & M. Hartmann, Progress in enzyme immobilization in ordered mesoporous materials and related applications. *Chemical Society Reviews*, **42** (2013) 3894. https://doi.org/10.1039/c3cs60059a

[52]   A. Vinu & K. Ariga, New Ideas for Mesoporous Materials. *Advanced Porous Materials*, **1** (2013) 63–71. https://doi.org/10.1166/apm.2013.1005

[53]   J.-R. Li, J. Sculley, & H.-C. Zhou, Metal–Organic Frameworks for Separations. *Chemical Reviews*, **112** (2012) 869–932. https://doi.org/10.1021/cr200190s

[54]   I. Ahmed & S. H. Jhung, Composites of metal–organic frameworks: Preparation and application in adsorption. *Materials Today*, **17** (2014) 136–146. https://doi.org/10.1016/J.MATTOD.2014.03.002

[55]   N. A. Khan, Z. Hasan, & S. H. Jhung, Adsorption and Removal of Sulfur or Nitrogen-Containing Compounds with Metal-Organic Frameworks (MOFs). *Advanced Porous Materials*, **1** (2013) 91–102. https://doi.org/10.1166/apm.2013.1002

[56]   Heavy Metals In The Environment - Bibudhendra Sarkar - Google Kitaplar. (n.d.). https://books.google.com.tr/books?hl=tr&lr=&id=OJboWGzbq1EC&oi=fnd&pg=PP1 &ots=lRv1MU1-p2&sig=z_t1IzMYFyNnkB4ZA9fVe- 0fQuE&redir_esc=y#v=onepage&q&f=false (accessed December 28, 2018)

[57]   Toxicity and source of Pb, Cd, Hg, Cr, As, and Radionuclides in the Environment. (n.d.). https://www.researchgate.net/publication/291159112_Toxicity_and_source_of_Pb_Cd _Hg_Cr_As_and_Radionuclides_in_the_Environment (accessed December 28, 2018)

[58]   † Mohamed Eddaoudi, † David B. Moler, † Hailian Li, † Banglin Chen, † Theresa M. Reineke, ‡ and Michael O'Keeffe, & † Omar M. Yaghi*, Modular Chemistry: Secondary Building Units as a Basis for the Design of Highly Porous and Robust Metal–Organic Carboxylate Frameworks. (2001). https://doi.org/10.1021/AR000034B

[59]   H. Furukawa, N. Ko, Y. B. Go, N. Aratani, S. B. Choi, E. Choi, A. O. Yazaydin, R. Q. Snurr, M. O'Keeffe, J. Kim, & O. M. Yaghi, Ultrahigh porosity in metal-organic frameworks. *Science (New York, N.Y.)*, **329** (2010) 424–8. https://doi.org/10.1126/science.1192160

[60]   * O. M. Yaghi, and Hailian Li, & T. L. Groy, Construction of Porous Solids from Hydrogen-Bonded Metal Complexes of 1,3,5-Benzenetricarboxylic Acid. (1996). https://doi.org/10.1021/JA960746Q

[61]   E. E. Moushi, T. C. Stamatatos, W. Wernsdorfer, V. Nastopoulos, G. Christou, & A. J. Tasiopoulos, A Family of 3D Coordination Polymers Composed of Mn19 Magnetic Units. *Angewandte Chemie International Edition*, **45** (2006) 7722–7725. https://doi.org/10.1002/anie.200603498

[62]   T. J. Prior & M. J. Rosseinsky, Crystal engineering of a 3-D coordination polymer from 2-D building blocks. *Chemical Communications*, **0** (2001) 495–496. https://doi.org/10.1039/b009455m

[63] M. E. Kosal, J.-H. Chou, S. R. Wilson, & K. S. Suslick, A functional zeolite analogue assembled from metalloporphyrins. *Nature Materials*, **1** (2002) 118–121. https://doi.org/10.1038/nmat730

[64] †,‡ Hitoshi Kumagai, § and Cameron J. Kepert, & † Mohamedally Kurmoo*, Construction of Hydrogen-Bonded and Coordination-Bonded Networks of Cobalt(II) with Pyromellitate: Synthesis, Structures, and Magnetic Properties. (2002). https://doi.org/10.1021/IC020065Y

[65] G. Férey, Hybrid porous solids: past, present, future. *Chem. Soc. Rev.*, **37** (2008) 191–214. https://doi.org/10.1039/B618320B

[66] P. Kumar, A. Deep, & K.-H. Kim, Metal organic frameworks for sensing applications. *TrAC Trends in Analytical Chemistry*, **73** (2015) 39–53. https://doi.org/10.1016/J.TRAC.2015.04.009

[67] K. K. Tanabe & S. M. Cohen, Postsynthetic modification of metal–organic frameworks—a progress report. *Chem. Soc. Rev.*, **40** (2011) 498–519. https://doi.org/10.1039/C0CS00031K

[68] G. Férey, C. Mellot-Draznieks, C. Serre, F. Millange, J. Dutour, S. Surblé, I. Margiolaki, N. G. Berry, Y. Z. Khimyak, A. Y. Ganin, P. Wiper, J. B. Claridge, M. J. Rosseinsky, J. F. Stoddart, & O. M. Yaghi, A chromium terephthalate-based solid with unusually large pore volumes and surface area. *Science (New York, N.Y.)*, **309** (2005) 2040–2. https://doi.org/10.1126/science.1116275

[69] P. Kumar, K.-H. Kim, & A. Deep, Recent advancements in sensing techniques based on functional materials for organophosphate pesticides. *Biosensors and Bioelectronics*, **70** (2015) 469–481. https://doi.org/10.1016/J.BIOS.2015.03.066

[70] S. Horike, S. Shimomura, & S. Kitagawa, Soft porous crystals. *Nature Chemistry*, **1** (2009) 695–704. https://doi.org/10.1038/nchem.444

[71] D. Bradshaw, J. B. Claridge, E. J. Cussen, and T. J. Prior, & M. J. Rosseinsky*, Design, Chirality, and Flexibility in Nanoporous Molecule-Based Materials. (2005). https://doi.org/10.1021/AR0401606

[72] O. K. Farha & J. T. Hupp, Rational Design, Synthesis, Purification, and Activation of Metal–Organic Framework Materials. *Accounts of Chemical Research*, **43** (2010) 1166–1175. https://doi.org/10.1021/ar1000617

[73] J. M. Taylor, T. Komatsu, S. Dekura, K. Otsubo, M. Takata, & H. Kitagawa, The Role of a Three Dimensionally Ordered Defect Sublattice on the Acidity of a Sulfonated Metal–Organic Framework. *Journal of the American Chemical Society*, **137** (2015) 11498–11506. https://doi.org/10.1021/jacs.5b07267

[74]   A. Douvali, A. C. Tsipis, S. V. Eliseeva, S. Petoud, G. S. Papaefstathiou, C. D. Malliakas, I. Papadas, G. S. Armatas, I. Margiolaki, M. G. Kanatzidis, T. Lazarides, & M. J. Manos, Turn-On Luminescence Sensing and Real-Time Detection of Traces of Water in Organic Solvents by a Flexible Metal-Organic Framework. *Angewandte Chemie International Edition*, **54** (2015) 1651–1656.
https://doi.org/10.1002/anie.201410612

[75]   K. K. Barnes, D. W. Kolpin, E. T. Furlong, S. D. Zaugg, M. T. Meyer, & L. B. Barber, A national reconnaissance of pharmaceuticals and other organic wastewater contaminants in the United States — I) Groundwater. *Science of The Total Environment*, **402** (2008) 192–200.
https://doi.org/10.1016/J.SCITOTENV.2008.04.028

[76]   V. K. Gupta, R. Jain, A. Mittal, T. A. Saleh, A. Nayak, S. Agarwal, & S. Sikarwar, Photo-catalytic degradation of toxic dye amaranth on TiO2/UV in aqueous suspensions. *Materials Science and Engineering: C*, **32** (2012) 12–17.
https://doi.org/10.1016/J.MSEC.2011.08.018

[77]   D. W. Kolpin, M. Skopec, M. T. Meyer, E. T. Furlong, & S. D. Zaugg, Urban contribution of pharmaceuticals and other organic wastewater contaminants to streams during differing flow conditions. *Science of The Total Environment*, **328** (2004) 119–130. https://doi.org/10.1016/J.SCITOTENV.2004.01.015

[78]   M. J. M. Bueno, M. J. Gomez, S. Herrera, M. D. Hernando, A. Agüera, & A. R. Fernández-Alba, Occurrence and persistence of organic emerging contaminants and priority pollutants in five sewage treatment plants of Spain: Two years pilot survey monitoring. *Environmental Pollution*, **164** (2012) 267–273.
https://doi.org/10.1016/J.ENVPOL.2012.01.038

[79]   S. Khan, Q. Cao, Y. M. Zheng, Y. Z. Huang, & Y. G. Zhu, Health risks of heavy metals in contaminated soils and food crops irrigated with wastewater in Beijing, China. *Environmental Pollution*, **152** (2008) 686–692.
https://doi.org/10.1016/J.ENVPOL.2007.06.056

[80]   N. J. Ashbolt, Microbial contamination of drinking water and disease outcomes in developing regions. *Toxicology*, **198** (2004) 229–238.
https://doi.org/10.1016/J.TOX.2004.01.030

[81]   S. Naidoo, A. Olaniran, S. Naidoo, & A. O. Olaniran, Treated Wastewater Effluent as a Source of Microbial Pollution of Surface Water Resources. *International Journal of Environmental Research and Public Health*, **11** (2013) 249–270.
https://doi.org/10.3390/ijerph110100249

[82] M. . Wong, Ecological restoration of mine degraded soils, with emphasis on metal contaminated soils. *Chemosphere*, **50** (2003) 775–780. https://doi.org/10.1016/S0045-6535(02)00232-1

[83] S. Liu, C. Gunawan, N. Barraud, S. A. Rice, E. J. Harry, & R. Amal, Understanding, Monitoring, and Controlling Biofilm Growth in Drinking Water Distribution Systems. *Environmental Science & Technology*, **50** (2016) 8954–8976. https://doi.org/10.1021/acs.est.6b00835

[84] H. R. Rogers, Sources, behaviour and fate of organic contaminants during sewage treatment and in sewage sludges. *Science of The Total Environment*, **185** (1996) 3–26. https://doi.org/10.1016/0048-9697(96)05039-5

[85] C. Hignite & D. L. Azarnoff, Drugs and drug metabolites as environmental contaminants: Chlorophenoxyisobutyrate and salicylic acid in sewage water effluent. *Life Sciences*, **20** (1977) 337–341. https://doi.org/10.1016/0024-3205(77)90329-0

[86] F. Mapanda, E. N. Mangwayana, J. Nyamangara, & K. E. Giller, The effect of long-term irrigation using wastewater on heavy metal contents of soils under vegetables in Harare, Zimbabwe. *Agriculture, Ecosystems & Environment*, **107** (2005) 151–165. https://doi.org/10.1016/J.AGEE.2004.11.005

[87] P. E. Stackelberg, E. T. Furlong, M. T. Meyer, S. D. Zaugg, A. K. Henderson, & D. B. Reissman, Persistence of pharmaceutical compounds and other organic wastewater contaminants in a conventional drinking-water-treatment plant. *Science of The Total Environment*, **329** (2004) 99–113. https://doi.org/10.1016/J.SCITOTENV.2004.03.015

[88] A. Leung, Z. W. Cai, & M. H. Wong, Environmental contamination from electronic waste recycling at Guiyu, southeast China. *Journal of Material Cycles and Waste Management*, **8** (2006) 21–33. https://doi.org/10.1007/s10163-005-0141-6

[89] T. Heberer, Occurrence, fate, and removal of pharmaceutical residues in the aquatic environment: a review of recent research data. *Toxicology Letters*, **131** (2002) 5–17. https://doi.org/10.1016/S0378-4274(02)00041-3

[90] Scopus - Document details. (n.d.). https://www.scopus.com/record/display.uri?eid=2-s2.0-34848883573&origin=inward (accessed December 28, 2018)

[91] excelwater.com. (n.d.). http://www.excelwater.com/thp/filters/Water-Purification.htm (accessed December 28, 2018)

[92] Xiu-Sheng Miao, and Jian-Jun Yang, & C. D. Metcalfe*, Carbamazepine and Its Metabolites in Wastewater and in Biosolids in a Municipal Wastewater Treatment Plant. (2005). https://doi.org/10.1021/ES050261E

[93]   Y. Zhang, S.-U. Geißen, & C. Gal, Carbamazepine and diclofenac: Removal in wastewater treatment plants and occurrence in water bodies. *Chemosphere*, **73** (2008) 1151–1161. https://doi.org/10.1016/J.CHEMOSPHERE.2008.07.086

[94]   N. Nakada, T. Tanishima, H. Shinohara, K. Kiri, & H. Takada, Pharmaceutical chemicals and endocrine disrupters in municipal wastewater in Tokyo and their removal during activated sludge treatment. *Water Research*, **40** (2006) 3297–3303. https://doi.org/10.1016/J.WATRES.2006.06.039

[95]   M. Petrović, S. Gonzalez, & D. Barceló, Analysis and removal of emerging contaminants in wastewater and drinking water. *TrAC Trends in Analytical Chemistry*, **22** (2003) 685–696. https://doi.org/10.1016/S0165-9936(03)01105-1

[96]   K. V Gernaey, M. C. . van Loosdrecht, M. Henze, M. Lind, & S. B. Jørgensen, Activated sludge wastewater treatment plant modelling and simulation: state of the art. *Environmental Modelling & Software*, **19** (2004) 763–783. https://doi.org/10.1016/J.ENVSOFT.2003.03.005

[97]   S. D. Kim, J. Cho, I. S. Kim, B. J. Vanderford, & S. A. Snyder, Occurrence and removal of pharmaceuticals and endocrine disruptors in South Korean surface, drinking, and waste waters. *Water Research*, **41** (2007) 1013–1021. https://doi.org/10.1016/J.WATRES.2006.06.034

[98]   O. A. H. Jones, N. Voulvoulis, & J. N. Lester, The occurrence and removal of selected pharmaceutical compounds in a sewage treatment works utilising activated sludge treatment. *Environmental Pollution*, **145** (2007) 738–744. https://doi.org/10.1016/J.ENVPOL.2005.08.077

[99]   Y. Luo, W. Guo, H. H. Ngo, L. D. Nghiem, F. I. Hai, J. Zhang, S. Liang, & X. C. Wang, A review on the occurrence of micropollutants in the aquatic environment and their fate and removal during wastewater treatment. *Science of The Total Environment*, **473–474** (2014) 619–641. https://doi.org/10.1016/J.SCITOTENV.2013.12.065

[100] S. Parsons, Advanced oxidation processes for water and wastewater treatment. (2004)

[101] E. M. Dias & C. Petit, Towards the use of metal–organic frameworks for water reuse: a review of the recent advances in the field of organic pollutants removal and degradation and the next steps in the field. *Journal of Materials Chemistry A*, **3** (2015) 22484–22506. https://doi.org/10.1039/C5TA05440K

[102] A. Oyelami, B. Elegbede, K. Semple, A. Oyelami, B. Elegbede, & K. Semple, Impact of Different Types of Activated Carbon on the Bioaccessibility of 14C-phenanthrene in Sterile and Non-Sterile Soils. *Environments*, **1** (2014) 137–156. https://doi.org/10.3390/environments1020137

Metal-Organic Framework Composites - Volume I
Materials Research Foundations 53 (2019) 73-104

Materials Research Forum LLC
https://doi.org/10.21741/9781644900291-4

[103] A. Dąbrowski, Z. Hubicki, P. Podkościelny, & E. Robens, Selective removal of the heavy metal ions from waters and industrial wastewaters by ion-exchange method. *Chemosphere*, **56** (2004) 91–106. https://doi.org/10.1016/J.CHEMOSPHERE.2004.03.006

[104] P. Kumar, A. Pournara, K.-H. Kim, V. Bansal, S. Rapti, & M. J. Manos, Metal-organic frameworks: Challenges and opportunities for ion-exchange/sorption applications. *Progress in Materials Science*, **86** (2017) 25–74. https://doi.org/10.1016/J.PMATSCI.2017.01.002

[105] A. U. Czaja, N. Trukhan, & U. Müller, Industrial applications of metal–organic frameworks. *Chemical Society Reviews*, **38** (2009) 1284. https://doi.org/10.1039/b804680h

[106] P. Kumar, K.-H. Kim, E. E. Kwon, & J. E. Szulejko, Metal–organic frameworks for the control and management of air quality: advances and future direction. *Journal of Materials Chemistry A*, **4** (2016) 345–361. https://doi.org/10.1039/C5TA07068F

[107] O. M. Yaghi & H. Li, Hydrothermal Synthesis of a Metal-Organic Framework Containing Large Rectangular Channels. *Journal of the American Chemical Society*, **117** (1995) 10401–10402. https://doi.org/10.1021/ja00146a033

[108] G. Férey, C. Mellot-Draznieks, C. Serre, F. Millange, J. Dutour, S. Surblé, I. Margiolaki, N. G. Berry, Y. Z. Khimyak, A. Y. Ganin, P. Wiper, J. B. Claridge, M. J. Rosseinsky, J. F. Stoddart, & O. M. Yaghi, A Chromium Terephthalate-Based Solid with Unusually Large Pore Volumes and Surface Area. *Science*, **309** (2013) 2040–2042. https://doi.org/10.1126/science.1116275

[109] Synthesis of Metal-Organic Frameworks (MOFs): routes to various MOF topologies. (n.d.). https://www.scopus.com/record/display.uri?eid=2-s2.0-84899527159&origin=inward (accessed December 28, 2018)

[110] D. Britt, D. Tranchemontagne, & O. M. Yaghi, Metal-organic frameworks with high capacity and selectivity for harmful gases. *Proceedings of the National Academy of Sciences of the United States of America*, **105** (2008) 11623–7. https://doi.org/10.1073/pnas.0804900105

[111] Z. Hu, B. J. Deibert, & J. Li, Luminescent metal–organic frameworks for chemical sensing and explosive detection. *Chem. Soc. Rev.*, **43** (2014) 5815–5840. https://doi.org/10.1039/C4CS00010B

[112] N. A. Khan & S. H. Jhung, Scandium-Triflate/Metal–Organic Frameworks: Remarkable Adsorbents for Desulfurization and Denitrogenation. *Inorganic Chemistry*, **54** (2015) 11498–11504. https://doi.org/10.1021/acs.inorgchem.5b02118

[113] S. R. Caskey, A. G. Wong-Foy, & A. J. Matzger, Dramatic Tuning of Carbon Dioxide Uptake via Metal Substitution in a Coordination Polymer with Cylindrical Pores. *Journal of the American Chemical Society*, **130** (2008) 10870–10871. https://doi.org/10.1021/ja8036096

[114] L. Hamon, C. Serre, T. Devic, T. Loiseau, F. Millange, G. Férey, & G. De Weireld, Comparative Study of Hydrogen Sulfide Adsorption in the MIL-53(Al, Cr, Fe), MIL-47(V), MIL-100(Cr), and MIL-101(Cr) Metal−Organic Frameworks at Room Temperature. *Journal of the American Chemical Society*, **131** (2009) 8775–8777. https://doi.org/10.1021/ja901587t

[115] H. Furukawa, F. Gándara, Y.-B. Zhang, J. Jiang, W. L. Queen, M. R. Hudson, & O. M. Yaghi, Water Adsorption in Porous Metal–Organic Frameworks and Related Materials. *Journal of the American Chemical Society*, **136** (2014) 4369–4381. https://doi.org/10.1021/ja500330a

[116] B. Arstad, H. Fjellvåg, K. O. Kongshaug, O. Swang, & R. Blom, Amine functionalised metal organic frameworks (MOFs) as adsorbents for carbon dioxide. *Adsorption*, **14** (2008) 755–762. https://doi.org/10.1007/s10450-008-9137-6

[117] J. R. Karra & K. S. Walton, Effect of Open Metal Sites on Adsorption of Polar and Nonpolar Molecules in Metal−Organic Framework Cu-BTC. *Langmuir*, **24** (2008) 8620–8626. https://doi.org/10.1021/la800803w

[118] G. Blanco-Brieva, J. M. Campos-Martin, S. M. Al-Zahrani, & J. L. G. Fierro, Effectiveness of metal–organic frameworks for removal of refractory organo-sulfur compound present in liquid fuels. *Fuel*, **90** (2011) 190–197. https://doi.org/10.1016/J.FUEL.2010.08.008

[119] F. Glover & J.-K. Hao, The case for strategic oscillation. *Annals of Operations Research*, **183** (2011) 163–173. https://doi.org/10.1007/s10479-009-0597-1

[120] E. Haque, J. W. Jun, & S. H. Jhung, Adsorptive removal of methyl orange and methylene blue from aqueous solution with a metal-organic framework material, iron terephthalate (MOF-235). *Journal of Hazardous Materials*, **185** (2011) 507–511. https://doi.org/10.1016/J.JHAZMAT.2010.09.035

[121] L. Hamon, H. Leclerc, A. Ghoufi, L. Oliviero, A. Travert, J.-C. Lavalley, T. Devic, C. Serre, G. Férey, G. De Weireld, A. Vimont, & G. Maurin, Molecular Insight into the Adsorption of H_2S in the Flexible MIL-53(Cr) and Rigid MIL-47(V) MOFs: Infrared Spectroscopy Combined to Molecular Simulations. *The Journal of Physical Chemistry C*, **115** (2011) 2047–2056. https://doi.org/10.1021/jp1092724

Materials Research Forum LLC
https://doi.org/10.21741/9781644900291-4

[122] S.-H. Huo & X.-P. Yan, Metal–organic framework MIL-100(Fe) for the adsorption of malachite green from aqueous solution. *Journal of Materials Chemistry*, **22** (2012) 7449. https://doi.org/10.1039/c2jm16513a

[123] N. A. Khan & S. H. Jhung, Remarkable Adsorption Capacity of CuCl2-Loaded Porous Vanadium Benzenedicarboxylate for Benzothiophene. *Angewandte Chemie*, **124** (2012) 1224–1227. https://doi.org/10.1002/ange.201105113

[124] M. R. Azhar, H. R. Abid, H. Sun, V. Periasamy, M. O. Tadé, & S. Wang, Excellent performance of copper based metal organic framework in adsorptive removal of toxic sulfonamide antibiotics from wastewater. *Journal of Colloid and Interface Science*, **478** (2016) 344–352. https://doi.org/10.1016/J.JCIS.2016.06.032

[125] B. Seoane, J. Coronas, I. Gascon, M. E. Benavides, O. Karvan, J. Caro, F. Kapteijn, & J. Gascon, Metal–organic framework based mixed matrix membranes: a solution for highly efficient CO$_2$ capture? *Chemical Society Reviews*, **44** (2015) 2421–2454. https://doi.org/10.1039/C4CS00437J

[126] N. L. Torad, M. Hu, S. Ishihara, H. Sukegawa, A. A. Belik, M. Imura, K. Ariga, Y. Sakka, & Y. Yamauchi, Direct Synthesis of MOF-Derived Nanoporous Carbon with Magnetic Co Nanoparticles toward Efficient Water Treatment. *Small*, **10** (2014) 2096–2107. https://doi.org/10.1002/smll.201302910

[127] N. Rangnekar, N. Mittal, B. Elyassi, J. Caro, & M. Tsapatsis, Zeolite membranes – a review and comparison with MOFs. *Chemical Society Reviews*, **44** (2015) 7128–7154. https://doi.org/10.1039/C5CS00292C

[128] C. Wang, X. Liu, N. Keser Demir, J. P. Chen, & K. Li, Applications of water stable metal–organic frameworks. *Chemical Society Reviews*, **45** (2016) 5107–5134. https://doi.org/10.1039/C6CS00362A

[129] D. F. Sava, T. J. Garino, & T. M. Nenoff, Iodine Confinement into Metal–Organic Frameworks (MOFs): Low-Temperature Sintering Glasses To Form Novel Glass Composite Material (GCM) Alternative Waste Forms. *Industrial & Engineering Chemistry Research*, **51** (2012) 614–620. https://doi.org/10.1021/ie200248g

[130] J. A. Prince, S. Bhuvana, V. Anbharasi, N. Ayyanar, K. V. K. Boodhoo, & G. Singh, Self-cleaning Metal Organic Framework (MOF) based ultra filtration membranes--a solution to bio-fouling in membrane separation processes. *Scientific reports*, **4** (2014) 6555. https://doi.org/10.1038/srep06555

[131] Y.-X. Tan, Y.-P. He, M. Wang, & J. Zhang, A water-stable zeolite-like metal–organic framework for selective separation of organic dyes. *RSC Adv.*, **4** (2014) 1480–1483. https://doi.org/10.1039/C3RA41627E

Materials Research Forum LLC
https://doi.org/10.21741/9781644900291-4

[132] H. Wang, X. Yuan, Y. Wu, G. Zeng, H. Dong, X. Chen, L. Leng, Z. Wu, & L. Peng, In situ synthesis of In2S3@MIL-125(Ti) core–shell microparticle for the removal of tetracycline from wastewater by integrated adsorption and visible-light-driven photocatalysis. *Applied Catalysis B: Environmental*, **186** (2016) 19–29. https://doi.org/10.1016/J.APCATB.2015.12.041

[133] X.-J. Liu, Y.-H. Zhang, Z. Chang, A.-L. Li, D. Tian, Z.-Q. Yao, Y.-Y. Jia, & X.-H. Bu, A Water-Stable Metal–Organic Framework with a Double-Helical Structure for Fluorescent Sensing. *Inorganic Chemistry*, **55** (2016) 7326–7328. https://doi.org/10.1021/acs.inorgchem.6b00935

[134] S.-H. Huo, J. Yu, Y.-Y. Fu, & P.-X. Zhou, In situ hydrothermal growth of a dual-ligand metal–organic framework film on a stainless steel fiber for solid-phase microextraction of polycyclic aromatic hydrocarbons in environmental water samples. *RSC Advances*, **6** (2016) 14042–14048. https://doi.org/10.1039/C5RA26656D

[135] Y. Yu, X.-M. Zhang, J.-P. Ma, Q.-K. Liu, P. Wang, & Y.-B. Dong, Cu(i)-MOF: naked-eye colorimetric sensor for humidity and formaldehyde in single-crystal-to-single-crystal fashion. *Chem. Commun.*, **50** (2014) 1444–1446. https://doi.org/10.1039/C3CC47723A

[136] Activated Carbon. *Adsorbents Fundam. Appl.* (Hoboken, NJ, USA: John Wiley & Sons, Inc.), pp. 79–130. https://doi.org/10.1002/047144409X.ch5

[137] J. Weitkamp, M. Schwark, & S. Ernst, Removal of thiophene impurities from benzene by selective adsorption in zeolite ZSM-5. *Journal of the Chemical Society, Chemical Communications*, **0** (1991) 1133. https://doi.org/10.1039/c39910001133

[138] K. A. Cychosz, A. G. Wong-Foy, & A. J. Matzger, Liquid Phase Adsorption by Microporous Coordination Polymers: Removal of Organosulfur Compounds. *Journal of the American Chemical Society*, **130** (2008) 6938–6939. https://doi.org/10.1021/ja802121u

[139] M. Zhang, M. Bosch, T. Gentle III, & H.-C. Zhou, Rational design of metal–organic frameworks with anticipated porosities and functionalities. *CrystEngComm*, **16** (2014) 4069–4083. https://doi.org/10.1039/C4CE00321G

[140] S. Rapti, A. Pournara, D. Sarma, I. T. Papadas, G. S. Armatas, A. C. Tsipis, T. Lazarides, M. G. Kanatzidis, & M. J. Manos, Selective capture of hexavalent chromium from an anion-exchange column of metal organic resin–alginic acid composite. *Chemical Science*, **7** (2016) 2427–2436. https://doi.org/10.1039/C5SC03732H

[141] A. D. Levine, G. Tchobanoglous, & T. Asano, Size distributions of particulate contaminants in wastewater and their impact on treatability. *Water Research*, **25** (1991) 911–922. https://doi.org/10.1016/0043-1354(91)90138-G

[142] Y. Peng, H. Huang, D. Liu, & C. Zhong, Radioactive Barium Ion Trap Based on Metal–Organic Framework for Efficient and Irreversible Removal of Barium from Nuclear Wastewater. *ACS Applied Materials & Interfaces*, **8** (2016) 8527–8535. https://doi.org/10.1021/acsami.6b00900

[143] Z. Hasan, E.-J. Choi, & S. H. Jhung, Adsorption of naproxen and clofibric acid over a metal–organic framework MIL-101 functionalized with acidic and basic groups. *Chemical Engineering Journal*, **219** (2013) 537–544. https://doi.org/10.1016/J.CEJ.2013.01.002

[144] K. Wang, C. Li, Y. Liang, T. Han, H. Huang, Q. Yang, D. Liu, & C. Zhong, Rational construction of defects in a metal–organic framework for highly efficient adsorption and separation of dyes. *Chemical Engineering Journal*, **289** (2016) 486–493. https://doi.org/10.1016/J.CEJ.2016.01.019

[145] X. Zhao, K. Wang, Z. Gao, H. Gao, Z. Xie, X. Du, & H. Huang, Reversing the Dye Adsorption and Separation Performance of Metal–Organic Frameworks via Introduction of $-SO_3H$ Groups. *Industrial & Engineering Chemistry Research*, **56** (2017) 4496–4501. https://doi.org/10.1021/acs.iecr.7b00128.

Metal-Organic Framework Composites - Volume I
Materials Research Foundations **53** (2019) 105-121

Materials Research Forum LLC
https://doi.org/10.21741/9781644900291-5

Chapter 5

High Performance Polymer Fibre-Metal Matrix Composites of Metal-Organic Frameworks - Metallization, Processing, Properties and Applications

D. Kumaran[1], K. Padmanabhan*[2], and A. Rajadurai[3]

[1] SRM Institute of Science and Technology, India

[2] Vellore Institute of Technology, India

[3] Madras Institute of Technology, India

* padmanabhan.k@vit.ac.in

Abstract

Metal-Organic Frameworks (MOFs) are increasingly being applied in space, automotive, electronic and aerospace applications. In this chapter, high performance engineering polymeric fibres are demonstrated to not only reinforce low melting metals and alloys due to their superior mechanical properties, high thermal resistance up to about 650 °C and resistance to corrosive environment but also light weight resulting composites by reducing the densities. High-performance polymer fibres have superior mechanical properties like specific strength and stiffness that enable them for structural applications. Though high-performance polymer fibres like Kevlar® and Zylon® were developed two to three decades back, they were still not used as a reinforcement in metal matrix composites, but were used only with other polymer matrices, which generally have good wettability and low processing temperature which are sufficiently low enough not to degrade the reinforcing polymers but also provide cost effective product level solutions with superior performance. Some of the reasons for not using high-performance polymer fibres as a potential reinforcement in metallic materials are their low wettability, poor interfacial properties, and low degradation or heat distortion temperatures in relation to the metal matrix, when processed by conventional manufacturing processes. This chapter presents techniques of metallizing the high-performance polymer fibre surfaces for compatibility with the metallic matrices, using techniques such as electroless coatings, Radio Frequency (RF) ion sputtering and High Velocity Oxygen Fuel (HVOF) thermal spray processes. The metallic coatings improve the wettability of the polymer fibre with the metal matrix and protect the fibre from high thermal gradients that arise during

Materials Research Forum LLC
https://doi.org/10.21741/9781644900291-5

processing. Thus superior thermo-mechanical properties are seen to result from the novel method. The need to process such metal-polymer composites are that the evaluated mechanical properties are comparatively superior to ceramic-metal or glass-metal composites with considerable weight savings over these materials. The corrosion and hygrothermal resistance can be accurately designed to suit the expected durability. This chapter also presents metal-polymer composites made by the HVOF thermal spray technique with low melting metallic matrices. This technique employing additive manufacturing by incorporating high-performance polymers as reinforcements, is the first of its kind in reported literature. It was found that electroless coatings, and RF ion sputtering of copper metal on high-performance polymers, HVOF thermal spray of aluminium on Zylon® fibres are effective ways of metallizing organic high-performance polymer fibres. However, the coating thickness optimization for the desired properties are case specific and method specific. The metallized polymers can be used as reinforcements for thermal-structural applications or as function members in radiation shielding. The high-performance polymer reinforced metal matrix composites can thus be used for structural, thermal, electrically conducting and lightning applications in aerospace, automotive and marine domains with considerable weight savings and performance superiority that save energy consumption.

Keywords

Metal-Organic Frameworks (MOFs), High-Performance Polymers, Metal Matrix, Reinforcement, Metallization, Thermal Spray, HVOF

Contents

1.   Introduction

High-performance polymers (HPP) [1] are engineered synthetic polymers developed for special structural and thermal applications. They are used as reinforcements in polymer matrix composites, and in fibre metal laminates (FMLs) owing to their high tensile strength and/or high tensile modulus. The tensile strength is in the range 3 GPa to 6 GPa, and tensile modulus between 50 GPa to 600 GPa inclusively. Some polymers, in addition to their high strength and stiffness also exhibit high temperature resistance and high thermal stability, whose degradation temperature is greater than 723.15 K (450 °C) or more for short duration, and can perform satisfactorily around 450.15 K (177 °C) for longer duration [2]. The advantages of such high-performance polymers are evident from their high specific strength and specific modulus owing to their low density. PBO (Poly P Phenylene 2,6 Benzo Bis Oxazole), poly aramid, PBT (Poly P Phenylnene 2,6 Benzo Bis Thiazole), and PBI (Poly Benzimidazole) are some such high performance polymers. Such high-performance polymers were not used as a reinforcement in metal matrices (MMCs) due to drastic variation in the properties of the reinforcing polymers and the matrix material, which leads to poor wettability and poor interfacial adhesion when processed by conventional processing techniques like liquid metallurgy techniques. If successfully reinforced, such HPP reinforced MMCs offer wide range of applications in automobile and aero-space and marine industries which necessitates the development of some novel technique to utilize such HPPs as a potential reinforcement in metal matrices without degrading the properties of the polymers.

This novel work confines itself mainly to the design and development of high-performance polymer fibre reinforced metal matrix composite materials by HVOF (High Velocity Oxygen fuel) thermal spray technique and to study the properties like yield strength, ultimate tensile strength, tensile modulus, specific tensile modulus and bulk density of HPP reinforced MMCs. Some successful experiments on metallization of HPP were also presented. This work reports the use of such polymer reinforced composites

Metal-Organic Framework Composites - Volume I   Materials Research Forum LLC
Materials Research Foundations **53** (2019) 105-121   https://doi.org/10.21741/9781644900291-5

developed by HVOF thermal spray technique, with PBO fibres as a reinforcement and with thermal sprayed aluminium as a matrix material. Other metals and alloys like zinc, magnesium its alloys can also be used in place of aluminium, and any other HPP or its combination can be used in place of PBO filament fibres to develop the HPP reinforced MMCs. Magnesium which is inflammable under open air liquid casting can be successfully used as a matrix material through novel processes.

2.   Materials

2.1   Poly-benzobisoxazole (PBO)

Poly-benzobisoxazole (PBO), (IUPAC name: Poly(p-phenylene-2,6-BenzOxazole)), is a super fibre with high temperature resistance that degrades fully at about 650 °C, and has very high resistance to organic solvents [3,4]. It has high resistance to heat and flame and has excellent mechanical properties [1]. The limiting oxygen index (LOI) is 68, which is the highest among organic super fibres making it highly flame retardent. The PBO fibres used for the present study were PBO-AS (PBO As Spun) supplied by the Toyobo Co. Ltd. Japan, in trade name Zylon AS whose properties are given in Table I.

Table 1: The properties of high-performance reinforcing fibre, PBO

Property	Tensile Strength (GPa)	Tensile Modulus (GPa)	Density ($\times 10^3$ kg/m^3)	LOI %	Heat Resistance (K) Up to 20 minutes
Value	5.8	180	1.54	68	923.15

2.2   Aluminium

Pure aluminium metal wire, 1050 grade, 99.5 % purity and of 4 mm in diameter, supplied by MetcoIndia Pvt. Ltd., was used as a feedstock for HVOF thermal spray method.

3.   Metallization of HPP

Bonding at the interface of the reinforcing phase and that of the surrounding matrix phases should have some interaction that keeps them bound to together. The continuity of these constituents could be maintained by means of three mechanisms namely: i) Mechanical interlocking or mechanical coupling, which occurs when the interfacial surfaces are rough enough to form a lock and key mechanism, which physically prevents the relative movement of the constituent materials. ii) Intra-molecular forces like Van der Waal's forces, which arise due to atomic or molecular level interactions at the interface surface. Such intermolecular forces are weak and so a physical separation of the constituents is easily possible. iii) Chemical interactions between the constituent materials results in formation of new phases, called interphase. Such interphases have properties different form each of the reacting constituent materials, which might influence positively and/or as well as negatively, depending upon the nature and properties of the interphase. In some cases, the interactions between the constituents might be poor owing to the chemical nature of the constituents or the conditions of processing. In such cases, additional material, which improves the interfacial property with that of the reinforcement and matrix, should be introduced which necessitates coating the polymer fibres with metals or alloys.

Reinforcing HPP fibres are usually surface cleaned to remove contaminants like grease, oil, dirt and thus to effect good interface with the matrix material, otherwise slipping or de-bonding of the fibres might occur at low stress or load levels. Clean surface of fibres ensures good chemical bonding, Van der Waal's bonding and mechanical friction. At times, clean surfaces are not enough to form better interface and/or interphase between the reinforcing material and the matrix material. Hence, they are coated with appropriate coating material to improve the overall properties of the composites. Thus, a coating might be used to improve few aspects like wettability of the reinforcement and the matrix [5], prevent damage to the reinforcements during handling and processing, form good mechanical interlocking adhesion, form metallurgical or chemical bonds, and totally improve the adhesion at the interfaces. Metallization of reinforcements is done for this purpose. Kumaran et al. [6] have demonstrated the need to metallize the surface of PBO fibres with electroless copper coating before liquid infiltration casting with LM6 Al-Si Alloys. A number of techniques are available for metallizing the surface of polymeric materials, three of which were attempted in this work, which are described below.

4. Electroless coating

Electroless coating is a process of coating metal by chemical means without the application of electric driving force, in which the substrate material (polymer) to be coated is immersed in a suitable chemical solution containing a reducing agent, so that metal ions present in the solution are deposited on the activated substrate surface [7]. Electroless coating can be used for metallizing non-conductors including polymers and ceramics. Since polymer fibres are non-conductors of electricity, this method was resorted to for metallizing the polymer fibres to improve the wettability of the fibres with the metal matrix for the development of HPP reinforced MMC. It was already reported [6,8] that PBO fibres can be electroless copper coated, and that copper is the most suitable material that has very good adhesion with the PBO fibres when compared with

Metal-Organic Framework Composites - Volume I   Materials Research Forum LLC
Materials Research Foundations **53** (2019) 105-121   https://doi.org/10.21741/9781644900291-5

other materials like nickel, silver, and gold. Copper metal also has higher melting point of 1358.15 K, which is much higher than the processing temperature of the metal matrix materials considered here, and also forms chemical compounds with that of suitable matrix materials which can be inferred with the help of phase diagrams. The high melting temperature of copper coating on the polymer fibres also forms a protective shielding against oxidation of the fibres during processing along with the metal matrices.

5.   Pre-treatment of PBO fibres

The PBO fibres were pre-treated before the actual electroless copper coating process [8]. This pre-treatment prepares the surface of the polymer fibres for electroless copper coating. This process starts with washing the PBO fibres with ethanol for surface cleaning, followed by etching with a solution of mixture of H_2SO_4 and HCL, to create an increase in the surface area by forming a rough undulated surface. Such rough surface also improves mechanical interlocking of the metal coating. The etched fibres were treated further by a solution of NaOH for acid neutralization, and further neutralization with CH_3COOH. The fibres were further treated with a mixture of solution of $SnCl_2$, $PdCl_2$ and HCl for forming the active sites for copper deposition. Finally, H_2SO_4 was used as an accelerator to speed up the electroless copper coating process.

6.   Copper coating of PBO fibres

The surface activated PBO fibres from the pre-treatment were immersed in an electroless copper coating bath solution containing copper sulphate-penta-hydrate, sodium salt of ethylenediamine-tetraacetic-acid, formaldehyde, sodium sulphate, sodium acetate, and Potassium Cyanide, for 30 min at 303.15 K, in basic pH range, maintained by ammonia buffer [8]. From the prepared electroless copper coating bath solution the copper ions present in the electrolyte solution get coated on the activated PBO polymer fibres as metal copper. The copper coated fibres were then washed with distilled water and dried in a hot air oven for 120 min at 333.15 K. Then electroless copper coated PBO fibres were characterized using SEM, XRD, and TGA. Experiments were also conducted with solution baths at temperatures other than room temperature and the results of copper coating was only slightly better but could not offset the effort.

SEM image of pre-treated PBO fibres reveals a slightly rough surface morphology relative to the un-treated or as received PBO fibres shown in Figure 1 (a) and Figure 1 (b). This might be due to acid etching of PBO fibres. Some palladium and tin deposits on the active sites on the surface of the PBO fibres were found [9]. The average diameter of the pre-treated fibres was estimated by taking the mean of nine measurements, with three

Metal-Organic Framework Composites - Volume I Materials Research Forum LLC
Materials Research Foundations **53** (2019) 105-121 https://doi.org/10.21741/9781644900291-5

measurements on each filament fibre, measured perpendicular to the longitudinal axis of the fibres, and was found to be 11.65 µm. The reduction in the average diameter was attributed to the uniform etching caused by the etchants during the pre-treatment.

The average diameter of the electroless copper-coated polymer filament fibres was estimated to be 13.55 µm. It was inferred from the SEM image by comparing with the pre-treated PBO fibres, that copper was electroless deposited on the active surface of the pre-treated PBO fibres as shown in Figure 1 (c). The surface morphology also reveals that copper was uniformly coated on the entire surface of the PBO polymer fibres. However, it was also observed that there were some local depositions of copper at the active sites, than the rest of the non-active sites. This is because of the preferential deposition of copper ions at the active sites, which continue to deposit until the entire PBO fibre is coated.

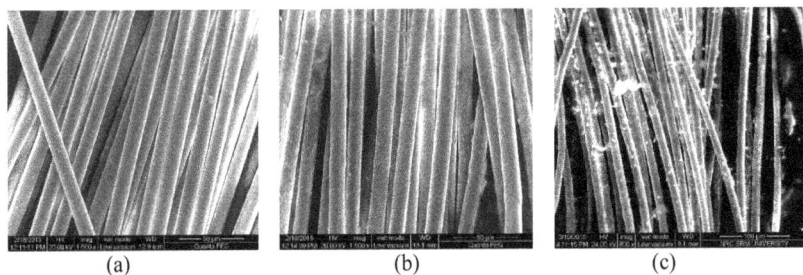

(a) (b) (c)

Figure 1: (a) Un-treated PBO fibres, (b) Acid treated PBO fibres, (c) Electroless Copper Coated Fibres

7. TGA characterization

The Thermogravimetric analysis (TGA) curve for the pre-treated PBO fibres is shown in Figure 2 (a) reveals that the degradation temperature of the pre-treated fibres was 916.52 K (643.37 °C), which was slightly less than the un-treated or as-received fibres. The decrease in the degradation temperature of the PBO fibres is possibly attributed to the increase in the surface area caused by the etched surface. A relatively steeper slope for the pre-treated PBO fibres was observed when compared to the un-treated PBO fibres, until degradation temperature was reached, was attributed to the presence of more volatile substances deposited during the pre-treatment process and to the increase in the surface area formed due to etching.

The TGA curve of the electroless copper coated PBO shown in Figure 2 (b) reveals that the degradation of electroless copper coated PBO fibres starts at 951.22 K (678.07 °C), which was slightly higher than the 916.52 K of the pre-treated PBO fibres. The relatively higher rate of weight loss before the degradation temperature was attributed to the presence of volatile substances present in the electroless deposited PBO fibres and due to moisture. The slight increase in the degradation temperature was attributed to the thin layer of electroless copper coating on the PBO surface. It was inferred that the coated fibres could not reach a temperature as high as that of copper metal for the reason that possibly the PBO fibres starts degrading around 650 °C and starts building up partial pressure from inside on the copper coating to rupture the same, which eventually exposes the PBO fibres to degradation.

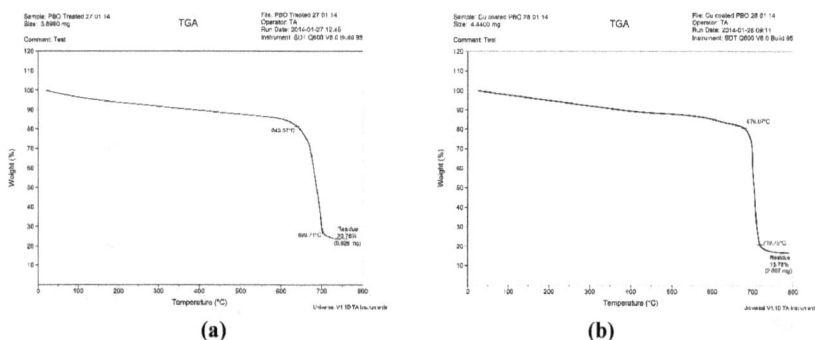

(a)   (b)

Figure 2: (a) Pre-treated PBO fibres, (b) Electroless Copper coated PBO fibres

8.   HVOF thermal spray coating

High Velocity Oxy-fuel Flame (HVOF) thermal spray is a technique in which feedstock of metal or alloy in the form of powder or in the form of wire is fed through a torch where they are instantaneously melted by the high temperature flame produced by a mixture of combustion gases, oxygen and LPG. The molten liquid metal or alloy is then atomized at the nozzle tip by high pressure expanding gases and then accelerated towards the substrate for deposition. The schematic diagram of HVOF thermal spray process is shown in Figure 3.

In metallizing the reinforcing HPP (PBO) by HVOF thermal spraying, the PBO fibres were first wound on rectangular steel frames of SS304 of 0.5 mm thickness, and 80 mm x 50 mm inner dimensions, and 10 mm width on all sides or on C frames of 2 mm diameter. The polymer fibres were wound manually, and barely tight enough (approximately 1 Newton) not to get loosened during the process of thermal spraying. The fibres were wound such that there is no overlapping of fibres that prevents the deposition of thermal spray material. The wound fibres were cleaned with acetone to remove surface contaminants like dust, grease, oil etc. that might have deposited in the due course of handling and winding. The cleaned framework of polymer fibres was dried in hot-air oven for 120 min. at 333.15 K to remove volatile substances and moisture present at the surface of the polymer fibres. The wound, cleaned and dried fibres were placed on a flat horizontal surface and then thermal sprayed with the material of choice. The PBO fibres were HVOF thermal sprayed with aluminium metal perpendicular to the plane of windings, first, on one side of the wound fibres and then on the other side after flipping the frame. This ensures that the fibres were well coated on both the sides of the wound bundle. The mean diameter of the aluminium wire feedstock used was 4.0 mm. The other parameters of thermal spray were LPG, with 84 % CH_4 at 48 Litres/min., with an oxygen flow rate of 2 Litres/min., feedstock feed rate of 100 mm/min., spraying distance or stand-off distance of 300 mm and a spray angle of 60° to the horizontal.

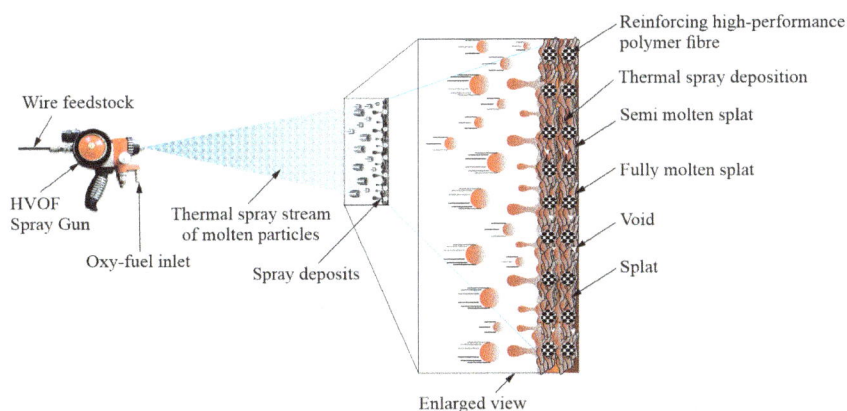

Figure 3: Schematic diagram of HVOF thermal spray process

HVOF thermal spray aluminium coated PBO fibre morphology and the nature of initial coating made on a single supporting wire frame is shown in Figure 4.

It was also inferred from Figure 4 that the wound PBO fibres were in place, as-wound, and without defibrillation. There was no visible colour change despite of the exposure of PBO fibres to the temperature of HVOF thermal spray stream. It was also observed that the aluminium coating was adhering well and did not peel off during handling. The external surface appeared to be rough due to the aluminium HVOF coating.

Figure 4 : Aluminium HVOF thermal sprayed PBO fibres

SEM images of HVOF thermal sprayed aluminium coated PBO fibres shown in Figure 5 (a) and Figure 5 (b) reveal that the thermally sprayed aluminium has bonded the individual filaments in the bundle together. Such a binding is predicted to increase the transverse stiffness to flexure. Aluminium thermal spray coating on PBO reveals that the atomized particles of aluminium were deposited on single filaments with a sound or a splat. The diameter of the aluminium splashes deposited on the fibre were in the range of 1 μm to 3 μm and were smaller than the mean diameter of individual PBO filament fibres used. The SEM image also reveals that the rough surfaces are formed as a result of randomness of the thermal spray deposit, and because of the random joining of the filament fibres into a single strand.

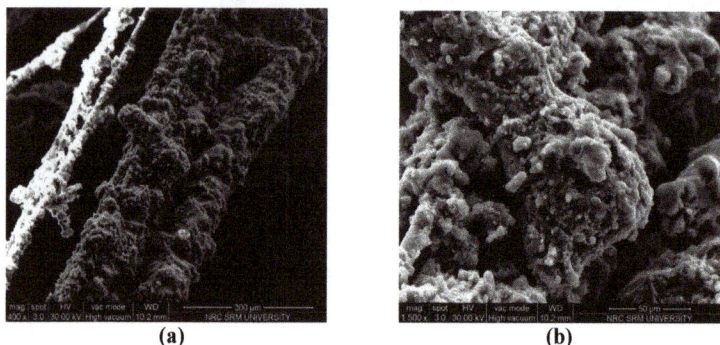

(a)   (b)

Figure 5 : SEM micrograph of Aluminium HVOF thermal sprayed PBO fibres

It was inferred that the increase in the surface roughness on micrometer scale and waviness in micrometer to millimeter scale could possibly contribute to the improved mechanical interlocking of the fibres along their longitudinal direction [10]. It was also expected that the aluminium coating might improve the wettability and adhesion [5] of the matrix material with the aluminium thermal spray coated PBO fibres.

The TGA curve of HVOF aluminium thermal spray coated PBO fibres is shown in Figure 6 . It reveals that the degradation of HVOF thermal spray aluminium coated PBO fibres starts at 938.69 K (665.54 °C). There is a steady weight loss recorded before the degradation temperature due to the evaporation of volatile substances present in the HVOF thermal spray aluminium coated PBO fibres. The slight increase in the degradation temperature was attributed to the melting of relatively thick aluminium thermal spray coating of about 100 μm, at 933.15 K until the PBO fibres were exposed and degrade further, in addition to an existing weight loss of about 5 %.

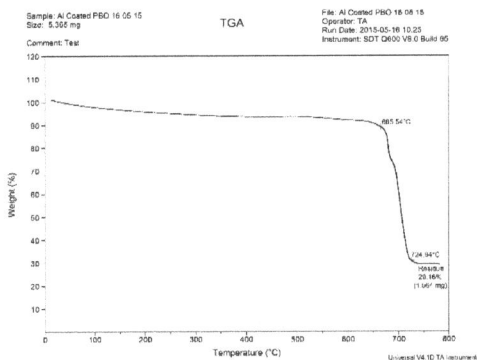

Figure 6 : TGA curve of Aluminium HVOF thermal sprayed PBO fibres

9. HVOF thermal spray high performance polymer reinforced MMCs

The development of HPP reinforced MMCs by HVOF thermal spray technique is similar to that outlined in the HVOF thermal spray coating processes. The changes effected for the development after thermal spray coating of the frame wound reinforcing HPP were as follows: HVOF Thermal spray coated fibres were further thermally sprayed by stacking them one over the other, inside a die, until the required matrix material was built-up on each of the stacked layers by the thermal spray processes. Once one layer of matrix was built up on the previous frame-wound fibres, then the next frame was placed and thermally sprayed. The process was repeated until a required volume fraction or required thickness was achieved. The thermal sprayed HPP reinforced MMCs were trimmed by water jet machining, to remove the stainless steel frame, and further surface finished to make the specimens for tensile testing and characterization. Some physical properties of the developed composite are presented in Table 2.

Materials Research Forum LLC

https://doi.org/10.21741/9781644900291-5

Table 2: Physical Properties of PBO reinforced Aluminium thermal sprayed composites

Reinforcement Volume (%)	Actual Density ($\times 10^3$ kg/m^3)	Theoretical Density ($\times 10^3$ kg/m^3)	Porosity (%)
0	2.56	2.7	5.19
1	2.52	2.69	6.32
2	2.48	2.68	7.46

The results of density indicate that the density of the HVOF thermal sprayed composite was lesser than the theoretical density computed from the properties of the constituent materials. It was observed that the actual density of the thermal sprayed PRMMC decrease with increase in volume fraction for all thermal sprayed HPP reinforced MMCs. This is due to the increased volume fraction of reinforcing polymers of relatively low density than the matrix material; and due to the increase in void volume fraction with increase in volume content of reinforcement during the thermal spray process.

Porosity measurements were made by calculating the difference in actual and theoretical bulk densities in the thermal sprayed composites. Total porosity of 5.19 % was obtained for a baseline aluminium thermal spray composite. This may be due to a higher amount of voids present in the thermal spray of bulk material, which is higher than what was reported for general HVOF thermal spray coating process [11]. This higher value of porosity is mainly attributed to the deviation of flatness of the surface during the HVOF thermal spray process, which is usually obtained when thermal-sprayed on a flat surface. Thus, an uneven substrate surface increases the chances of formation of voids, when compared with a thin layer of coating on an even substrate. The porosity of the reinforced MMCs slightly increases compared to the unreinforced thermal spray matrix, with increase in volume fractions of the HPP fibres. This is due to the formation of voids at the fibre-matrix interfaces, at the down side of the reinforcing fibres masked by an upper layer of the same reinforcement.

Optical Micrographs of cross-sections of PBO reinforced aluminium HVOF thermal sprayed MMCs is shown in Figure 7 (a) and Figure 7 (b). The images re

veal a uniform deposition of wet aluminium particles with a splat forming a dried up matrix portion around the HPP reinforcing fibres. The splats were randomly deposited. Micro-voids of cross sectional dimensions from 1 μm to 10 μm were found on the micrograph of the developed aluminium HVOF thermal sprayed MMC. These voids were distributed randomly and were formed as a result of deposition of aluminium with a splat like sound.

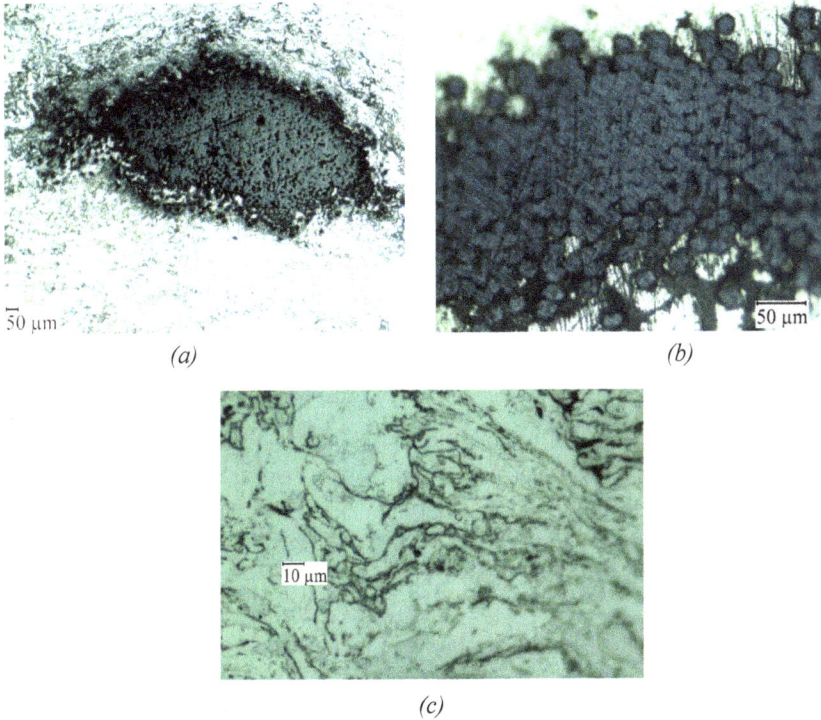

(a)   *(b)*

(c)

Figure 7: (a) Optical Micrograph of cross sectional view of PBO reinforced aluminium thermal sprayed MMC, (b) magnified cross sectional view showing adhesion of matrix material to the filament fibres in a single strand of reinforcing PBO, (c) flow of matrix material as a result of HVOF thermal spray forming.

The micrograph of PBO cross-section reinforcements in aluminium HVOF thermal sprayed MMC also reveals that the aluminium spray splats fill the space between the fibres adequately. However, some micro-voids were also observed at a few polymer fibre-matrix interfaces as well as in the thermal sprayed matrix. It is clearly visible that the aluminium thermal spray has even filled the space between the reinforcing polymer fibres of lesser diameters; say 10 μm. It is also evident from Figure 7 (c) that molten aluminium stream has completely covered even the surface of single filament fibres. However, filament fibres grouped together in a strand were also observed.

The test results of the tensile test is presented in Table 3. It was observed that the HVOF thermal sprayed un-reinforced aluminium material, which was used as a baseline reference, possess a tensile strength of 83 MPa, which was less than the tensile strength of bulk aluminium metal developed by casting. This reduction in the tensile strength is mainly attributed to the presence of pores in the thermally sprayed aluminium that was formed from 1050 Grade commercially pure 99.5 % wires of 4 mm diameter. The pores act as stress concentrators and possess a stress concentration factor of up to 3 near their edges that reduce the near field yield strength. With 1 % reinforcement volume of PBO fibres, the tensile strength of the PBO reinforced HVOF aluminium thermal sprayed MMC exhibits a tensile strength of 141.2 MPa, which is 70.12 % higher than the unreinforced one, and with 2 % reinforcement volume of PBO fibres the tensile strength of 193.4 MPa was obtained which was 133.01 % higher than the unreinforced HVOF aluminium thermal sprayed material. It was generally observed that there was an increase in the tensile strength of the HVOF aluminium thermal sprayed MMCs, with increase in the volume percentage of PBO fibre reinforcement, which was contributed by the high tensile strength of the reinforcing PBO fibres. The increase in the tensile strength of the MMC with fibre additions was not in line with the predictions of the rule of mixtures. Unreinforced aluminium thermal spray specimen has a yield strength of 12.4 MPa. Reinforcement with treated PBO fibres has improved the yield strength as 35.9 MPa and 55.2 MPa for 1 % and 2 % volume additions respectively which was 189 % and 377 % higher than the unreinforced aluminium specimens.

The high-performance polymer reinforced metal matrix composites can thus be used for structural, thermal, electrically conducting and lightning applications in aerospace, automotive and marine domains with considerable weight savings and performance superiority that save energy consumption. As metal matrices are used that have lightning conducting capabilities, the high performance polymer reinforcement in MOFs would save weight and significantly improve the mechanical properties like strength whereas the metal alloy matrices would conduct electricity and lightning which is impossible with plastic matrices.

Table 3:   *Mechanical Properties of PBO reinforced Aluminium thermal sprayed composites*

Reinforcement Volume (%)	Yield Strength (MPa)	Tensile Strength (MPa)	Tensile Modulus (GPa)	Percentage Increase in Tensile strength (%)
0	12.4	83	57	-
1	35.9	141.2	61.2	70.12
2	59.2	193.4	62.4	133.01

Metal-Organic Framework Composites - Volume I   Materials Research Forum LLC
Materials Research Foundations **53** (2019) 105-121   https://doi.org/10.21741/9781644900291-5

Conclusions

This chapter presents the conclusions drawn based on the experimentation made on MOFs during the process of development of high-performance Polymer (HPP) reinforced Metal Matrix Composites (MMCs). Based on the qualitative and quantitative evidence captured and estimated during the characterization and tensile testing of the developed MMCs of various reinforcing fibre volume fractions by HVOF thermal spray techniques, the following deliverables are documented;

1. The HVOF thermal spray technique was found to be a method for developing directionally reinforced Polymer Reinforced MMCs.

2. Aluminium alloys can be used as a matrix material for PBO reinforced HVOF thermal sprayed MMCs.

3. Porosity of around 4 % was observed for aluminium HVOF thermal spray composites.

4. The density of the HVOF thermal sprayed MMCs decreases due to presence of voids and due to low density reinforcement polymer fibres as the volume fraction of reinforcement increases.

5. The tensile strengths of the thermal spray HPP reinforced MMCs were found to be higher than the unreinforced base-line matrix materials. However, the base-line matrix have lower strength than bulk material property due to the presence of pores in the HVOF thermal sprayed matrix.

6. The tensile modulus of the HVOF thermal sprayed Polymer Reinforced MMCs showed a slight increase when compared with the unreinforced as-sprayed matrix material due to the impregnation of a small volume fraction of the high modulus polymer fibre reinforcement.

Limitations and further research

The topic of high-performance polymer reinforced MMCs offers a novel and promising area in which materials of widely differing properties can be combined in a way to produce synergistic materials with in the topic of MOFs. Though the present study is a good beginning for the development of HPP reinforced MMCs, it also identifies good amount of opportunities for further research in-line with the current research work and its findings, which are as follows:

- A closer study on reinforcement-matrix interface could be made, including the study on interfacial adhesion that could be investigated.

Materials Research Forum LLC
https://doi.org/10.21741/9781644900291-5

- Detailed studies on wettability could be made.

- Characterization techniques such as Micro Computed Tomography / Transmission Electron Microscopy could be made to study the nature of bonding between high-performance reinforcing polymer fibres subjected to HVOF thermal spray metallic coatings.

- New methodologies could be devised to incorporate higher volume content of reinforcing high-performance polymers without degradation of properties.

Acknowledgement

The authors thank the managements of SRM, VIT and MIT-Anna University, for the support and encouragement.

References

[1] J.E. Mark, Polymer Data Handbook, Oxford University Press, 1999.

[2] P.M. Hergenrother, 'The Use, Design, Synthesis, and Properties of High Performance/High Temperature: an overview', High Performance Polymers, 15 (2003) 3-45. https://doi.org/10.1177/095400830301500101

[3] T. Kitagawa and K. Yabuki, Jl. Polym. Sci: Part B, Polymer Physics, 38(2000) 2901-2911.

[4] T. Kitagawa and K. Yabuki, Jl. Polym. Sci: Part B, Polymer Physics, 38 (2000) 2937-2942.

[5] S. Pourhosseini, H. Beygi, & S.A. Sajjadi, 'Effect of metal coating of reinforcements on the microstructure and mechanical properties of Al-Al2O3 nanocomposites, Materials Science and Technology,34 (2018) 145-152. https://doi.org/10.1080/02670836.2017.1366708

[6] D. Kumaran, K. Padmanabhan, A. Rajadurai, Design and Fabrication of High Strength, High modulus Polymer Reinforced Metal matrix Composites For Light Weight Applications, Indian journal of Science and technology, 9 (2016) 42. https://doi.org/10.17485/ijst/2016/v9i42/103304

[7] M. Radoeva, M. M. Monev, I.T. Ivanov, G.S. Georgiev, B. Radoev, 'Adhesion improvement of electroless copper coatings by polymer additives', Colloids and Surfaces A: Physicochemical and Engineering Aspects, 460 (2014) 441-447. https://doi.org/10.1016/j.colsurfa.2014.02.003

[8] M. Matsuo, H. Ishikawa, Y. Xi, Y. Bin, 'High modulus and high strength fibers with high electric conductivity prepared by copper electroless plating on the surface of poly (p- phenylenebenzobisoxazole) (PBO)', Polymer Journal, 39.4 (2007) 389-396. https://doi.org/10.1295/polymj.pj2006161

[9] H.K. Lee, J.Y. Hur, H.N. Lee, C.M Lee, 'Sn-Ag two step activation process for electroless copper plating. 2. Experimental Procedures', Latest Trends in Environmental and Manufacturing Engineering, (2012) 135-137. https://doi.org/10.1080/00218469008030185

[10] A.N. Gent, C.W. Lin, 'Model Studies of the Effect of Surface Roughness and Mechanical Interlocking on Adhesion', The Journal of Adhesion, 32 (1990) 113-125.

[11] P. Fauchais , 'Understanding plasma spraying: an invited review', Journal of Physics D: Applied Physics, 37(2004) 86-108.

Metal-Organic Framework Composites - Volume I
Materials Research Foundations **53** (2019) 122-139

Materials Research Forum LLC
https://doi.org/10.21741/9781644900291-6

Chapter 6

Condensation of WO$_4^{2-}$ Polyhedra Units on Layered Rare Earth Hydroxides Nanosheets: Hierarchical Channels and Heavy Metal Adsorption

Solomon Omwoma

Department of Physical Sciences, Jaramogi Oginga Odinga University of Science and Technology, P.O. Box 210-40604, Bondo, Kenya

solomwoma@yahoo.com

Abstract

Porous materials exhibit advantages in adsorption applications due to their ability to increase accessibility, surface area and enhanced liquid/molecular/ion transport. Condensation of WO$_4^{2-}$ polyhedra units on delaminated europium containing layered rare earth hydroxide nanosheets, $[Eu_8(OH)_{20}(H_2O)_n]^{4+}$, at pH 5 under nitrogen environment is herein reported to form a porous material with hierarchical channels characterized as $Eu_2(OH)_5[H_2W_{12}O_{40}]_{0.17} \cdot 7H_2O$. Heavy metal adsorption by this material was found to be 1.8, 2.4 and 4.1mmol/g for Cd^{2+}, Pb^{2+} and Cr^{6+} within contact times of 18, 25 and 20 minutes respectively. The sorption mechanism was pseudo-second order and obeyed the Langmuir sorption model.

Keywords

Layered Rare Earth Hydroxides, Heavy Metals, Adsorption

Contents

1.   Introduction

Toxic heavy metals such as cadmium, lead, and hexavalent chromium are ubiquitous, have no beneficial role in human homeostasis, and contribute to non-communicable chronic diseases [1, 2]. Their presence in aquatic environment, even at trace levels, is detrimental to aquatic life [3]. There exist various adsorbents for removal of this heavy metals from aqueous solutions such as activated carbon, zeolites, chitosan, waste slurry, lignin, agricultural wastes, *etc.* [4]. However, low adsorption efficiency, long contact times, high production cost and availability limit their applications [2, 5-7]. Consequently, there is need to develop synthetic, highly efficient and readily available heavy metal adsorbents. Described herein is a technique that involves condensation of WO_4^{2-} polyhedra units on delaminated europium containing layered rare earth hydroxide nanosheets to form a material with hierarchical channels, and is presumably affordable and highly efficient adsorbent for heavy metal removal from wastewater within short contact times.

Layered rare-earth hydroxides (LRHs) are anion-exchangeable lamellar compounds similar in structure to layered double hydroxides (LDHs). They consist of positively charged rare-earth hydroxide layers with exchangeable charge-balancing anions in the interlayer space [8, 9]. Intercalation of polyoxometalate (POMs) anions into LRHs/LDHs has resulted into versatile materials such as heterogeneous catalysts and adsorbents [8-10]. POMs are a class of discrete anionic metal oxides [11].

Various techniques such as ion exchange, calcination, reconstitution, self-assembly, delamination, *etc.* can be used for intercalation of POMs into LRHs/LDHs [9]. In this work, the simple aqueous delamination method was employed to generate single layer positively charged LRH nanosheets that provided a plat-form for condensation of WO_4^{2-} polyhedra units into $[H_2W_{12}O_{40}]^{6-}$ POM anions. Finally, POM anions and LRH nanosheets underwent self-assembly process to produce an intercalated LRH-POM porous material (Fig. 1).

The LRH-POM porous material was characterized and utilized in heavy metal (Cd^{2+}, Pb^{2+} and Cr^{6+}) adsorption experiments. The sorption mechanism was best described by Langmuir and pseudo second order kinetics.

Metal-Organic Framework Composites - Volume I
Materials Research Foundations **53** (2019) 122-139

Materials Research Forum LLC
https://doi.org/10.21741/9781644900291-6

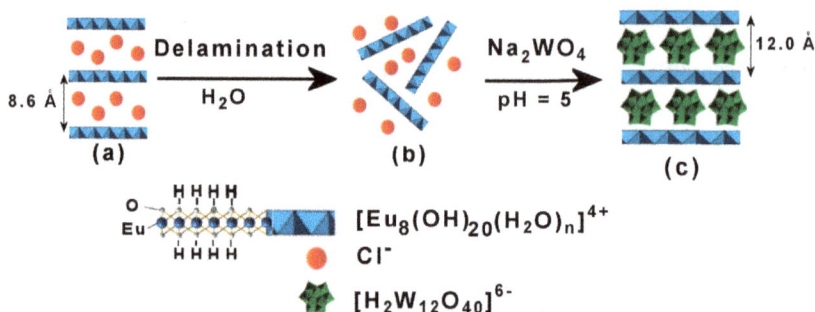

Figure 1. Proposed condensation mechanism of tungstate polyhedra units on layered rare earth hydroxides nanosheets in nitrogen environment at 70 °C: a) $Eu_2(OH)_5Cl \cdot nH_2O$ b) $[Eu_8(OH)_{20}(H_2O)_n]^{4+}$ nanosheets c) $Eu_2(OH)_5[H_2W_{12}O_{40}]_{0.17} \cdot 7H_2O$; water molecules avoided for clarity.

2. Experimental

2.1 Materials and methods

Analytically pure $K_2Cr_2O_7$, $PbCl_3$, $CdCl_3$, NaOH, HCl, H_2SO_4, KOH, KBr, Na_2WO_4 and Eu_2O_3 were purchased from Alfa Aesar-China and used as supplied without further purification. The synthesis of Eu-containing layered rare earth hydroxide, $Eu_2(OH)_5Cl \cdot nH_2O$ abbreviated as LEuH-Cl, was synthesized and characterized according to the literature methods [12-14].

2.2 Methods

EuH nanosheets ($[Eu_8(OH)_{20}(H_2O)_n]^{4+}$) were prepared by ultra-sonication of fresh wet LEuH-Cl material (0.2 g) in aqueous media for 10 minutes to delaminate the positively charged nanosheets followed by centrifugation at 2000 rpm to remove the un-delaminated material [12]. The resultant delaminated nanosheets (0.05 g) were dispersed in de-.carbonated deionized water (75 ml) at 70 °C, under vigorous stirring in nitrogen environment. The pH of the nanosheets was gradually adjusted to 5 using 0.5N HCl, and $Na_2WO_4.2H_2O$ (6.6 mM) dissolved in 150 ml of de-carbonated deionized water added while maintaining the same pH. The mixture was continuously stirred for 3 hours at 70 °C, and finally left undisturbed for 12 hours to yield a white solid which was separated, washed several times with de-carbonated deionized water and dried in vacuum at 40 °C for 12 hours. The elemental analysis (%) of the white solid revealed: Eu 32.54 %, OH 8.5 %, W 42.32 %; H_2O 12.6 %); FT-IR (KBr, cm^{-1}): 3375 (vs), 1625 (vs), 946 (vs), 864 (vs), 740 (vs), 620 (vs), 660 (vs), and 546 (s). It is important to note that preparation of LEuH-Cl in air leads to a little contamination with CO_2 due to the high affinity of LRHs to CO_2. However, it is difficult to delaminate LEuH-CO_3 hence this contaminated samples are removed during the 2000 rpm centrifugation step.

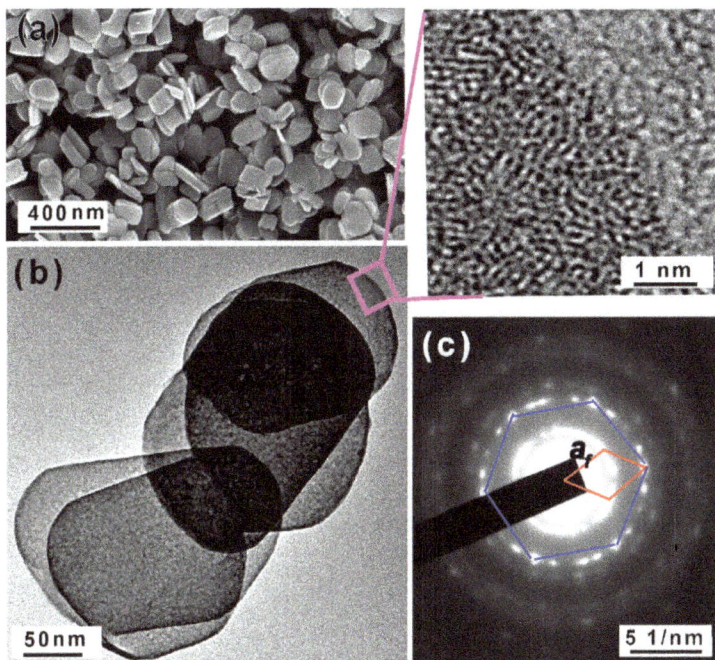

Figure 2. a) SEM image of LEuH-Cl, b) HRTEM of delaminated LEuH nanosheets, Inset: a higher magnification at the edge of a single plate lying on a copper grid. c) SAED pattern taken from an individual plate lying on a copper grid.

2.3   Characterization of the Sorbent materials

Powder X-ray diffraction (XRD) patterns were recorded on a Rigaku XRD-6000 diffractometer under the following conditions: 40 kV, 30 mA, Cu-Ka radiation (λ = 0.154 nm), scan step of 0.01° and scan range between 3° and 80°. Fourier transform infrared (FT-IR) spectra were recorded on a Bruker Vector 22 infrared spectrometer, using the KBr pellet method. Scanning electron microscopy (SEM) images and energy dispersive

X-ray spectroscopy (EDX) analytical data were obtained using a Zeiss Supra 55 SEM equipped with an EDX detector. High resolution transmission electron microscopy (HRTEM) micrographs were recorded using a Hitachi H-800 instrument. Inductively coupled plasma-atomic emission spectroscopy (ICP-AES) analysis was performed using a Shimadzu ICPS-7500 spectrometer after dissolution of a given amount of the sorbent material in *aqua-regia* solution (1:3 HNO_3:HCl). The specific surface area determination and pore volume and size analysis were performed by Brunauer−Emmett−Teller (BET) and Barrett−Joyner−Halenda (BJH) methods, respectively, using a Quantachrome Autosorb-1C-VP analyzer. Prior to the measurements, the samples were degassed at 100 °C for 6 h. The OH content was obtained by neutralization back-titration after dissolving the sample in 0.1N standard H_2SO_4. Thermo gravimetric analysis was carried out on a locally-produced HCT thermal analysis system in flowing N_2 and a heating rate of 10 °C min^{-1}.

2.4 Adsorption experiment

The as-prepared sorbent material (0.2 g) was added to different aqueous solutions (50 ml) containing known concentrations of heavy metals: Pb^{2+}, Cr^{6+} and Cd^{2+}. The test solution was then continuously stirred for 60 minutes, after which the sorbent was separated by centrifugation and the concentration of the heavy metal ions in the supernatant tested using inductively coupled plasma-atomic emission spectroscopy (ICP-AES).

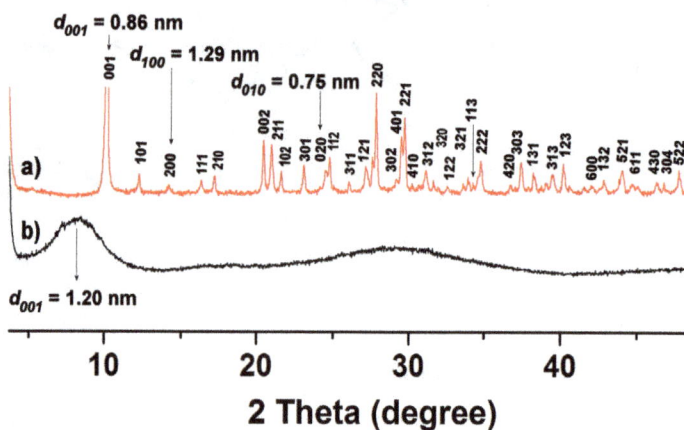

Figure 3. XRD diffraction patterns of: a) LEuH-Cl and b) LEuH-$H_2W_{12}O_{40}$ ($\lambda = 0.154$ nm).

2.5   Kinetic studies

In order to determine the adsorption rate and equilibrium constant of the reaction, ten samples (50 ml) of the test solution containing 500 ppm of the heavy metal ions were treated with 0.2 g of the sorbent material for various time periods of: 0, 2, 4, 8, 10, 12, 15, 20, 30 and 60 minutes, and then measuring the final concentrations of the heavy metal ions at the end of each period. The optimum reaction temperature and pH for the sorption experiment was established by the reaction of 0.2 g of the sorbent material with 50 ml of the test ion solution (500 ppm) at different pH and temperatures.

3.   Results and discussion

Chemical analysis of the starting material, LEuH-Cl, showed its elemental composition to be $Eu(OH)_{2.45}Cl_{0.46}(CO_3)_{0.045} \cdot 0.8H_2O$, which can simply be written as $Eu_8(OH)_{20}Cl_4 \cdot 6.4H_2O$. The SEM images of this material exhibit stacked layers of approximately 200 by 100 nm (Fig. 2a). HRTEM images of a single nanosheet lying on a cupper grid (Fig. 2b) also show the layer morphology and its selected-area electron diffraction pattern (SAED) displays crystal arrangement of atoms in a pseudo-hexagonal symmetry with a unit fundamental cell of $a_f = 3.7$ Å (Fig. 2c). It can be proved that $d_{100} = 2\sqrt{3}a_f = 12.8$ Å and $d_{010} = 2a_f = 7.4$ Å. The XRD diffraction patterns of LEuH-Cl (Fig. 3a) can be indexed in a single orthorhombic unit cell with lattice parameters of a = 12.90 Å (d_{100}) b = 7.52 Å (d_{010}) and c = 8.63 Å (d_{001}), which is in agreement with the calculated values from SAED patterns in Fig. 2c. It is worth noting that these results are in agreement with the literature on characterization of LEuH-Cl [15].

The LEuH-Cl material is easily delaminated into LEuH nanosheets by sonication [12]. The nanosheets form a colloidal solution of pH 7 in water. This pH was adjusted to 5 using 0.5 M HCl. Addition of WO_4^{2-} polyhedra anions to the colloidal solution of pH 5 led to their condensation to form $[H_2W_{12}O_{40}]^{6-}$ POM cluster. Self-assembly of the resultant condensed POM cluster and the LEuH nanosheets under nitrogen environment over a period of 12 hours yields a nanocomposite material, LEuH-$[H_2W_{12}O_{40}]$, which exhibits hierarchical channels with an interior rectangular shape (Fig. 4a). XRD diffraction patterns of this material (Fig. 3b) display a broad spectrum with a maximum intensity at $d_{001} = 12.0$ Å.

Condensation of WO_4^{2-} polyhedra units to form large POM clusters is governed by the solution pH [16]. At pH 5, $[H_2W_{12}O_{40}]^{6-}$, with Td symmetry of nearly spherical shape, is the most stable POM cluster [16, 17]. Therefore, the WO_4^{2-} polyhedra units condense on the surface of the plate-like nanosheets of LEuH to form the metatungtsate cluster $[H_2W_{12}O_{40}]^{6-}$ which is used as interlayer charge balancing anions in LEuH nanosheets

Metal-Organic Framework Composites - Volume I
Materials Research Forum LLC
Materials Research Foundations **53** (2019) 122-139
https://doi.org/10.21741/9781644900291-6

stacking during the self-assembly process. The observed morphology is suspected to be as a result of the two processes taking place simultaneously.

SAED pattern of the resultant LEuH-[$H_2W_{12}O_{40}$] material show LEuH to maintain its Eu-Eu pseudo-hexagonal super lattice structure (a_f = 3.7Å: Fig. 4e). However, the intensity of the sports corresponding to this Eu-Eu arrangement decrease, signifying the loss of LEuH layered structure arrangement.

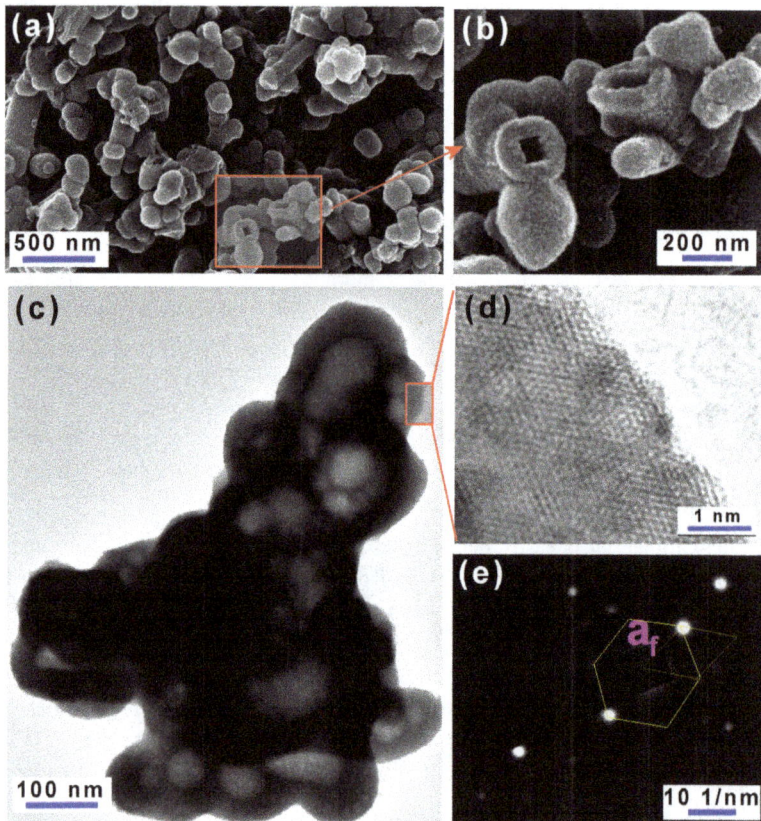

Figure 4. a) SEM images of LEuH-[$H_2W_{12}O_{40}$], Inset is a higher magnification of the hierarchical macro pore channels of the material, b) HRTEM of LEuH-[$H_2W_{12}O_{40}$], inset is a higher magnification of the edge of the material, c) SAED pattern taken from a section of the material.

However, LEuH-$[H_2W_{12}O_{40}]$ material is highly crystalline as is shown in Fig. 4d. The appearance of large sharp bright dots in the material may be attributed to the crystallinity of large POM anions in the structure (Fig. 4e).

The POM anion crystals show FTIR vibrations at 946, 740 and 664 cm^{-1} (Fig. 5). This is ascribed to asymmetric stretching of the W$-O_d$ bond, W$-O_b-$W bridge and W$-O_c-$W bridge (Table 1). These vibrations and pH controlled scale of tungstate condensation[16] point to the characterization of the POM species as $[H_2W_{12}O_{40}]^{6-}$. Energy dispersive X-ray spectroscopy metal mapping experiments (Fig. 6), TGA analysis (Fig. 7) and elemental analysis confirmed the structure characterization of the synthesized material to be $Eu_2(OH)_5[H_2W_{12}O_{40}]_{0.17}.7H_2O$.

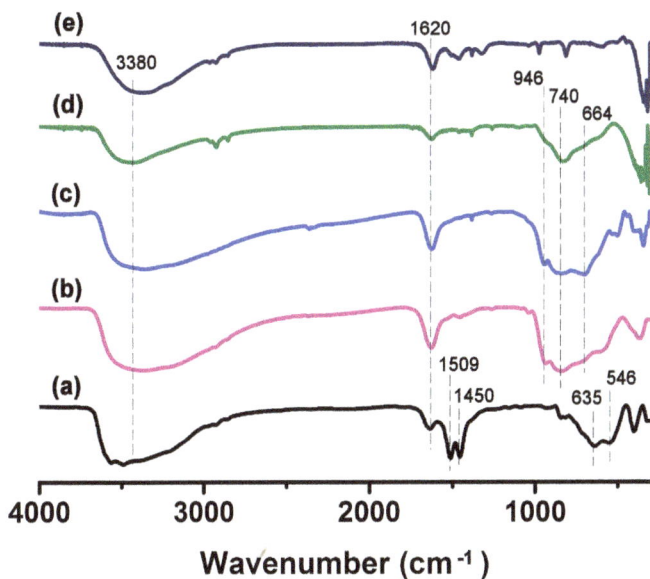

Figure 5. FT-IR vibrations of: a) LEuH-Cl, b) $Na_6[H_2W_{12}O_{40}]$, c) LEuH-$[H_2W_{12}O_{40}]$, d) LEuH-$[H_2W_{12}O_{40}]@MX_2$ e) Physical mixture of LEuH-Cl, MX_2, $M = Pb^{2+}$, Cr^{6+} or Cd^{2+}, $X = Cl$.

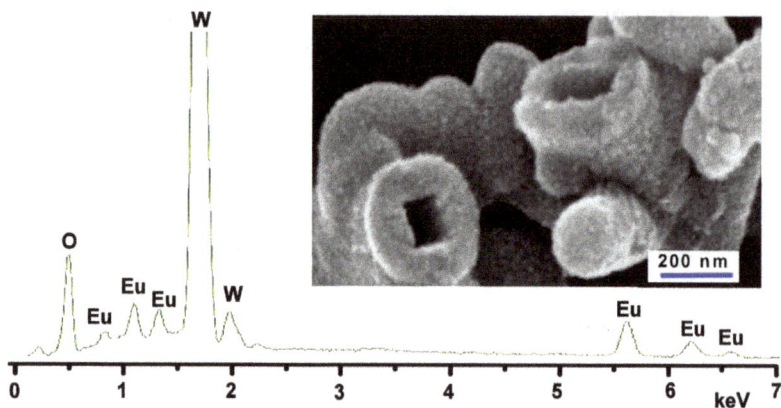

Figure 6: Energy dispersive X-ray spectroscopy of metal mapping for LEuH-[$H_2W_{12}O_{40}$]; inset is the SEM image.

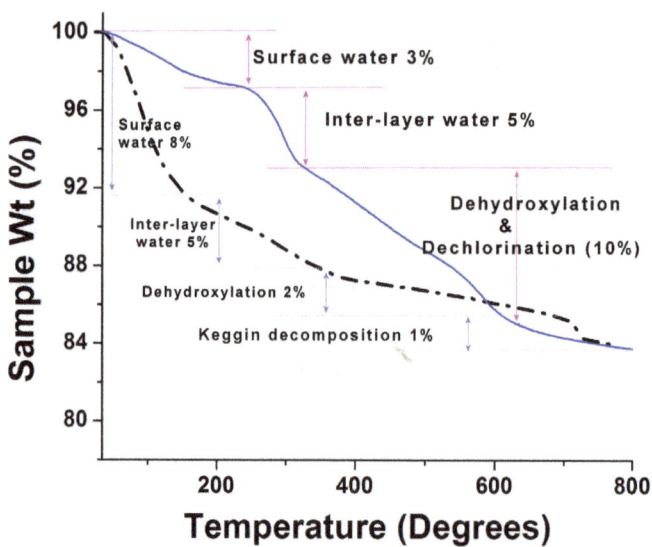

Figure 7: Thermogravimetric analysis for: a) LEuH-Cl and b) LEuH-[$H_2W_{12}O_{40}$].

Table 1. FTIR vibrations assignment

Vibration Bands (cm^{-1})	Assignment
3380	O-H stretching
1620	H_2O bending mode
1509 and 1450	CO_3^{2-} vibrations*
946, 740, 664	asymmetric stretching of $W-O_d$ bond, $W-O_b-W$ bridge and $W-O_c-W$ respectively
635 and 546	Eu-O vibrations
Below 400	Vibrations due to Pb^{2+}, Cd^{2+} and Cr^{2+}

*LEuH-Cl was prepared in air hence contaminated by CO_3^{2-}. This contamination is subsequently removed during the delamination stage as they are centrifuged off at 2000 rpm hence the subsequent materials do not have CO_3^{2-} vibrations.

Table 2. Comparison of adsorption parameters for toxic heavy metals by LEuH-[$H_2W_{12}O_{40}$] with other sorbent materials

Sorbent material	Toxic metal	Optimal sorbed amounts q_m (mmol/g)	Contact time (hours)	pH	Temp. (K)	Ref.
LEuH-[$H_2W_{12}O_{40}$]	Cr^{6+}	4.1	0.33			
	Cd^{2+}	1.8	0.30	5-7	298	This study
	Pb^{2+}	2.4	0.42			
Hazelnut shell	Cr^{6+}	0.32	72	1	323	[18]
HNO$_3$ Treated Activated Carbon	Cr^{6+}	0.30	3	5.2	298	[19, 20]
HNO$_3$ Treated Coconut Shell Charcoal	Cr^{6+}	0.21	3	5.2	298	[19]
Iron slags	Pb^{2+}	0.46	24	3.5-8.5	291	[21]
Coconut Shell Charcoal	Pb^{2+}	0.13	2	4.5	318	[22]
HNO3-treated Activated Carbon	Cd^{2+}	1.3	168	4-6	298	[23]
Jackfruit	Cd^{2+}	0.47	12	5.0	305	[24]
Sugar-cane bagasse pith	Cd^{2+}	0.22	5	5-9	301	[25]

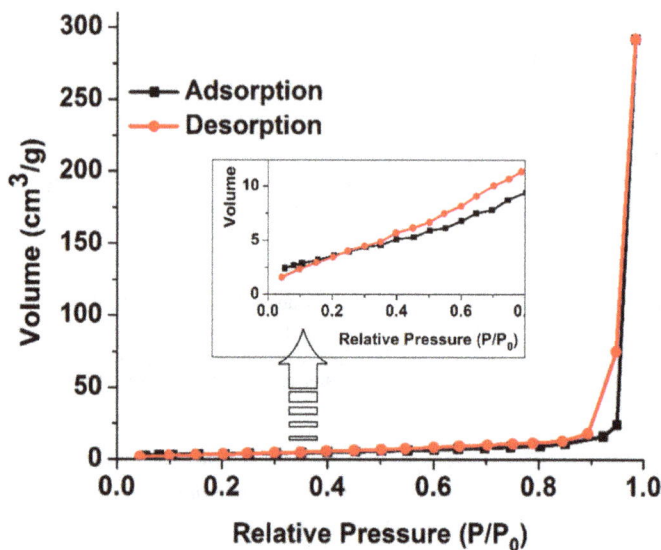

Figure 8. Brunauer–Emmett–Teller (BET) isotherms for LEuH-[H₂W₁₂O₄₀].

The formation of macro-sized hierarchical channels within the resultant material was verified with BET measurements that showed an isotherm of type ii with a rapid vertical rise near $p/p° = 1$ (Fig. 8) [26]. However, with an average pore volume of 0.452cm^3/g, BET surface area of 13.67m^2/g and average pore radius of 66.1nm, the as-synthesised material is considered as a poor physico-sorption adsorbent. Nevertheless, the experimental results on heavy metal sorption ability of LEuH-[H₂W₁₂O₄₀] show its effectiveness as a sorbent material (Fig. 5 and 9).The sorbent has a loading rate for Cd^{2+}, Cr^{6+} and Pb^{2+-} of 1.8, 4.1 and 2.4 mmol/g within contact times of 18, 20 and 25 minutes respectively (Fig. 9 and 10). Its adsorption rates qualify it as the best among the top three best sorbent materials reported so far for heavy metal adsorption from aqueous solutions (Table 2). In addition to short conduct times reported herein, the sorption efficiency of LEuH-[H₂W₁₂O₄₀] was not affected by low or high levels of initial sorbate (Cr^{6+}, Pb^{2+}, Cd^{2+}) concentration.

The sorption mechanism is assumed to be majorly chemisorption since the experimental data obeys Langmuir isotherms and pseudo-second order kinetics (Fig. 10 and 11).The presence of hierarchical channels is noted to be responsible for the increased sorption capacity of LEuH-[H₂W₁₂O₄₀] as they function as transport channels that provide free motion of sorbate within the porous body.

Metal-Organic Framework Composites - Volume I   Materials Research Forum LLC
Materials Research Foundations **53** (2019) 122-139   https://doi.org/10.21741/9781644900291-6

Under alkaline conditions (pH 8~12), the interlayer metatungtsate ($[H_2W_{12}O_{40}]^{6-}$) probably rearranges to the stable tungstate ion (WO_4^{2-}), and subsequently leads to leaching/decomposition of the sorbent material. This phenomena maybe responsible for the inability of LEuH-$[H_2W_{12}O_{40}]$ to sorb metal ions at pH = 8~12 (Fig. 12). For this reason, the sorbent material can only work at pH 5~7 where its sorption capacity is at maximum (Fig. 12). Below pH 4.5, the sorbent material decomposes. In terms of Gibbs energy, the adsorption systems have -1.00 x 10^5, -1.53 x 10^5 and -1.23 x 10^5 J for Cd^{2+}, Pb^{2+} and Cr^{6+} respectively, signifying their spontaneous character (Fig. 13).

Conclusion

In summary, the condensation of WO_4^{2-} polyhedra units on delaminated europium containing layered rare earth hydroxide nanosheets yields a porous material, $Eu_2(OH)_5[H_2W_{12}O_{40}]_{0.17} \cdot 7H_2O$, with hierarchical charnels. This material is demonstrated to be an efficient sorbent for heavy metals such as Cd^{2+}, Cr^{6+} and Pb^{2+} in aqueous solutions. Its sorption mechanism is mainly chemical in nature, obeying

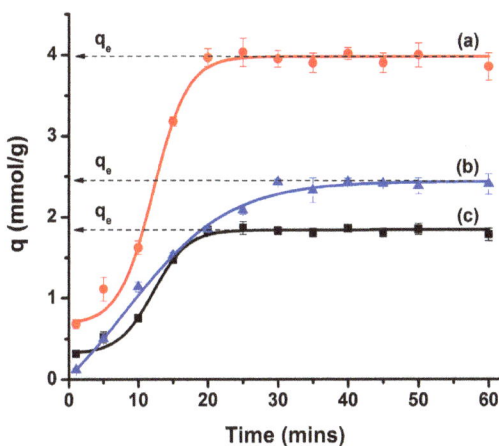

Figure 9: Kinetic studies of LEuH-[$H_2W_{12}O_{40}$] sorbent loading rate: a) Cr^{6+}, b) Pb^{2+} and c) Cd^{2+}; q_e maximum sorbed amount at equilibrium; pH = 4; Temperature = 298K; Mass of sorbent = 0.2g.

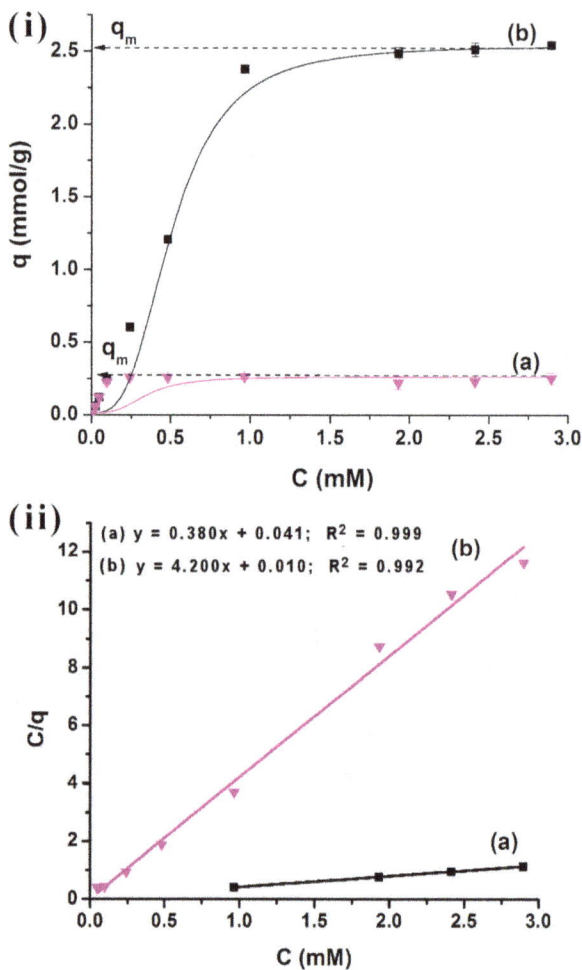

Figure 10. i) Optimal sorbed amounts (q_m) for: Pb^{2+} and ii) Langmuir fit for Pb^{2+}: a) = LEuH-Cl and b) = LEuH-[$H_2W_{12}O_{40}$].

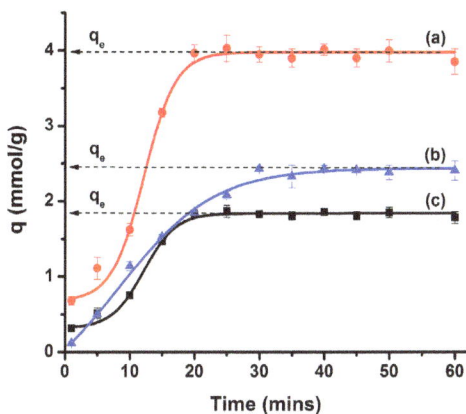

Figure 11: Kinetic studies of LEuH-[H$_2$W$_{12}$O$_{40}$] sorbent loading rate: a) Cr^{6+}, b) Pb^{2+} and c) Cd^{2+}; q$_e$ maximum sorbed amount at equilibrium; pH = 4; Temperature = 298K; Mass of sorbent = 0.2g.

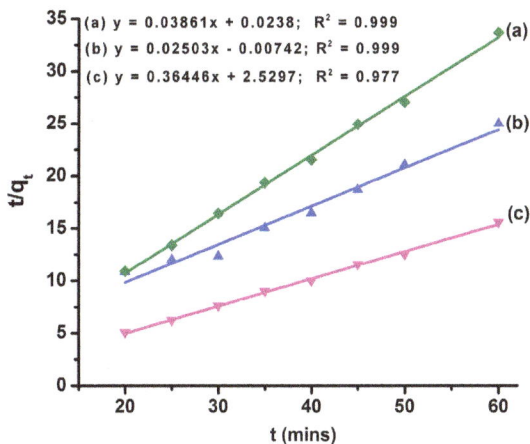

Figure 12: Pseudo-second order kinetic studies on sorption mechanism of LEuH-[H$_2$W$_{12}$O$_{40}$] nanocomposite material: a) Cd^{2+}, b) Pb^{2+} and c) Cr^{6+}; pH=5; Mass of sorbent = 0.2g .

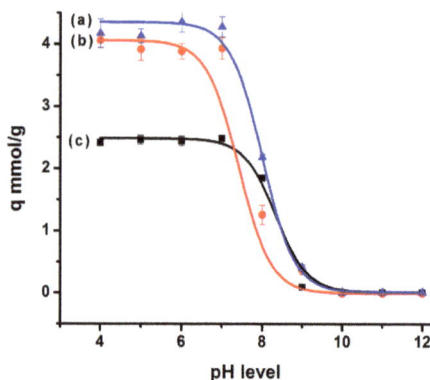

Figure 13: Effect of pH on sorption mechanism of LEuH-[$H_2W_{12}O_{40}$] nanocomposite material: a) Cd^{2+}, b) Cr^{6+} and c) Pb^{2+}; Temperature = 298K; Mass of sorbent = 0.2g.

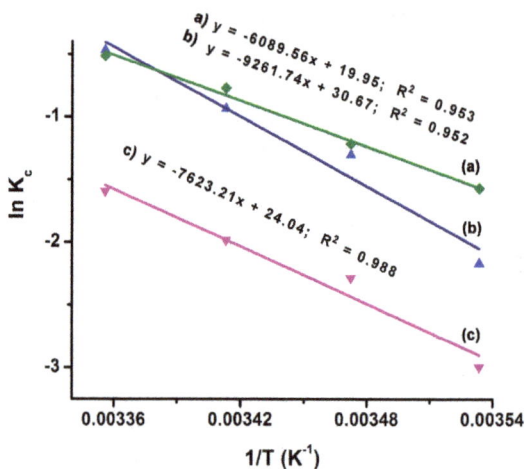

Figure 14. Thermodynamic studies on LEuH-[$H_2W_{12}O_{40}$] nanocomposite material sorption mechanisms: a) Cd^{2+}, b) Pb^{2+} and c) Cr^{6+}; pH = 5, Mass of sorbent = 0.2 g.

Langmuir sorption model and the sorption data fits pseudo-second order kinetics.

Metal-Organic Framework Composites - Volume I
Materials Research Foundations **53** (2019) 122-139

Materials Research Forum LLC
https://doi.org/10.21741/9781644900291-6

Acknowledgements

Financial support from National Research Fund, Kenya (2017/2018 FY) is highly appreciated.

References

[1] L. Hou, D. Wang, A. Baccarelli, Environmental chemicals and microRNAs, Mutat. Res. 714 (2011) 105– 112.

[2] T.A. Kurniawan, G.Y. Chan, W.H. Lo, S. Babel, Comparisons of low-cost adsorbents for treating wastewaters laden with heavy metals, The Science of the total environment 366 (2006) 409-426. https://doi.org/10.1016/j.scitotenv.2005.10.001

[3] L. Jarup, Hazards of heavy metal contamination, Brit. Med. Bull. 68 (2003) 167-182.

[4] S. Babel, T.A. Kurniawan, Low-cost adsorbents for heavy metals uptake from contaminated water: a review, J. Hazard. Mater. 97 (2003) 219-243. https://doi.org/10.1016/s0304-3894(02)00263-7

[5] S.E. Bailey, T.J. Olin, R.M. Bricka, D.D. Adrian, A review of potentially low-cost sorbents for heavy metals, Wat. Res. 33 (1999) 2469-2479. https://doi.org/10.1016/s0043-1354(98)00475-8

[6] W.S. Wan Ngah, M.A. Hanafiah, Removal of heavy metal ions from wastewater by chemically modified plant wastes as adsorbents: a review, Bioresource technology 99 (2008) 3935-3948. https://doi.org/10.1016/j.biortech.2007.06.011

[7] H. Marsh, F.R. Reinoso, Activated Carbon, Technology & Engineering , Elsevier, Britain, 2006.

[8] S. Miyata, Anion-Exchanged Properties of Hydrotalcite-like Compounds, Clay Clay Miner. 31 (1983) 305-311. https://doi.org/10.1346/ccmn.1983.0310409

[9] S. Omwoma, W. Chen, R. Tsunashima, Y.-F. Songa, Recent advances on polyoxometalates intercalated layered double hydroxides: From synthetic approaches to functional material applications, Coord. Chem. Rev. 258-259 (2014) 58-71. https://doi.org/10.1016/j.ccr.2013.08.039

[10] B. Sels, D. De Vos, M. Buntinx, F. Pierard, A.K.-D. Mesmaeker, P. Jacobs, Layered double hydroxides exchangedwith tungstate as biomimetic catalysts for mild oxidative bromination, Nature 400 (1999) 855-857. https://doi.org/10.1038/23674

[11] M.T. Pope, A. Müller, Polyoxometalate Chemistry: An Old Field with New Dimensions in Several Disciplines, Angew. Chem. Int. Ed. Engl. 30 (1991) 34-48. https://doi.org/10.1002/anie.199100341

[12] H. Jeong, B.I. Lee, S.H. Byeon, Directional self-assembly of rare-earth hydroxocation nanosheets and paradodecatungstate anions, Dalton transactions 41 (2012) 14055-14058. https://doi.org/10.1039/c2dt32421k

[13] J.B. Christian, M.S. Whittingham, Structural study of ammonium metatungstate, J. Solid State Chem. 181 (2008) 1782– 1791. https://doi.org/10.1016/j.jssc.2008.03.034

[14] F. Geng, H. Xin, Y. Matsushita, R. Ma, M. Tanaka, F. Izumi, N. Iyi, T. Sasaki, New Layered Rare-Earth Hydroxides with Anion-Exchange Properties, Chem. Eur. J. 14 (2008) 9255-9260. https://doi.org/10.1002/chem.200800127

[15] F. Geng, Y. Matsushita, R. Ma, H. Xin, M. Tanaka, F. Izumi, N. Iyi, T. Sasaki, General Synthesis and Structural Evolution of a Layered Family of Ln8(OH)20Cl4 ·nH2O (Ln) Nd, Sm, Eu, Gd, Tb, Dy, Ho, Er, Tm, and Y), J. AM. CHEM 130 (2008) 16344–16350. https://doi.org/10.1021/ja807050e

[16] M. Del-Arco, D. Carriazo, S. Gutie´rrez, C. Martı´n, V. Rives, Synthesis and Characterization of New Mg$_2$Al-Paratungstate Layered Double Hydroxides, Inorg. Chem. 43 (2004) 375-384. https://doi.org/10.1021/ic0347790

[17] S. Kyeong, P.T. J., Layered Double Hydroxides Intercalated by Polyoxometalate Anions with Keggin (A-H$_2$W$_{12}$O$_{40}$$^{6-}$), Dawson (A-P$_2W_{12}O_{62}$$^{6-}$), and Finke (Co$_4$(H$_2$O)$_2$(PW$_9O_{34}$)$_2$$^{10-}$ Structures, Inorg. Chem. 35 (1996) 6853-6860. https://doi.org/10.1021/ic960287u

[18] M. Kobya, Removal of Cr(VI) from aqueous solutions by adsorption onto hazelnut shell activated carbon: kinetic and equilibrium studies, Bioresource technology 91 (2004) 317-321. https://doi.org/10.1016/j.biortech.2003.07.001

[19] S. Babel, T.A. Kurniawan, Cr(VI) removal from synthetic wastewater using coconut shell charcoal and commercial activated carbon modified with oxidizing agents and/or chitosan, Chemosphere 54 (2004) 951-967. https://doi.org/10.1016/j.chemosphere.2003.10.001

[20] K. Selvi, Removal of Cr(VI) from aqueous solution by adsorption onto activated carbon, Bioresource technology 80 (2001) 87-89. https://doi.org/10.1016/s0960-8524(01)00068-2

[21] D. Feng, J.S.J. van Deventer, C. Aldrich, Removal of pollutants from acid mine wastewater using metallurgical by-product slags, Sep. Purif. Technol. 40 (2004) 61-67. https://doi.org/10.1016/j.seppur.2004.01.003

[22] M. Sekar, V. Sakthi, S. Rengaraj, Kinetics and equilibrium adsorption study of lead(II) onto activated carbon prepared from coconut shell, J. Colloid Interface Sci. 279 (2004) 307-313. https://doi.org/10.1016/j.jcis.2004.06.042

[23] J.R. Rangel-Mendez, M. Streat, Adsorption of cadmium by activated carbon cloth: influence of surface oxidation and solution pH, Wat. Res. 36 (2002) 1244-1252. https://doi.org/10.1016/s0043-1354(01)00343-8

[24] B.S. Inbaraj, N. Sulochana, Carbonised jackfruit peel as an adsorbent for the removal of Cd(II) from aqueous solution, Bioresource technology 94 (2004) 49-52. https://doi.org/10.1016/j.biortech.2003.11.018

[25] K.A. Krishnan, T.S. Anirudhan, Removal of cadmium(II) from aqueous solutions by steam-activated sulphurised carbon prepared from sugar-cane bagasse pith: kinetics and equilibrium studies, Water SA 29 (2003). https://doi.org/10.4314/wsa.v29i2.4849

[26] K.S.W. Sing, D.H. Everett, R.A.W. Haul, L. Moscou, R.A. Pierotti, J. Rouquero, T. Siemieniewska, Reporting Physisorption Data for Gas Solid Systems with Special Reference to the Determination of Surface Area and Porosity, Pure & Appl. Chem 57 (1985) 603—619. https://doi.org/10.1515/iupac.57.0013

Metal-Organic Framework Composites - Volume I
Materials Research Foundations **53** (2019) 140-169

Materials Research Forum LLC
https://doi.org/10.21741/9781644900291-7

Chapter 7

Designing Metal-Organic Frameworks for Clean Energy Applications

Vasi Uddin Siddiqui[1]*, Afzal Ansari[1], Irshad Ahmad[1], Imran Khan[2], M. Khursheed Akram[3], Weqar Ahmad Siddiqi[1], Anish Khan[4,5], Abdullah Mohamed Asiri[4,5]

[1]Department of Applied Sciences and Humanities, Faculty of Engineering and Technology, Jamia Millia Islamia, New Delhi-110025, India

[2]Applied Sciences and Humanities Section, University Polytechnic, Faculty of Engineering and Technology, Aligarh Muslim University, Aligarh-202002, India

[3]Applied Sciences and Humanities Section, University Polytechnic, Faculty of Engineering and Technology, Jamia Millia Islamia, New Delhi-110025, India

[4]Chemistry Department, Faculty of Science, King Abdulaziz University, Jeddah, Saudi Arabia

[5]Center of Excellence for Advanced Materials Research, King Abdulaziz University, Jeddah, Saudi Arabia

* Vasi.siddiqui@gmail.com

ORCID- https://orcid.org/0000-0003-3427-8943

Abstract

The high surface area metal organic-frameworks (MOFs) are highly porous structures made of distinct inorganic and organic building blocks, so their chemical and structural multiplicity is vast. While that variety has led to lots of potential applications, including fuel cell, solar cell, supercapacitor, lithium-ion batteries, gas storage and conversion, it also means that finding the optimal MOF for a specific application is hard to reach. This chapter discusses the design approach of MOFs based on clean energy applications in four categories. It starts with an overview of the MOFs approach for the giant problem of clean energy. It then explains how the MOFs material is a strong candidate to overcome this problem by comparing the different design approaches. By the end of the chapter, the challenges facing to commercialize this technique is discussed.

Metal-Organic Framework Composites - Volume I   Materials Research Forum LLC
Materials Research Foundations **53** (2019) 140-169   https://doi.org/10.21741/9781644900291-7

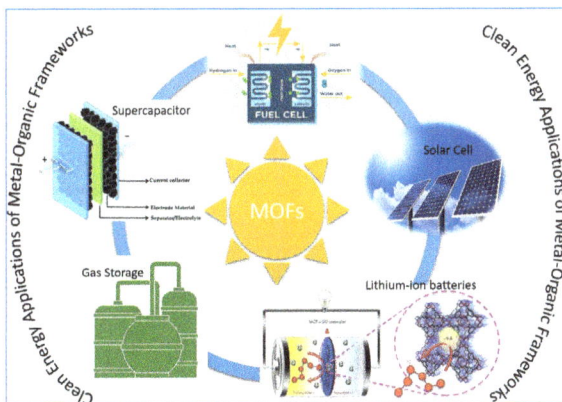

Clean energy application of metal organic-framework.

Keywords

Metal Organic-Framework, Fuel Cell, Solar Cell, Supercapacitor, Clean Energy

Contents

1.   Introduction

Statistics show the exponential increase in energy consumption globally and expected to 820 quadrillion British thermal units (Btu) in the year 2040. Presently, 80% of consumption is from non-renewable energy sources [1]. Fossil fuel/non-renewable energy sources are the primary challenges to overcome environmental pollution and shortage of energy to gain clean energy which comprises the different technologies for storage and conversion. Several approaches have been made so far to optimize many materials for getting high efficiency in storage and conversion until this 'wonder material' called Metal-Organic framework (MOFs) with low cost, high porosity and yield was proposed to the scientific community. Organic and inorganic linkers are connected in such a way that forms a porous, crystalline material and one-, two- or three-dimensional structure called MOFs also considered as co-ordination polymer or coordination network is a typical example of the interconnected organic-inorganic cluster having high potential in various applications. MOF-5 unfold the enormous possibility for the researcher as it had more than three times the internal surface area of the most porous zeolite [2]. Schematic representation of a typical MOFs is shown in figure 1.

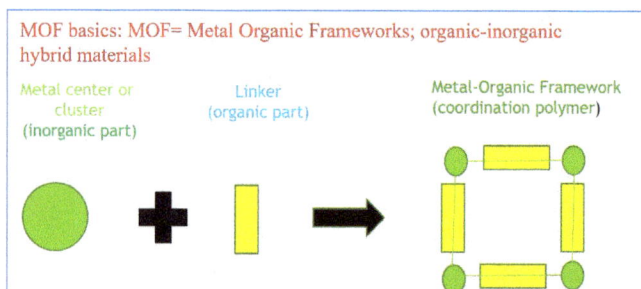

Figure 1- Schematic representation of typical MOFs.

MOFs provide a robust solution for storage and conversion systems of clean energy or renewable energy like fuel cells, lithium-ion batteries (LIBs), supercapacitor, solar energy and hydrogen energy [3]. As the non-renewable energy sources depleting significantly, renewable energy shows possibilities to overcome the demand in various energy sectors despite various challenging aspects hinder the use of its full potentials such as yielding the product with economically suitable and environment-friendly, low activity, poor stability and toxicity being the main hurdles to boost its usages. Zeolites, activated carbon, metal complex hybrids etc. are the class of porous material that are used in various traditional areas related to porous materials such as sensors, catalysis, drug delivery, gas separation and storage. Among these traditional porous materials, MOFs and MOFs derived nanomaterials are promising materials for energy storage and conversion technologies as they have the multifunctional ability by tuning the topology and functionality of MOFs via linker design or using the guest@MOF concept [4,5]. The large surface area provides enhanced accessibility to anchor the different chemical

Metal-Organic Framework Composites - Volume I   Materials Research Forum LLC
Materials Research Foundations **53** (2019) 140-169   https://doi.org/10.21741/9781644900291-7

functional groups at the surface and is an advantage of porous materials among many. Zn(II), Co(II), Cu(II), Mg(II), Ni(II), Al(III) etc. are the metal ions connecting units linking with bidentate to polyvalent carboxylate; an organic linker to form typical MOFs of a surface area ranging from 1000 to 10,000 m^2/g with tunable pore size of 9.8 nm by varying metal ions and organic linkers [6]. In 1995, Yaghi *et al.* [7] reported a MOF by hydrothermal synthesis process using hydrogen bonding interaction, metal-ligand coordination and metal-cluster co-polymerization reactions to link their molecular component that stable up to 180 °C under an inert atmosphere and for hours at 70 °C in water. MOFs deal in different studies as a functional material since various MOF-based design materials like MOF composite such as nanomaterial@MOF, Graphene/MOF and MOF derived materials showed immense potential in energy storage and conversion sector, even though different nanomaterial gained quite a familiarity either in pristine or composite form in this field. Metal-organic framework including pristine MOFs and derived materials can be synthesized at low cost and high yield from an inexpensive precursor that encourage the large-scale production for industrial application. MOF

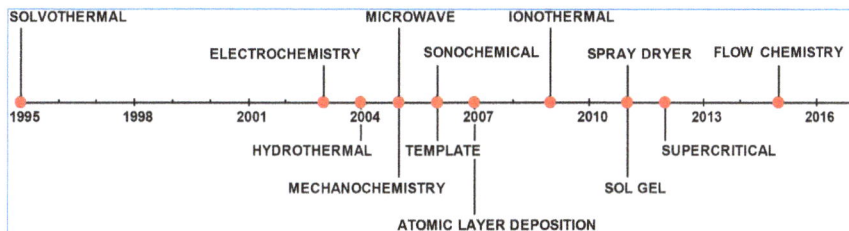

Figure 2 - Timeline of synthetic route for the synthesis of MOFs [10].

architecture has two key components, the one acts like a 'strut' is the organic linker and considered as organic secondary building unit (SBU) connecting the metal centers which considered as inorganic secondary building unit and act as the 'joint' in the desired metal organic-framework structure. Components are attached to each other by coordination bonds, together with other intermolecular interactions to form a network with a definite topology [8]. Generally, the term 'design' used for the synthesis of MOF. Apart from the traditional solvents used for the synthesis and designing for MOFs, new different solvents having properties of low vapour pressure, non-inflammability, high thermal stability and ionic conductivity have been used also and summarized by Li *et al.* [9] in the category of ionic liquid, deep eutectic solvent and surfactant ionic liquid (ILs) with melting point of less than 100 °C are shown in scheme 1 and the timeline of various synthetic route to synthesize MOF in figure 2 [10].

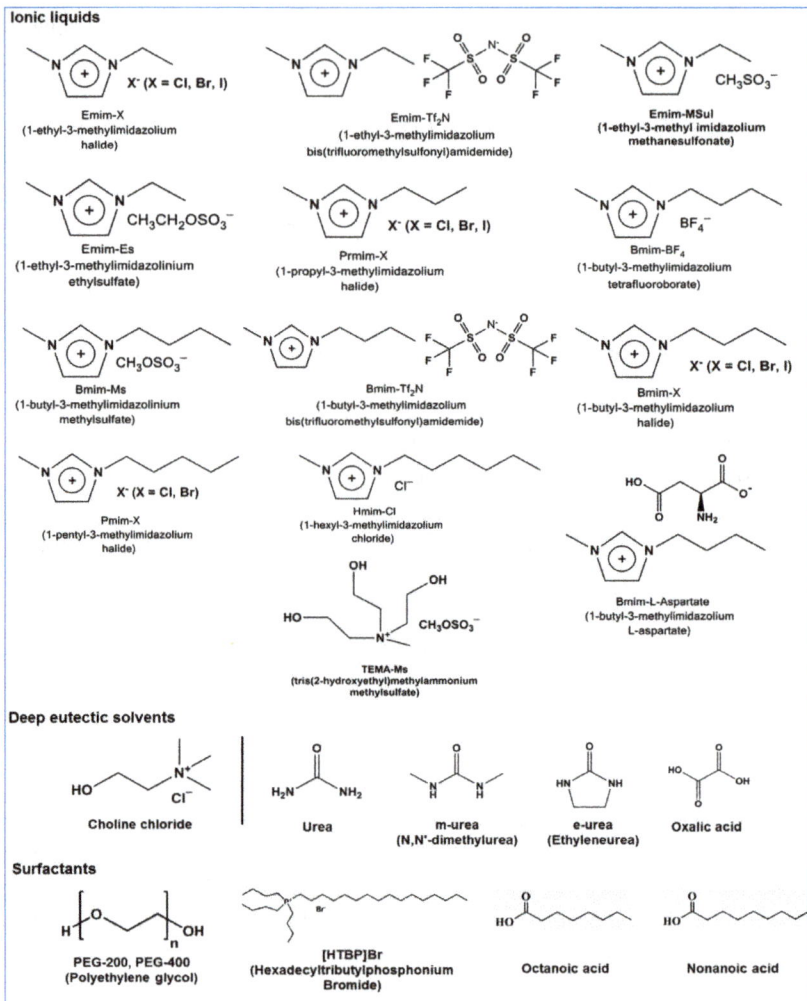

Scheme 1- ILs, deep eutectic solvents, and surfactants used for the synthesis of MOFs [9].

In general, the positive ions of metals are neutralized by organic linker hence form a neutral MOFs, but the positive and negative charge of ionic liquid behaves as a template in the synthesis of MOFs hence form ionic MOFs that open the path for host-guest interactions. Moreover, deep eutectic solvents are the mixture of quaternary ammonium halide salts and hydrogen bond donors having similar properties of ILs and are comparatively less costly. A green approach for designing MOFs could be free from surfactants that is mechanochemistry (MC). Since surfactant that could be acidic, basic, neutral, cationic and anionic but provide different approaches to the reaction system can damage the environment more and increase the cost of synthesis. The prime objective in MOF design is to set the synthesis parameters that govern to define inorganic building block for avoiding the decomposition of organic linkers. Meanwhile, the kinetics of crystallization must be optimized to make ease for nucleation and growth of the anticipated phase. Stock *et al.* [11] have reported some conventional synthesis method at room temperatures such as conventional electric (CE) heating, microwave (MW) heating, electrochemistry (EC), mechanochemistry (MC) and ultrasonic method (UC) as shown in figure 3.

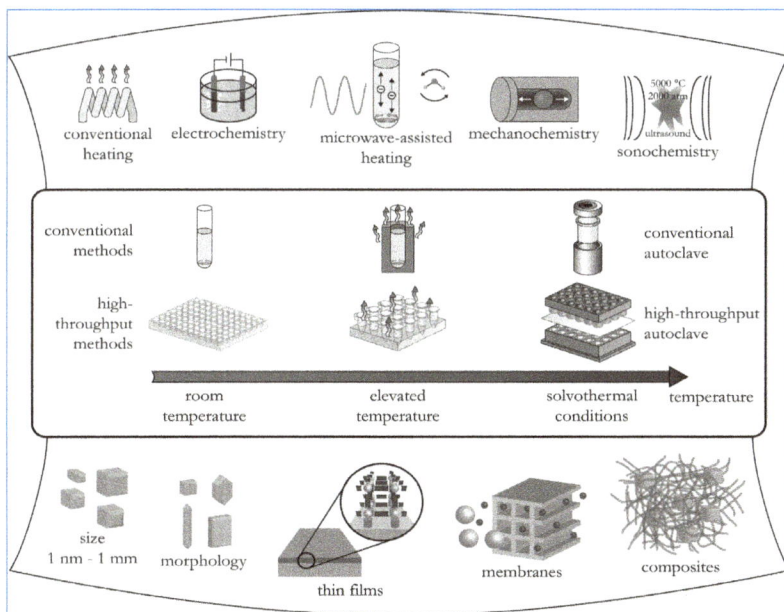

Figure 3-Overview of synthesis methods, possible reaction temperatures, and final reaction products in MOF synthesis [11].

These conventional and other than conventional methods have following challenges either in selecting the organic and inorganic unit for designing MOFs or in the modification of its component according to its application.

1. Designing 4(four) connected and porous MOF retaining expanded topology than zeolite. It could be by scheming inorganic and organic four connected nodes.

2. Designing MOFs from a large metal cluster as it replaces a vertex and size of inorganic SBUs resulting in large pore size and high porosity.

3. Designing MOFs with larger organic linkers. It expands the length in between inorganic SBUs leading to large pore size and high porosity.

Most commonly, the solvent evaporation method used for designing the MOFs although diffusion method, hydro/solvothermal method, microwave reaction and ultrasonic methods are other crystal growth methods [8]. The solvent evaporation method needs appropriate parameters such that the crystal will grow and appear in saturated solutions during the cooling step and increase in solubility with temperature. The principle of diffusion methods is to allow different species to be connected but gradually. For this, solvent liquid diffusion is one of the approaches. In which two layers of uneven densities are formed first; one has the solvent product, the second is the precipitant solvent and is separated with a solvent layer. The precipitant solvent slowly diffuses into the distinct layer and crystal growth occur at the interface. One more way could be the gradual diffusion of reactants by the division of physical barriers, such as two vessels of different sizes. Furthermore, gels are also worked as diffusion and crystallization media in some instances, primarily to slow down diffusion and to detour precipitation of bulk material. The diffusion method is preferred for getting a single crystal suitable for X-ray diffraction analysis rather than non- or poly-crystalline products, especially if the products are poorly soluble. In hydro/solvothermal method the products are self-assembled from the soluble precursors. Earlier, this method was used for the synthesis of zeolites but has been adapted to the synthesis of metal organic-frameworks in the temperature ranging 80–260°C inside the autoclave under autogenous pressure. This can be affected mainly by the rate of cooling speed at the end of the reaction. However, microwave and ultrasonic methods are not the first choice for the synthesis of crystalline MOFs but are used for rapid synthesis and to control the size and shape of the synthesized material [12]. Along with the synthesis method used for designing MOFs some other factors effect viz; solvent, pH, temperature and molar ratio of reactants is also important to architecture the MOFs structure [13] (Scheme 2). However, the desired structure with specific chemical functionality could be obtained by complementary approach is to modify known MOFs after synthesis.

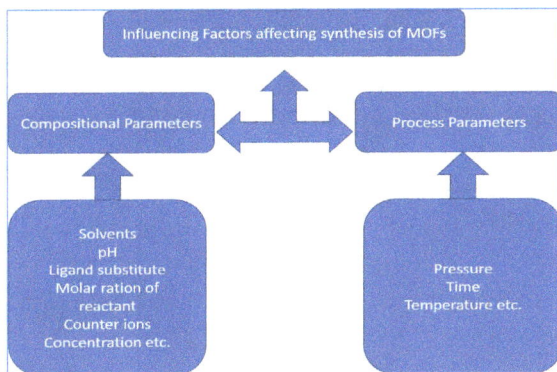

Scheme 2-Various parameters influencing the formation of MOFs [13].

For functionalizing MOF nodes Islamoglu et al., [14] discussed the solvent-assisted ligand incorporation (SALI) and atomic layer deposition in MOFs (AIM) as post-synthetic technique furthermore, to replace the structural linkers, termed solvent-assisted linker exchange (SALE), also known as a post-synthetic exchange (PSE).

In this chapter, we discuss the various design approaches of MOFs for clean energy applications and briefly categorize in four sections i.e. 1. MOFs for fuel cells 2. MOFs for solar cell 3. MOFs for Lithium-ion batteries 4. MOFs for supercapacitor. Finally, future prospective and conclusion have been discussed for MOF with clean energy perspective.

2. MOFs for clean energy applications

2.1 MOFs for fuel cells

Fuel cells (FCs) are defined as an electrochemical conversion device that contains a constant supply of fuel such as hydrogen, natural gas, or methanol and an oxidant like oxygen, air, or hydrogen peroxide. Anode (negative side), a cathode (positive side), electrolyte and the external circuit are the main part of each fuel cell, which allows ions amid the two sides of the fuel cell that provides an effective way to use various energy sources. Consequently, figure 4 clearly shows the consumption of less energy and emission than conventional techniques [14]. Oxidation of hydrogen into protons and electron occurs at anode while reduction of oxygen into oxide species occurs at the cathode and reacts to form water. In addition, depending on the electrolyte, either proton

Materials Research Forum LLC
https://doi.org/10.21741/9781644900291-7

or oxide ions are taken through the ion conductor electron-insulating electrolyte, while electrons travel to give electrical power through an external circuit. Nevertheless, FCs comes listed as renewable energy systems, if the fuel used is renewable (biofuel, hydrogen from the air or solar conversion of water) [15]. Furthermore, fuel cells are categorized into the following six groups based on the selection of fuel and electrolyte: [15]

- Alkaline fuel cells (AFC) utilize an aqueous potassium hydroxide (KOH) solution as the alkaline electrolyte;

- Proton exchange membrane fuel cells (PEMFC) and direct methanol fuel cells (DMFC) use a polymer membrane as the electrolyte;

- Phosphoric acid fuel cells (PAFC) use pure phosphoric acid as the electrolyte;

- Molten carbonate fuel cells (MCFC) use a molten mixture of lithium, sodium, and potassium carbonates as the electrolyte;

- Solid oxide fuel cells (SOFC) use a ceramic material as the electrolyte.

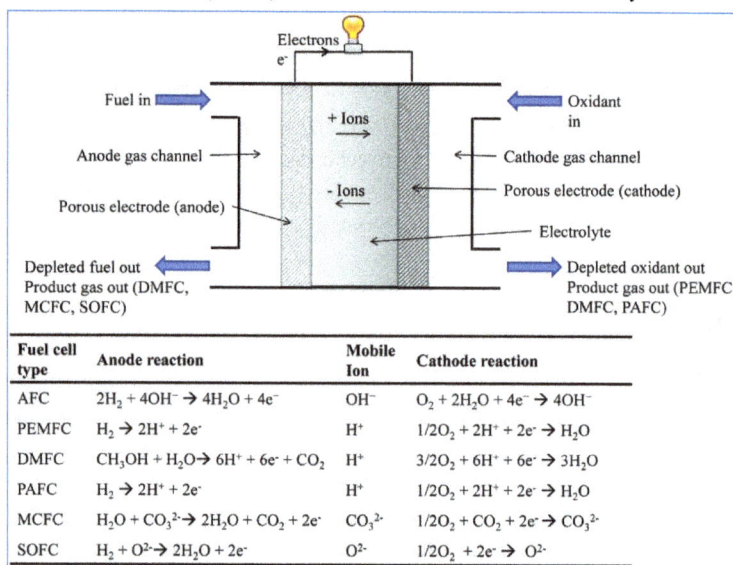

Fuel cell type	Anode reaction	Mobile Ion	Cathode reaction
AFC	$2H_2 + 4OH^- \rightarrow 4H_2O + 4e^-$	OH^-	$O_2 + 2H_2O + 4e^- \rightarrow 4OH^-$
PEMFC	$H_2 \rightarrow 2H^+ + 2e^-$	H^+	$1/2O_2 + 2H^+ + 2e^- \rightarrow H_2O$
DMFC	$CH_3OH + H_2O \rightarrow 6H^+ + 6e^- + CO_2$	H^+	$3/2O_2 + 6H^+ + 6e^- \rightarrow 3H_2O$
PAFC	$H_2 \rightarrow 2H^+ + 2e^-$	H^+	$1/2O_2 + 2H^+ + 2e^- \rightarrow H_2O$
MCFC	$H_2O + CO_3^{2-} \rightarrow 2H_2O + CO_2 + 2e^-$	CO_3^{2-}	$1/2O_2 + CO_2 + 2e^- \rightarrow CO_3^{2-}$
SOFC	$H_2 + O^{2-} \rightarrow 2H_2O + 2e^-$	O^{2-}	$1/2O_2 + 2e^- \rightarrow O^{2-}$

Figure 4-A schematic representation showing the general operating principles of a fuel cell and the principle anode and cathode reactions, as well as the mobile ions associated with the most common fuel cell types [15].

Metal-Organic Framework Composites - Volume I   Materials Research Forum LLC
Materials Research Foundations **53** (2019) 140-169   https://doi.org/10.21741/9781644900291-7

As the energy sources depleting conversely demand is constantly growing, clean energy sources will be the alternative; H_2 powered fuel cell is a promising candidate to match the need of alternative, clean and carbon-free method with almost twice efficient as fossil fuels (60% efficiency for fuel cell versus 34% for fossil fuel) since fuel cell process is not governed by carnot cycle laws [16]. Sharaf *et al.*[17] discussed the comparison of fuel cell technology with competing technologies and explain the principle, types and parts of FC structure at the design level

- – The unit cell; is the main part of FC where a basic electrochemical reaction takes place.

- – Fuel cell stack; is the connected series of several unit cell of typically between 0.5 and 0.8V as the single unit cell is too small for most practical applications.

- – The fuel cell system; made up of the fuel cell stack in addition to the balance of plant (BoP) subsystem that deliver the oxidant and fuel supply and storage, thermal management, water management, power conditioning and instrumentation and control of the fuel cell system.

Nevertheless, from decreasing the cost of fuel cell and increasing the efficiency, broadly MOFs act as – (a) for hydrogen evolution reaction (b) precursors for the oxygen reduction (c) proton conductive polymer for membranes in different fuel cell mechanism [3]. More importantly, high yields fabrication of MOFs from comparatively low-cost precursors make it feasible to use as precursors of catalyst, electrolytes and electrode catalysts for fuel cells [18]. Furthermore, MOF acts as an electrocatalyst in various types of reaction such as hydrogen evolution reaction (HER), oxygen evolution reaction (OER), oxygen reduction reaction (ORR), alcohol oxidation, nitrite and bromate reduction [19]. During the oxidation of hydrogen at the anode produced hydrogen ions and the lost electrons are combine with oxygen to produce water. This sort of reaction occurs at the cathode is called oxygen reduction reaction (ORR) which is usually catalyzed by Platinum (Pt)-based electrocatalysts. It is also a rate-determining step but slow kinetics causes the voltage loss in hydrogen FCs [19]. Since Pt costs high for fuel cell stack, a primary objective is to develop a new efficient and inexpensive cathode ORR catalyst to substitute costly Pt catalysts, however, Pt and Pt group metal (PGM) based materials are considered as the best catalysts for ORR. Alternatively, pristine MOF or non-platinum group metal (non-PGM), carbon-based, transition metal-based, and their composites MOF materials are performing for fuel cell applications [20].

Oxygen and hydrogen are used by the alkaline fuel cell to generate electricity and water but facing challenges of the lower mobility of hydroxyl ions (OH^-) than a proton (H^+). By increasing the electrolyte concentration OH^- ions cause to increase in corrosion,

Metal-Organic Framework Composites - Volume I
Materials Research Foundations **53** (2019) 140-169

Materials Research Forum LLC
https://doi.org/10.21741/9781644900291-7

wettability and handling difficulties at varying pressure. Since, the electrolyte plays a vital role in the advancement of this technology Nagarkar *et al.* [21] design a MOF electrolyte (Ni$_2$(µ-pymca)$_3$]OH·nH$_2$O) by Nickle (II) and 2-pyrimidinecarboxyllic acid under hydrothermal condition and obtained highest conductivity 0.8×10^{-4} Scm^{-1} from the Nyquist plot at room temperature with 99% relative humidity (RH). This illustrated in figure 5.

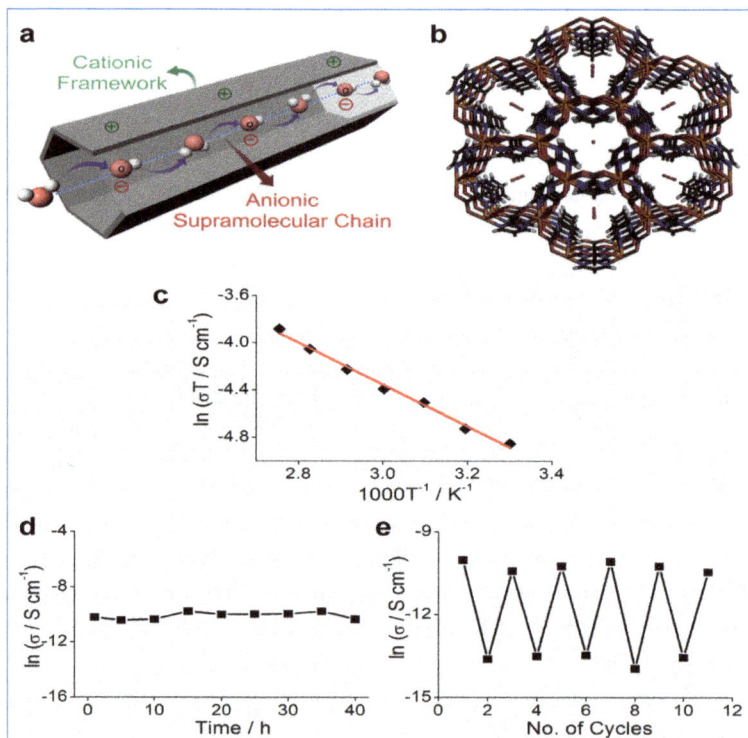

Figure 5- (a) Schematic representation of a cationic MOF with an anionic supramolecular chain of hydroxide anions with water molecules (b) 2D cationic honeycomb sheets of MOF (c) Arrhenius plot for the hydroxide conducting MOF (d) Time dependent conductivity measurements of MOF at 27 °C and 99% RH (e) Humidity cycling study of MOF at 27 °C between 50 and 99% RH [21].

Metal-Organic Framework Composites - Volume I   Materials Research Forum LLC
Materials Research Foundations **53** (2019) 140-169   https://doi.org/10.21741/9781644900291-7

For fuel cell membrane electrode assemblies conduction of proton in a hydrolytic condition is a challenge. Proton-conducting MOFs (PCMOF) that show the ability to conduct proton under a hydrated condition in both low (<100 °C) and high (<100 °C) temperature are needed. Ramaswamy *et al.* [22] reported magnesium MOF; $Mg_2(H_2O)_4(H_2L)\cdot H_2O$ (H_6L = 2,5-dicarboxy-1,4-benzene-diphosphonic acid), PCMOF10 that shows hydrolytic robustness and conduct proton at 3.55×10^{-2} Scm^{-1} at 70 °C and 95% RH. Electrocatalysis of oxygen is important to fuel cell technologies, but due to the lack of effective earth-abundant electrocatalysts and insufficient understanding of catalytic mechanisms challenges remain. Recently, Cheng *et al.* [23] reported the 50-100 times higher mass activities of 500 A g_{metal}^{-1} at 0.83 V for the ORR and 2,000 A g_{metal}^{-1} at 0.30 V for the OER than the pristine NiFe-MOF. This has been achieved by applying the lattice strain in noble metal free MOFs to activate the strong bifunctional ORR and OER activity. The Lattice-strained NiFe MOFs catalyst maintains ~97% of its initial activity after 200 h of continuous ORR/OER reaction at a high current density of 100–200 mA cm^{-2} and suggests a four-electron mechanistic pathway.

2.2 MOFs for solar cell

Industrial development and economic growth increase the energy demand globally and energy consumption will be increased by 25%, 33% and 30% of the world growing economic countries China, India and Africa by 2040 respectively, but China remains by far the world's largest producer of energy-intensive goods in 2040 [24]. MOFs have been studied as catalysts or support for oxidation and reduction to generate hydrogen, sensitizer, for conversion of CO_2 and water to hydrocarbon or alcohol in solar energy harvesting applications during the past decade. Conversion of solar energy is basically inspired by plants photosynthesis process and depends on the ability to absorb photons and stored the generate electricity as chemical energy until required. However, conversion needs significant improvement in efficiency and durability and all three fundamental steps: designing antenna/reaction center that collects sunlight for converting it to electrochemical potential (redox equivalent), water oxidation complex for the creation of redox equivalents in term of electrons and oxygen, and catalytic reactions system to store chemical energy in the products using redox equivalent [25] as shown in figure 6.

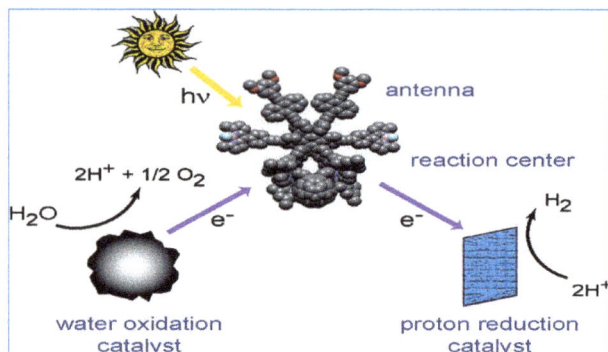

Figure 6-Schematic representation of solar fuel via artificial synthesis [25].

In the case of organic semiconductors, charge-carrier mobility and generation are not possible since their band gap are normally flat therefore band gap is direct. In 2015, Liu *et al.*[26] reported a pioneering study of a crystalline organic semiconductor an epitaxial MOF thin films that give high and efficient mobility of electron and holes and provides evidence for suppressing the charge-carrier recombination as in indirect band gap. This kind of MOF referred to as- surface grafted MOF (SURMOFs 2) that contained paddle-wheel units which stacked to yield an overall P4 symmetry. While free base porphyrin SURMOF 2 and saturated porphyrin molecule have similar band gap values of 1.58 eV and 1.60 eV respectively. However, introducing Pd and Zn (metal centres) increases the value to 1.94 eV and 1.88 eV respectively. Overall, from both the free-base porphyrin Zn-SURMOFs 2 and Pd-porphyrin Zn-SURMOFs 2, experimentally observed that the unexpected high performance of the MOF-based organic photovoltaic (OPV) devices are the porphyrin MOFs and are indirect band-gap semiconductors. These findings of MOFs give a remarkable direction for designing the flexible solar cell. Furthermore, Liu *et al.* [27] afresh used the layer by layer (LbL) fashion to fabricate Zn (II) porphyrin-based MOF thin films with an organic linker diphenylamine (DPA) that control the photophysical properties by donating electrons into the porphyrin skeleton and causes for a decrease in the band gap. However, it increases the absorption of solar light and feasible for higher photocurrents. To emphasize the increase in efficiency with changing the parameter for mechanical and structural of the device, dye-sensitized solar cell (DSSCs) usually used to deal with all conditions (even severe) of the environment. In an electrode system, a molecular dye produces photons afterwards transferred the excited electrons by photons into a high surface-area TiO_2 electrode framework. Maza *et al.*[28]

Materials Research Forum LLC

https://doi.org/10.21741/9781644900291-7

reported the first Ruthenium doped based film (RuDCBPY) which were encapsulated on MOF-TiO$_2$ and was highly efficient (>90%) for generation of photocurrent. Previously, RuDCBPY-doped UiO-67 films grown on fluorine-doped tin oxide (FTO) coated glass substrates [29], water stable zirconium (IV) biphenyldicarboxylic acid MOF but could not be achieved with much efficiency [30].

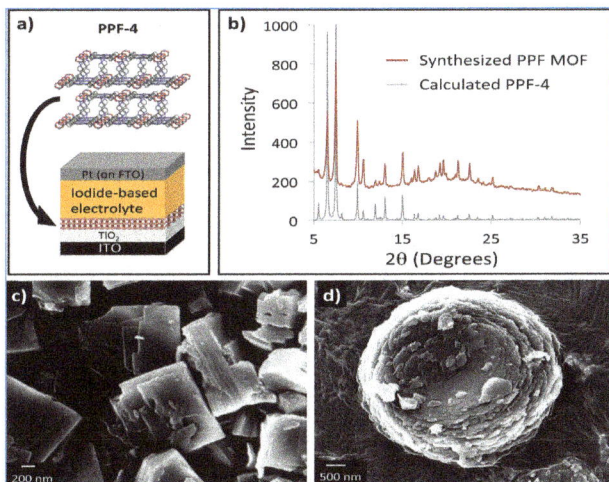

Figure 7-(a) Schematic representation of PPF-4 crystal structure incorporated into a basic MSSC assembly (b) XRD pattern of PPF-MOF and PPF-4 (c) SEM micrograph of PPF MOF crystal (d) SEM micrograph of Zn-TCPP [32].

In DSSCs porphyrin-based MOFs synthesized by LbL assembly (that used to control the thickness at single molecular level precision) and helps in varying the absorption-emission causing chromophores. In 2016, Goswami and co-workers [31] used solvent-assisted linker exchange (SALE) for designing MOF and change the MOF thin film 3D framework to 2D framework and high symmetry palladium porphyrin to lower symmetry dyes that reduce the contact area between light harvesting MOF and the phase containing the electrons and holes. The results increasing the excitons transport through 6-8 chromophores layer to 9-11 layers. Recently, Spoerke *et al.* [32] used MOF as a sensitizer by using pillared porphyrin framework (PPF) and reported the possibility of using MOF-DSSC (MSSC) without non-MOF species. This report established the fact that it is the MOF that acts as a sensitizer and not the organic linkers. Although the efficiency was low for photocurrent generated by the device since it was coated by atomic layer deposition (ALD) to avoid the difficulty from marking the measuring photoresponse from bound molecular components or to the assembled MOFs composition. This MOF architecture in figure 7 showing the Zinc-tetra(carboxyphenyl)porphyrin (Zn-TCPP) absorbing linker.

Organometal halide perovskite used for the very first time by Kojima *et al.* [33] as a light harvester in liquid-state sensitized solar cells and they reported the power conversion efficiency (PCE) of 3.8%. In perovskite solar cells (PSCs) Shen *et al.* [34] introduced MOFs as an interfacial layer for the first time, thru coating with a MOF ZIF-8 layer at the interface between mp-TiO_2 and the perovskite film. This modified interface with MOF ZIF-8 efficiently enhanced perovskite crystallinity, grain sizes and photovoltaic performance of 16.99% than that of the uncoated reference cell. The interface can efficiently repress the recombination of photogenerated carriers and improve charge extraction. Moreover, researchers published informative review articles and designed a variety of MOF and MOF based material for solar cell applications [1,3,35].

2.3   MOFs for lithium-ion batteries

Research for alternative fossil-fuel resources conceives the development of renewable energy sources and sustainable storage technologies. Production of energy from solar, wind, ocean, biomass, geothermal and other renewable sources are equally important as to store these energies in bulk with safety, ease of use, durability and recyclability for the best use that need significant improvement. Demand for the energy and power density of electrical energy storage (EES) devices in the field of portable electronics and transportation, hybrid electric vehicles (HEVs), plug-in hybrid vehicles (PHEVs) and full electric vehicles (EVs) is emerging [36,37].

Usually, batteries carrying high energy densities, but poor power is one of the major technologies that meet the current needs however not all. The development of new materials with high performance for EES is a big challenge [38]; enormous research efforts are being delivered to the exploration of next-generation high-performance rechargeable batteries. In this chapter we discussed Lithium-ion batteries (LIBs), having three primary components cathode (positive), an anode (negative) and electrolyte despite being light in weight, higher energy density, low cost, long life span are inefficient to meet the current requirements of energy storage [39] since commercially available LIBs using graphite as anode and Li-Cobalt oxide as cathode provides theoretically low capacities 372 mAh g^{-1} for graphite and 148 mAh g^{-1} for $LiCoO_2$ [20]. In LIBs, Li-ions transport to the cathode from the anode during the discharging process and migrate back

Materials Research Forum LLC
https://doi.org/10.21741/9781644900291-7

on charging, that generated through the oxidation of lithium-based materials as the positive electrode material [40]. For developing these components and to increase the storage capacity of batteries MOFs and MOFs derived materials are being used for the past few years as shown in table 1 [39].

Table 1. MOFs or MOF-derived materials for LIBs [38].

MOFs	Sample	CC/DC	RC/rate	CN	Vs/Li/Li$^+$
MOF-based anode materials					
MOF-177	MOF-177	110/425	-/50	2	0.1-1.6
Zn$_3$(HCOO)$_6$	Zn$_3$(HCOO)$_6$	693/1344	560/60	60	0.005-3
Mn-LCP	Mn(tfbdc)(4,40-bpy)(H$_2$O)$_2$	610/1807	390/50	50	0.1-3
Co$_2$(OH)$_2$BDC	Co$_2$(OH)$_2$BDC	1385/1005	650/50	100	0.02-3
Li/Ni-NTC	Li/Ni-NTC	601/1084	480/-	80	0.01-3
Co$_3$[Co(CN)$_6$]$_2$	Co$_3$[Co(CN)$_6$]$_2$	294.2/566.2	299.1/20	40	0.01-3
MONFs	Asp-Cu-nanofibers	334/1255	233/50	200	0.01-3
Ni-Me$_4$bpz	Ni-Me$_4$bpz	-/320	120/50	100	0.01-3
MOF-derived anode materials					
MOF-5	3D porous carbon	1084/2983	1015/100	100	0.01-3
ZIF-8	N-C-800	2037/3487	2132/100	50	0.01-3
Co(HO-BDC)(bbb)	Co$_3$O$_4$	1008/1186	852/500	200	0.01-3
ZIF-67	Co$_3$O$_4$	1083/1735	1335/100	150	0.01-3
MOF-71	Co$_3$O$_4$	879.5/1286.1	913/200	60	0.001-3
Cu-MOF	CuO	-/1208	470/100	100	0.05-3
MOF-199	CuO	538/1208	484.2/100	40	0.005-3
[Cu$_3$(btc)$_2$]n	CuO/Cu$_2$O	513/727	740/100	250	0.01-3
MIL-88	α-Fe$_2$O$_3$	940/1372	911/200	50	0.01-3
Fe-ZIF	Fe$_2$O$_3$@N-C	1368/1696	1573/100	50	0.01-3
Fe-MOF	α-Fe$_2$O$_3$	1024/1487	1024/100	40	0.005-3
MIL-125	TiO$_2$	-/-	166.2/1 C	500	1-3
MOF-5	ZnO/ZnFe$_2$O$_4$/C	1047/1385	1390/500	100	0.005-3
MIL-100	NG/Fe-Fe$_3$C	222/904	607/1000	100	0.01-3
MOF-5	550N	-/-	1200/75	50	0.02-3
ZIF-8	ZIF-8@chitosen-800N	-/-	750/50	50	0.02-3
MIL-88-Fe	Fe$_2$O$_3$/$_3$DGN	-/1133	864/200	50	0.01-3
MOF-5s	CZO@C	1024/1663	725/100	50	0.01-3
Cu-MOF	CuO-G	-/1350	700/60	40	0.005-3
ZIF-8	ZnO@ZnO QDs/C NRAs	712/785	699/500	100	0.01-2
MOF-based cathode materials					
MIL-53	MIL-53(Fe)	77/80	71/0.025C	50	1.5-3.5
MIL-53	MIL-53(Fe)_quinone 1	77.5/93	73/0.1 C	8	1.5-3.5
MIL-68	MIL-68(Fe)	31/40	32/0.02 C	12	1.5-3.5
MIL-136	MIL-136(Ni)	-/-	10/10 C	50	2-4
MOPOF	K$_{2.5}$(VO)$_2$(HPO$_4$)$_2$(C$_2$O$_4$)	81/62	70/40	60	2.5-4.6
MOPOF	Li$_2$[(VO)$_2$(HPO$_4$)$_{1.5}$(PO$_4$)$_{0.5}$(C$_2$O$_4$)]	78/55	80/12.5	25	2.5-4.5
MIL-101	MIL-101-Fe	65	-	30	2-3.5
Cu(2,7-AQDC)	Cu(2,7-AQDC)	147	105/1 Ma	50	1.7-4

CC: Charge capacity, DC: discharge capacity, RC: reversible capacity, (mA h g^{-1}) CN: cycle number

Metal-Organic Framework Composites - Volume I   Materials Research Forum LLC
Materials Research Foundations **53** (2019) 140-169   https://doi.org/10.21741/9781644900291-7

Various nanostructured materials are also been used as an alternative electrode material for the LIBs such as carbon nanotubes, transition metal oxide and graphite composite since graphite electrode provide the limited storage and energy density [41]. For LIBs cathode material Férey *et al.* [42] design very first a MIL53 ($Fe^{III}(OH)_{0.8}F_{0.2}O_2CC_6H_4CO_2.H_2O$) MOF which results in maintaining 70 mAhg^{-1} reversible capacity that initially was 80 mAhg^{-1}. The insertion amount of Li to x = 0.6 for $Li_xFe^{III}(OH)_{0.8}F_{0.2}O_2CC_6H_4CO_2.H_2O$ with a gravimetric capacity of 75 mAhg^{-1} corresponds to relatively low capacities. In 2010, Doublet and co-workers design the MOF based MIL53(Fe) as a positive electrode materials for LIBs that gives the direction for the use of MOFs in LIBs and given the detail explanation of the redox mechanism of lithium insertion by density functional theory (DFT) calculation and local chemical bond analyses [43]. The design material was antiferromagnetic at T = 0 K and paramagnetic at room temperature with iron ions in the high-spin S = 5/2 state. Wang *et al.* [39] presented an insight on this research that along with pristine MOFs MOF incorporated hybrid electrode materials and MOFs derived materials such as MOF derived carbons, MOF derived metal oxide, MOFs derived metal oxide/carbon composite used extensively for both positive and negative electrode materials. In 2013, Shun-Li Li and Qiang Xu present a view on the application of MOFs in LIBs as an electrode and as an electrolyte that MOFs used in LIBS are two types- the non-Li MOF and Li-MOF and for non-Li-MOFs as an electrolyte materials showed a high conductivity but Li-based MOFs is a promising contender for electrode materials [40].

Shin *et al.* [44] design Fe-based MOFs, the MIL-101(Fe) as a cathode material and since the low insertion amount found the unstable structure change of that redox chemistry of Fe^{2+}/Fe^{3+} is not completely reversible and led to the rapid decay of the capacity. Under controlled morphologies by the solvothermal process, Li *et al.*[45] have reported the very first polycrystalline MOF-177 ($Zn_4O(1,3,5$-benzenetribenzoate)$_2$) as an anode material for LIBs and results showed approximately 400 mAhg^{-1} for the first discharge and 105 mAhg^{-1} for charge capacities that involved reactions are:

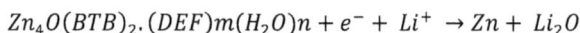

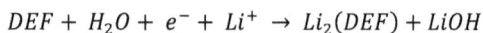

$$Zn_4O(BTB)_2.(DEF)m(H_2O)n + e^- + Li^+ \rightarrow Zn + Li_2O$$

$$DEF + H_2O + e^- + Li^+ \rightarrow Li_2(DEF) + LiOH$$

$$Zn + Li^+ + e^- \leftrightarrow LiZn$$

Furthermore, MOFs could also be used for the anode as a buffer matrix in the coating layer with carbon nanotubes, graphene, or polymer. Han *et al.* [46] reported a MOF coated silica nanoparticles through a one-pot in-situ mechanism that represents the remarkable storage capacity of 1050 mAhg^{-1} and more than 99 % recyclability at 500 cycles after pyrolysis at 700 °C for 1 h at nitrogen atmosphere. The schematic representation of the synthesis is shown in figure 8.

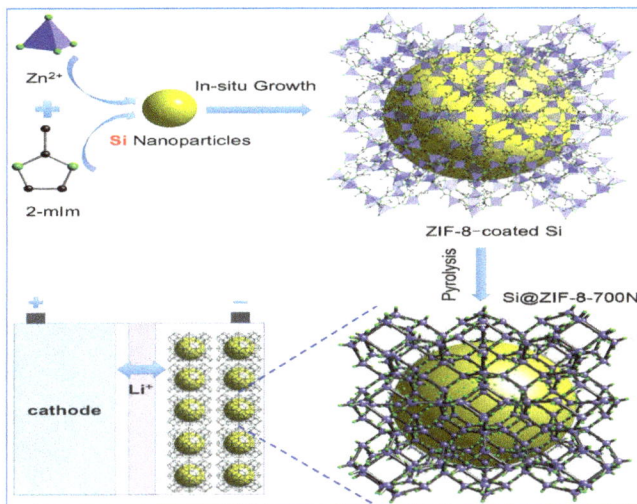

Figure 8-Schematic representation of the preparation of Si@ZIF-8-700N for anode material in LIBs. Color code: carbon, black; nitrogen, green; zinc, blue; silicon, yellow. After pyrolysis, ZIF-8 converts to amorphous carbon with monodispersed zinc ions [46].

Recently, Lin *et al*. [47] also designed and synthesized a cadmium-based MOF {[Cd(HTTPCA)]·2H$_2$O}$_n$ anode material under solvothermal method that stabilize after pyrolysis at 800 °C under nitrogen atmosphere and showed an enhanced capacity and cyclic stability as excellent electrochemical performance evaluated by cyclic voltammetry (CV) in the voltage range 0.1−3.0 V versus Li$^+$/Li. The first and after 100 cycles discharge capacity was 710 mAhg^{-1} and 302 mAhg^{-1} respectively before pyrolysis and increased remarkable with superior cyclic stability to 741 mAhg^{-1} and remains 741 mAhg^{-1} after the 100th cycle at a current density of 100 mAhg^{-1}. Moreover, researchers [1,18,39–41,47–50] usually are working on the different design approach of MOFs for LIBs such as an anode, cathode and electrolyte, in contrary Lagadec *et al*. [51] represent a different approach for the MOFs used in LIBs such as lithium-ion battery-separators plays a key role in ion transport and influences rate performance, cell life and safety and offer great prospects for developing separators to support encounter the exaction that new

Metal-Organic Framework Composites - Volume I   Materials Research Forum LLC
Materials Research Foundations **53** (2019) 140-169   https://doi.org/10.21741/9781644900291-7

applications place on LIB technology. In LIBs electrodes filled with electrolyte are isolated electronically by a polymeric membrane having pores that act as a separator and allows transfer of Li ions from the anode to cathode during discharge and reverse again during charge while preventing short circuits between the positive and negative electrodes. A complex 3-D structure of polyolefin membranes, usually less than 25µm thick prepared from either semi-crystalline polyethene and/or polypropylene are used in commercially available LIBs with a porosity, ε, approximately 40% as shown in figure 9.

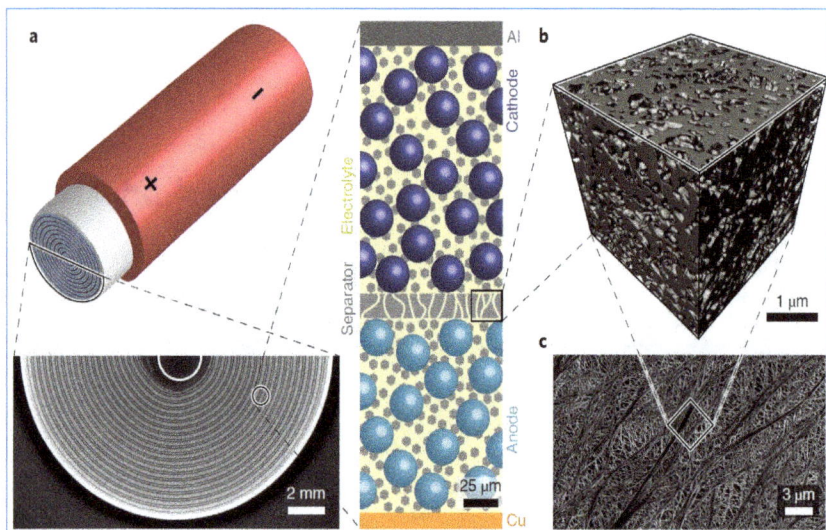

Figure 9- (a) Schematic and X-ray tomographic image showing the tightly wound layers of anode, separator and cathode of a cylindrical cell and a zoomed-in schematic of the layer structure (b) FIB-SEM tomographic rendering of a Targray PE16A separator(c) Top-view SEM image of the same separator [51].

2.4   MOFs for supercapacitor

Batteries possess high energy density, work the entire day and take much time to recharge when discharged. Electrochemical capacitors that are also known as supercapacitors (SCs) and ultracapacitor for accelerated power delivery with recharging, used in hybrid electric vehicles and fuel cell vehicles. Researchers studied the challenges for their high-power densities, long cycle life spans and speedy charging/discharging rates [52]. However, SCs have a low energy density and both batteries and SCs depend on electrochemical processes but distinct electrochemical mechanisms. Electrochemical double layer capacitor (EDLC) and pseudo-capacitors are the two main types of SCs depending on the energy storage mechanism [20]. In EDLCs, capacitance comes from the pure electrostatic charge stored at the interface of the electrode and electrolyte and the pseudo-capacitor is based on fast and reversible Faradic reaction process. Since a high surface area of the electrode material needed in EDLCs, porous carbonaceous materials like activated carbon, templated carbon, carbide-derived carbon, carbon nanotubes and graphene, are generally used as the electrode material for available to electrolyte ions for the good chemical and thermal properties with better electrochemical performance. Whereas many noble or cheap transition metal oxides or hydroxides, conducting polymers and carbon materials with oxygen- or nitrogen-containing functional groups on the surface are used for pseudo-capacitor, in which the size of the redox active species, the porosity and the specific surface area of the electrode material mainly determine the characteristics of it [53]. Due to adjustable post- and pre-synthetic changes in structure and highly porous nature with large specific surface area, MOFs showed great potential in the energy sector and has been widely investigated in the past decade. Wang *et al.* [39], reported three ways to apply MOFs for electrode materials of SCs in the category of pristine MOFs, MOFs derived metal oxide and MOFs derived carbons: (i) using pristine MOFs to store charges through physisorption of electrolyte ions on their internal surfaces or exploiting reversible redox reactions of the metal centres; (ii) destroying MOFs to obtain metal-oxides and preserving electrons through the charge transfer between the electrolyte and electrode; (iii) pyrolyzing MOFs to give porous carbons and improving the capacitance via enhancing the conductivity. MOFs a highly porous materials act as insulators typically and only a few exceptions are reported yet asserting the potential of MOFs for batteries and supercapacitors. Since for the synthesis of corresponding metal oxides, the MOFs are generally used as precursors and cannot be considered as a direct MOFs application [54]. Choi *et al.* [55] reported a series of 23 different nanocrystalline MOFs (nMOFs) with different functional groups and metal ions having discrete pore sizes and shapes and found the zirconium MOF ($Zr_6O_4(OH)_4$-$(BPYDC)_6$, nMOFs-867) performed remarkably high capacitance over at least 10000 charge/discharge cycles as shown in figure 10. A coin-shaped cell was fabricated by sandwiching the nMOF between two nHKUST-1 films with a separator and for charge/discharge profiles, CV and life cycle measurement that all have done by galvanostatic measurements. The energy density was 2.86×10^{-4} W cm_{stack}^{-3} at the power density of 0.386 W cm_{stack}^{-3} and was more than three times that of activated carbon which had the power density of 1.00×10^{-4} W cm_{stack}^{-3}. However, the insulating behaviour of MOFs is responsible for practical application since conductivity makes hurdle results in very low rate performance and demitting capacity over cycles.

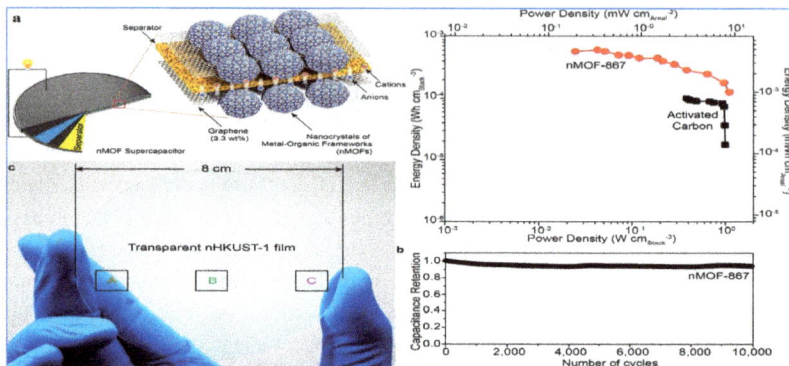

Figure 10- (a) Construct for nMOF Scs (b) Energy and power densities of nMOF-867 compared with activated carbon and life cycle for nMOF-867 (c) A large, transparent nHKUST-1 film on a quartz substrate 8 cm in diameter [55].

This work inspired many other and Wang *et al.* [56] minimize the insulating behaviour of MOFs by interweaving ZIF-67 crystals/carbon cloth with the conducting polymer polyaniline (PANI) in which the chains of PANI are interconnected and linked up and act as bridges for electrons transportation between the external circuit and the internal surface of MOFs as shown in scheme 3, results increased the conductivity and Faradaic reaction process. An extraordinary high areal capacitance of 2146 mF cm^{-2} at 10 mV s^{-1} was achieved.

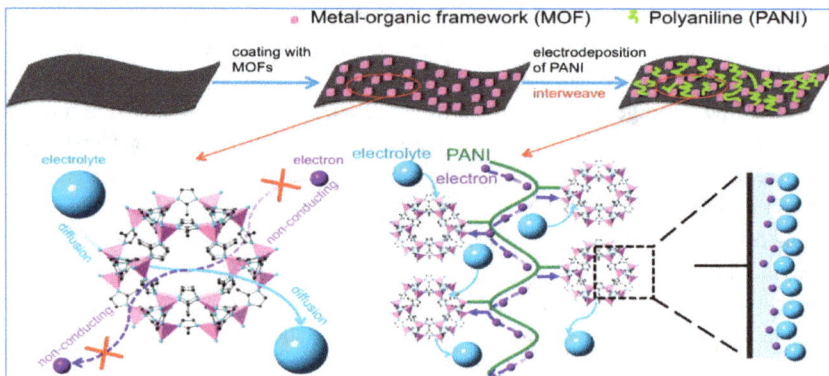

Scheme 3-The construct for PANI-ZIF-67-CC and the schematic representation of electron and electrolyte conduction in MOF and MOF interwoven by PANI [39].

Wen *et al.* [57] used the synergic effects of Ni-MOF structure and design Ni-MOF/CNT that achieved a specific capacitance of 1765 F g^{-1} at a current density of 0.5 A g^{-1}. Furthermore, reduced graphene oxide (rGO) concept of high electrical conductivity, surface area and electrochemical stability in the form of reduced graphene oxide/graphitic carbon nitride (rGO/g-C_3N_4) as the negative electrode and Ni-MOF/CNTs as the positive electrode performed notably with 95% specific capacitance retention after 5000 successive charge-discharge cycles and in a working voltage range of 0–1.6 V based on a complementary potential window in 6 M KOH aqueous electrolyte achieved a high energy density of 36.6 W h kg^{-1} at a power density of 480 W kg^{-1}. A summary of the electrochemical performance like Electrolyte, rate, initial capacitance, cycle number and capacitance after cycles or the percentage of initial capacity of pristine MOFs in SCs are listed in table 2 [20].

Table 2. Summary of the performance of pristine MOFs in SCs [20].

MOFs	Electrolyte	Rate	IC	CN	AC
Ni-based MOFs	6 M KOH	50 mV s^{-1}	~125	2000	~100
Co-based	1 M LiOH	0.6 A g^{-1}	206	1000	~203
439-MOFs	6 M KOH	4 A g^{-1}	64	6000	~64
Ni-MOF-24	6 M KOH	10 A g^{-1}	668	-	-
Zn-doped Ni-MOF	6 M KOH	2 A g^{-1}	~1250	3000	~1200
nMOF-867	1 M $(C_2H_5)_4NBF_4$	12.7 mA cm^{-3}	726	10,000	~700
UiO-66	6 M KOH	10 mV s^{-1}	~900	2000	654
UiO-66	Polymer-gel	80 mV s^{-1}	-	1000	~89%
Ni-based MOFs	6 M KOH	0.5 A g^{-1}	1765	-	-
$Ni_3(HITP)_2$	1 M TEABF$_4$/CAN	2 A g^{-1}	-	10,000	~90%
Cu-CAT NWAs	3 M KCl	800 mV s^{-1}	-	5000	80%
Co-LMOFs	1 M KOH	2 A g^{-1}	1978	2000	94.3%
Layered Co-MOFs	5 M KOH	2 A g^{-1}	-	3000	95.8%
DABCO-MOFs	2 M KOH	10 A g^{-1}	-	16,000	98%
PANI-ZIF-67-CC	3 M KCl	0.05 mA cm^{-2}	35 mF cm^{-2}	2000	80%
ZIF-67	1 M KOH	2 A g^{-1}	188	3000	105%
MnOx-MHCF	1 M Na_2SO_4	5 mA cm^{-2}	127	10,000	94.5%
Accordion-like Ni-MOFs	3 M KOH	1.4 A g^{-1}	-	5000	90.7%
ZIF-LDH/GO	6 M KOH	1.4 A g^{-1}	-	5000	96.5%
$V^{IV}(O)(bdc)$	1 M Na_2SO_4	1 A g^{-1}	521	10,000	92.8%
CoNi-MOFs	1 M KOH	-	-	5000	94%
CoMn-MOFs	2 M KOH	100 mV s^{-1}	-	1500	96%

IC: Initial capacitance (F g^{-1}), CN: cycle number, AC: capacitance after cycle (F g^{-1})

Comparatively, MOFs as the only electrode provide high surface area that leads to increasing the capacitance for EDLCs than the addition of carbonaceous materials results reduces the surface area of composites. Recently, Sheberla *et al.* [58] used a single MOFs in EDLCs as electrode without any conductive binders or additive and achieved more than 90% capacitance retention after 10,000 cycle and give ~18 μF cm^{-2} capacitance that exceeding most of carbon-based materials. Many researchers used the concept and design conductive MOF nanowire arrays (NWAs) [59], 2D-layered cobalt based MOFs [60], Ni-based pillared DABCO-MOFs (DMOFs) (DABCO = 1,4-diazabicyclo[2.2.2]-octane) [61], zinc and cobalt based MOFs (ZIF-4, ZIF-7, ZIF-8, ZIF-14 and ZIF-67) [62] and got exciting results.

For pseudo-capacitance materials, metal oxide used as electrode materials to increase the capacitance for SCs, Wang and Chen [63] design a hierarchical double layered nickel hexacyanoferrate/MnO_2 (NiHCF)@MnO_2) nanostructure composite cathode material that give a high cyclic stability and capacitance of 224 F g^{-1} at 50 mV s^{-1} without fading over 500 cycles in the presence of aqueous electrolyte of Na_2SO_4 due to the synergic effect named interlayer concentration enhancement effect (ICE). Figure 11 shows the schematic diagram of charging and discharging mechanism of NiHCF@MnO_2, the different Na^+ ion concentrations in the space between the NiHCF and MnO_2 layers (C_{01}) at charged (right) and discharged (left) states.

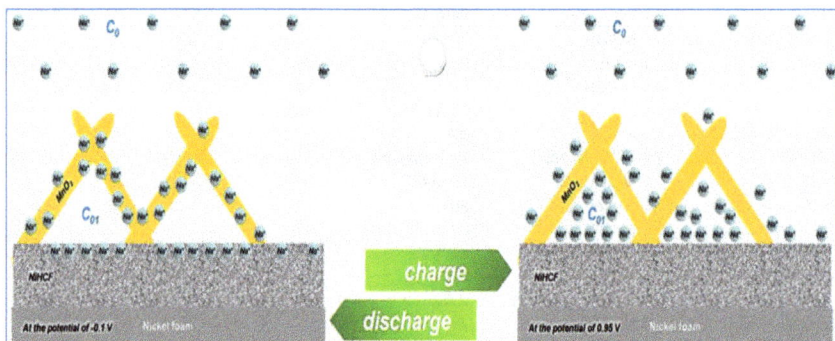

Figure 11-Schematic diagram showing the different Na^+ ion concentrations at charged (right) and discharged (left) states [63].

Nevertheless, the full potential of the synergic effect could not be achieved of this dual-layer structure fabricated by instrument dependent electrodeposition method. To overcome this, Zhang *et al.* used a one-step chemically induced self-transformation process for MnO_x into MOFs (MOF–manganese hexacyanoferrate hydrate (MHCF)), the added NH_4F helps the manganese to grew nanoflowers morphology into manganese oxides that decorated the surfaces of the smooth MHCF cubes and achieved the specific capacitance $\approx$1200 F g^{-1} that was three times higher than the pristine MOFs $\approx$300 F g^{-1} at 10 A g^{-1} that because of the more intimate connection between the pseudocapacitive manganese oxide and MOFs.

Conclusion and future perspective

In summary, this chapter deals with recent development and design approaches of MOFs in solar cells for light-harvesting material, fuel cell, lithium-ion batteries and supercapacitors as clean energy applications. MOFs and MOFs derived material has been widely investigated and extensive attention is giving to meet the demand of this material that shows a healthy approach for various clean energy applications and is difficult to summarize all the analysis in a single platform. For fuel cell, MOFs plays an important role from enhancing the efficiency to cut the cost of fuel cells through a different mode of catalytic action in HER, ORR and membrane material by tuning pore size, pore volume, surface area, active sites and electrical conductivity. From storage and conversion of energy to the mobility of ions in a different solar cell like DSSCs, the application of MOFs has significant effect to enhance the light harvesting results in terms of weight, size and design flexibility in various flexible solar cell approaches. Advanced chemistry and new properties of MOFs will mark new pathways for storage material. Being an insulator pristine MOFs used in batteries and supercapacitor as an electrode material has been proven to be one of the most suitable electrode materials carrying high power density, energy density, capacitance and cyclability. In addition, low stability with non-conductivity are key properties for the direct applications of MOFs. Although, introducing metal or metal oxide and carbon material on MOFs fabricating a double-layered structure enhanced the charge/discharge cyclic stability with electrical conductivity. However, an optimize designed approach with new synthesis approach is needed covering all the criteria like environmentally-friendly, low-cost, thermal, mechanical, chemical stability and high-yield MOFs for using MOFs as a stabilizing material for the practical applications in a wider perspective.

References

[1]   H. Zhang, J. Nai, L. Yu, X.W. (David) Lou, Metal-Organic-Framework-Based Materials as Platforms for Renewable Energy and Environmental Applications, Joule. 1 (2017) 77–107. https://doi.org/10.1016/j.joule.2017.08.008

[2]   D. Liu, J.J. Purewal, J. Yang, A. Sudik, S. Maurer, U. Mueller, J. Ni, D.J. Siegel, MOF-5 composites exhibiting improved thermal conductivity, Int. J. Hydrogen Energy. 37 (2012) 6109–6117. https://doi.org/10.1016/j.ijhydene.2011.12.129

[3]   V. Bon, Metal-organic frameworks for energy-related applications, Curr. Opin. Green Sustain. Chem. 4 (2017) 44–49. https://doi.org/10.1016/j.cogsc.2017.02.005

[4]   W. Lu, Z. Wei, Z.-Y. Gu, T.-F. Liu, J. Park, J. Park, J. Tian, M. Zhang, Q. Zhang, T. Gentle III, M. Bosch, H.-C. Zhou, Tuning the structure and function of metal–

organic frameworks via linker design, Chem. Soc. Rev. 43 (2014) 5561–5593.
https://doi.org/10.1039/C4CS00003J

[5]   D. Zhao, D.J. Timmons, D. Yuan, H.-C. Zhou, Tuning the Topology and
Functionality of Metal−Organic Frameworks by Ligand Design, Acc. Chem. Res. 44
(2011) 123–133. https://doi.org/10.1021/ar100112y

[6]   H. Furukawa, N. Ko, Y.B. Go, N. Aratani, S.B. Choi, E. Choi, A.O. Yazaydin,
R.Q. Snurr, M. O'Keeffe, J. Kim, O.M. Yaghi, Ultrahigh Porosity in Metal-Organic
Frameworks, Science (80-.). 329 (2010) 424–428.
https://doi.org/10.1126/science.1192160

[7]   O.M. Yaghi, H. Li, Hydrothermal Synthesis of a Metal-Organic Framework
Containing Large Rectangular Channels, J. Am. Chem. Soc. 117 (1995) 10401–10402.
https://doi.org/10.1021/ja00146a033

[8]   S. Qiu, G. Zhu, Molecular engineering for synthesizing novel structures of metal–
organic frameworks with multifunctional properties, Coord. Chem. Rev. 253 (2009)
2891–2911. https://doi.org/10.1016/j.ccr.2009.07.020

[9]   P. Li, F.-F. Cheng, W.-W. Xiong, Q. Zhang, New synthetic strategies to prepare
metal–organic frameworks, Inorg. Chem. Front. 5 (2018) 2693–2708.
https://doi.org/10.1039/C8QI00543E

[10]   M. Rubio-Martinez, C. Avci-Camur, A.W. Thornton, I. Imaz, D. Maspoch, M.R.
Hill, New synthetic routes towards MOF production at scale, Chem. Soc. Rev. 46
(2017) 3453–3480. https://doi.org/10.1039/c7cs00109f

[11]   N. Stock, S. Biswas, Synthesis of Metal-Organic Frameworks (MOFs): Routes to
Various MOF Topologies, Morphologies, and Composites, Chem. Rev. 112 (2012)
933–969. https://doi.org/10.1021/cr200304e

[12]   K.K. Gangu, S. Maddila, S.B. Mukkamala, S.B. Jonnalagadda, A review on
contemporary Metal–Organic Framework materials, Inorganica Chim. Acta. 446
(2016) 61–74. https://doi.org/10.1016/j.ica.2016.02.062

[13]   R. Seetharaj, P.V. Vandana, P. Arya, S. Mathew, Dependence of solvents, pH,
molar ratio and temperature in tuning metal organic framework architecture, Arab. J.
Chem. (2016). https://doi.org/10.1016/j.arabjc.2016.01.003

[14]   T. Islamoglu, S. Goswami, Z. Li, A.J. Howarth, O.K. Farha, J.T. Hupp,
Postsynthetic Tuning of Metal–Organic Frameworks for Targeted Applications, Acc.
Chem. Res. 50 (2017) 805–813. https://doi.org/10.1021/acs.accounts.6b00577

[15] N. Linares, A.M. Silvestre-Albero, E. Serrano, J. Silvestre-Albero, J. García-Martínez, Mesoporous materials for clean energy technologies, Chem. Soc. Rev. 43 (2014) 7681–7717. https://doi.org/10.1039/C3CS60435G

[16] Y. Ren, G.H. Chia, Z. Gao, Metal–organic frameworks in fuel cell technologies, Nano Today. 8 (2013) 577–597. https://doi.org/10.1016/j.nantod.2013.11.004

[17] O.Z. Sharaf, M.F. Orhan, An overview of fuel cell technology: Fundamentals and applications, Renew. Sustain. Energy Rev. 32 (2014) 810–853. https://doi.org/10.1016/j.rser.2014.01.012

[18] Y. Zhao, Z. Song, X. Li, Q. Sun, N. Cheng, S. Lawes, X. Sun, Metal organic frameworks for energy storage and conversion, Energy Storage Mater. 2 (2016) 35–62. https://doi.org/10.1016/j.ensm.2015.11.005

[19] A. Morozan, F. Jaouen, Metal organic frameworks for electrochemical applications, Energy Environ. Sci. 5 (2012) 9269. https://doi.org/10.1039/c2ee22989g

[20] X. Zhang, A. Chen, M. Zhong, Z. Zhang, X. Zhang, Z. Zhou, X.-H. Bu, Metal–Organic Frameworks (MOFs) and MOF-Derived Materials for Energy Storage and Conversion, Electrochem. Energy Rev. (2018). https://doi.org/10.1007/s41918-018-0024-x

[21] S.S. Nagarkar, B. Anothumakkool, A. V Desai, M.M. Shirolkar, S. Kurungot, S.K. Ghosh, High hydroxide conductivity in a chemically stable crystalline metal–organic framework containing a water-hydroxide supramolecular chain, Chem. Commun. 52 (2016) 8459–8462. https://doi.org/10.1039/C6CC04436K

[22] P. Ramaswamy, N.E. Wong, B.S. Gelfand, G.K.H. Shimizu, A Water Stable Magnesium MOF That Conducts Protons over 10 −2 S cm −1, J. Am. Chem. Soc. 137 (2015) 7640–7643. https://doi.org/10.1021/jacs.5b04399

[23] W. Cheng, X. Zhao, H. Su, F. Tang, W. Che, H. Zhang, Q. Liu, Lattice-strained metal–organic-framework arrays for bifunctional oxygen electrocatalysis, Nat. Energy. (2019). https://doi.org/10.1038/s41560-018-0308-8

[24] L. Capuano, International Energy Outlook 2018 (IEO2018), 2018. www.eia.gov

[25] D. Gust, T.A. Moore, A.L. Moore, Solar fuels via artificial photosynthesis, Acc. Chem. Res. 42 (2009) 1890–1898. https://doi.org/10.1021/ar900209b

[26] J. Liu, W. Zhou, J. Liu, I. Howard, G. Kilibarda, S. Schlabach, D. Coupry, M. Addicoat, S. Yoneda, Y. Tsutsui, T. Sakurai, S. Seki, Z. Wang, P. Lindemann, E. Redel, T. Heine, C. Wöll, Photoinduced Charge-Carrier Generation in Epitaxial MOF

Thin Films: High Efficiency as a Result of an Indirect Electronic Band Gap?, Angew. Chemie Int. Ed. 54 (2015) 7441–7445. https://doi.org/10.1002/anie.201501862

[27]   J. Liu, W. Zhou, J. Liu, Y. Fujimori, T. Higashino, H. Imahori, X. Jiang, J. Zhao, T. Sakurai, Y. Hattori, W. Matsuda, S. Seki, S.K. Garlapati, S. Dasgupta, E. Redel, L. Sun, C. Wöll, A new class of epitaxial porphyrin metal–organic framework thin films with extremely high photocarrier generation efficiency: promising materials for all-solid-state solar cells, J. Mater. Chem. A. 4 (2016) 12739–12747. https://doi.org/10.1039/C6TA04898F

[28]   W.A. Maza, A.J. Haring, S.R. Ahrenholtz, C.C. Epley, S.Y. Lin, A.J. Morris, Ruthenium(II)-polypyridyl zirconium(IV) metal– organic frameworks as a new class of sensitized solar cells, Chem. Sci. 7 (2016) 719–727. https://doi.org/10.1039/C5SC01565K

[29]   D.Y. Lee, C.Y. Shin, S.J. Yoon, H.Y. Lee, W. Lee, N.K. Shrestha, J.K. Lee, S.-H. Han, Enhanced photovoltaic performance of Cu-based metal-organic frameworks sensitized solar cell by addition of carbon nanotubes, Sci. Rep. 4 (2015) 3930. https://doi.org/10.1038/srep03930

[30]   W.A. Maza, S.R. Ahrenholtz, C.C. Epley, C.S. Day, A.J. Morris, Solvothermal Growth and Photophysical Characterization of a Ruthenium(II) Tris(2,2′-Bipyridine)-Doped Zirconium UiO-67 Metal Organic Framework Thin Film, J. Phys. Chem. C. 118 (2014) 14200–14210. https://doi.org/10.1021/jp5034195

[31]   S. Goswami, L. Ma, A.B.F. Martinson, M.R. Wasielewski, O.K. Farha, J.T. Hupp, Toward Metal–Organic Framework-Based Solar Cells: Enhancing Directional Exciton Transport by Collapsing Three-Dimensional Film Structures, ACS Appl. Mater. Interfaces. 8 (2016) 30863–30870. https://doi.org/10.1021/acsami.6b08552

[32]   E.D. Spoerke, L.J. Small, M.E. Foster, J. Wheeler, A.M. Ullman, V. Stavila, M. Rodriguez, M.D. Allendorf, MOF-Sensitized Solar Cells Enabled by a Pillared Porphyrin Framework, J. Phys. Chem. C. 121 (2017) 4816–4824. https://doi.org/10.1021/acs.jpcc.6b11251

[33]   A. Kojima, K. Teshima, Y. Shirai, T. Miyasaka, Organometal Halide Perovskites as Visible-Light Sensitizers for Photovoltaic Cells, J. Am. Chem. Soc. 131 (2009) 6050–6051. https://doi.org/10.1021/ja809598r

[34]   D. Shen, A. Pang, Y. Li, J. Dou, M. Wei, Metal–organic frameworks at interfaces of hybrid perovskite solar cells for enhanced photovoltaic properties, Chem. Commun. 54 (2018) 1253–1256. https://doi.org/10.1039/C7CC09452C

[35] M.B. Majewski, A.W. Peters, M.R. Wasielewski, J.T. Hupp, O.K. Farha, Metal–Organic Frameworks as Platform Materials for Solar Fuels Catalysis, ACS Energy Lett. 3 (2018) 598–611. https://doi.org/10.1021/acsenergylett.8b00010

[36] J.-M. Tarascon, M. Armand, Issues and challenges facing rechargeable lithium batteries, Nature. 414 (2001) 359–367. https://doi.org/10.1038/35104644

[37] P. Simon, Y. Gogotsi, Materials for electrochemical capacitors, Nat. Mater. 7 (2008) 845. https://doi.org/10.1038/nmat2297

[38] P.G. Bruce, S.A. Freunberger, L.J. Hardwick, J.-M. Tarascon, Li–O2 and Li–S batteries with high energy storage, Nat. Mater. 11 (2011) 19. https://doi.org/10.1038/nmat3191

[39] L. Wang, Y. Han, X. Feng, J. Zhou, P. Qi, B. Wang, Metal–organic frameworks for energy storage: Batteries and supercapacitors, Coord. Chem. Rev. 307 (2016) 361–381. https://doi.org/10.1016/j.ccr.2015.09.002

[40] S.-L. Li, Q. Xu, Metal–organic frameworks as platforms for clean energy, Energy Environ. Sci. 6 (2013) 1656. https://doi.org/10.1039/c3ee40507a

[41] M.H. Yap, K.L. Fow, G.Z. Chen, Synthesis and applications of MOF-derived porous nanostructures, Green Energy Environ. 2 (2017) 218–245. https://doi.org/10.1016/j.gee.2017.05.003

[42] G. Férey, F. Millange, M. Morcrette, C. Serre, M.-L. Doublet, J.-M. Grenèche, J.-M. Tarascon, Mixed-Valence Li/Fe-Based Metal–Organic Frameworks with Both Reversible Redox and Sorption Properties, Angew. Chemie Int. Ed. 46 (2007) 3259–3263. https://doi.org/10.1002/anie.200605163

[43] C. Combelles, M. Ben Yahia, L. Pedesseau, M.-L. Doublet, Design of Electrode Materials for Lithium-Ion Batteries: The Example of Metal−Organic Frameworks, J. Phys. Chem. C. 114 (2010) 9518–9527. https://doi.org/10.1021/jp1016455

[44] J. Shin, M. Kim, J. Cirera, S. Chen, G.J. Halder, T.A. Yersak, F. Paesani, S.M. Cohen, Y.S. Meng, MIL-101(Fe) as a lithium-ion battery electrode material: a relaxation and intercalation mechanism during lithium insertion, J. Mater. Chem. A. 3 (2015) 4738–4744. https://doi.org/10.1039/C4TA06694D

[45] X. Li, F. Cheng, S. Zhang, J. Chen, Shape-controlled synthesis and lithium-storage study of metal-organic frameworks Zn4O(1,3,5-benzenetribenzoate)2, J. Power Sources. 160 (2006) 542–547. https://doi.org/https://doi.org/10.1016/j.jpowsour.2006.01.015

[46] Y. Han, P. Qi, X. Feng, S. Li, X. Fu, H. Li, Y. Chen, J. Zhou, X. Li, B. Wang, In Situ Growth of MOFs on the Surface of Si Nanoparticles for Highly Efficient Lithium Storage: Si@MOF Nanocomposites as Anode Materials for Lithium-Ion Batteries, ACS Appl. Mater. Interfaces. 7 (2015) 2178–2182. https://doi.org/10.1021/am5081937

[47] X. Lin, J. Niu, J. Lin, L. Wei, L. Hu, G. Zhang, Y. Cai, Lithium-Ion-Battery Anode Materials with Improved Capacity from a Metal − Organic Framework, (2016) 1–4. https://doi.org/10.1021/acs.inorgchem.6b01123

[48] G. Xu, P. Nie, H. Dou, B. Ding, L. Li, X. Zhang, Exploring metal organic frameworks for energy storage in batteries and supercapacitors, Mater. Today. 20 (2017) 191–209. https://doi.org/10.1016/j.mattod.2016.10.003

[49] R. Zhao, Z. Liang, R. Zou, Q. Xu, Metal-Organic Frameworks for Batteries, Joule. 2 (2018) 2235–2259. https://doi.org/10.1016/j.joule.2018.09.019

[50] Y. Lin, Q. Zhang, C. Zhao, H. Li, C. Kong, C. Shen, L. Chen, An exceptionally stable functionalized metal-organic framework for lithium storage, Chem. Commun. 51 (2015) 697–699. https://doi.org/10.1039/c4cc07149b

[51] M.F. Lagadec, R. Zahn, V. Wood, Characterization and performance evaluation of lithium-ion battery separators, Nat. Energy. 4 (2019) 16–25. https://doi.org/10.1038/s41560-018-0295-9

[52] L.L. Zhang, X.S. Zhao, Carbon-based materials as supercapacitor electrodes, Chem. Soc. Rev. 38 (2009) 2520. https://doi.org/10.1039/b813846j

[53] Y. Zhao, J. Liu, M. Horn, N. Motta, M. Hu, Y. Li, Recent advancements in metal organic framework based electrodes for supercapacitors, Sci. China Mater. 61 (2018) 159–184. https://doi.org/10.1007/s40843-017-9153-x

[54] V. Bon, Metal-organic frameworks for energy-related applications, Curr. Opin. Green Sustain. Chem. 4 (2017) 44–49. https://doi.org/10.1016/j.cogsc.2017.02.005

[55] K.M. Choi, H.M. Jeong, J.H. Park, Y.-B. Zhang, J.K. Kang, O.M. Yaghi, Supercapacitors of Nanocrystalline Metal–Organic Frameworks, ACS Nano. 8 (2014) 7451–7457. https://doi.org/10.1021/nn5027092

[56] L. Wang, X. Feng, L. Ren, Q. Piao, J. Zhong, Y. Wang, H. Li, Y. Chen, B. Wang, Flexible Solid-State Supercapacitor Based on a Metal–Organic Framework Interwoven by Electrochemically-Deposited PANI, J. Am. Chem. Soc. 137 (2015) 4920–4923. https://doi.org/10.1021/jacs.5b01613

[57] P. Wen, P. Gong, J. Sun, J. Wang, S. Yang, Design and synthesis of Ni-MOF/CNT composites and rGO/carbon nitride composites for an asymmetric supercapacitor with

high energy and power density, J. Mater. Chem. A. 3 (2015) 13874–13883.
https://doi.org/10.1039/C5TA02461G

[58] D. Sheberla, J.C. Bachman, J.S. Elias, C.-J. Sun, Y. Shao-Horn, M. Dincă,
Conductive MOF electrodes for stable supercapacitors with high areal capacitance,
Nat. Mater. 16 (2016) 220. https://doi.org/10.1038/nmat4766

[59] W.-H. Li, K. Ding, H.-R. Tian, M.-S. Yao, B. Nath, W.-H. Deng, Y. Wang, G. Xu,
Conductive Metal–Organic Framework Nanowire Array Electrodes for High-
Performance Solid-State Supercapacitors, Adv. Funct. Mater. 27 (2017) 1702067.
https://doi.org/10.1002/adfm.201702067

[60] X. Liu, C. Shi, C. Zhai, M. Cheng, Q. Liu, G. Wang, Cobalt-Based Layered
Metal–Organic Framework as an Ultrahigh Capacity Supercapacitor Electrode
Material, ACS Appl. Mater. Interfaces. 8 (2016) 4585–4591.
https://doi.org/10.1021/acsami.5b10781

[61] C. Qu, Y. Jiao, B. Zhao, D. Chen, R. Zou, K.S. Walton, M. Liu, Nickel-based
pillared MOFs for high-performance supercapacitors: Design, synthesis and stability
study, Nano Energy. 26 (2016) 66–73.
https://doi.org/10.1016/J.NANOEN.2016.04.003

[62] S.D. Worrall, H. Mann, A. Rogers, M.A. Bissett, M.P. Attfield, R.A.W. Dryfe,
Electrochemical deposition of zeolitic imidazolate framework electrode coatings for
supercapacitor electrodes, Electrochim. Acta. 197 (2016) 228–240.
https://doi.org/10.1016/j.electacta.2016.02.145

[63] Y. Wang, Q. Chen, Dual-layer-structured nickel hexacyanoferrate/MnO2
composite as a high-energy supercapacitive material based on the complementarity
and interlayer concentration enhancement effect, ACS Appl. Mater. Interfaces. 6
(2014) 6196–6201. https://doi.org/10.1021/am5011173

Metal-Organic Framework Composites - Volume I
Materials Research Foundations **53** (2019) 170-176

Materials Research Forum LLC
https://doi.org/10.21741/9781644900291-8

Chapter 8

Metal-Organic Frameworks for Water Adsorption

Zhiqian Jia*

Lab for Membrane Technology, College of Chemistry, Beijing Normal University, Beijing 100875, China.

zhqjia@bnu.edu.cn

Abstract

The adsorption of water by porous materials is important for many applications. This chapter reviews the recent progress in MOFs for water adsorption, including adsorption isotherm, mechanisms, water harvesting processes, criteria for MOFs, and affecting factors.

Keywords

Water Adsorption, MOFs, Water Harvesting, Adsorption Isotherm, Adsorption Mechanisms

Contents

Metal-Organic Framework Composites - Volume I   Materials Research Forum LLC
Materials Research Foundations 53 (2019) 170-176   https://doi.org/10.21741/9781644900291-8

1. Introduction

The adsorption of water by porous materials is important for many applications, such as harvesting water from air, adsorption-driven heat exchangers, dehumidification, etc. It was reported that only 2.5% of the water resources on earth is fresh water[1]. Water pollution further aggravates the water stress. About 2 billion people in more than 80 countries are experiencing the freshwater crisis. On the other hand, in the air, water vapor and droplets (humidity) add up to about 10% of all other fresh water resources. Therefore, collecting water from the air plays an important role in alleviating the water shortages especially in desert, and has been paid more attentions in recent years [2]. In adsorption-driven heat exchangers, heating and cooling are respectively achieved upon adsorption and desorption of water into/from porous materials, and can be used as air-conditioning units (e.g. in vehicles).

Traditional porous materials, such as charcoal, silica gel, zeolites, hygroscopic salts[3], etc., display either low adsorption capacity or high energy consuming for water releasing due to the too weak or too strong interactions between water and the materials (e.g. adsorption enthalpy of more than 70 KJ mol^{-1} for zeolite) [4]. Therefore, scientists are striving to search novel materials with high water uptake and low desorption energy. Metal-organic frameworks (MOFs) possess adjustable pores size, large surface area and excellent thermal stability, and have found potential in water adsorption. This chapter reviewed the recent progress in MOFs for water adsorption, including adsorption isotherm, mechanisms, water harvesting processes, criteria for MOFs, and factors affecting the adsorption of MOFs.

2. Water adsorption isotherm of MOFs

Water adsorption isotherm of MOFs provides useful information about the adsorption mechanism and the MOFs/water interaction. Four quantities are commonly used to describe the water adsorption isotherm[5]:

(i) The water adsorption capacity (q_{max}, g g^{-1}). This value is related to the pore volume of MOFs.

(ii) The inflection point of the isotherm (α, i.e. the relative pressure at which half of the total capacity is reached). α correlates with the hydrophilicity of the pores, and decreases with increasing hydrophilicity. The relative hydrophobicity of isostructural MOFs can be compared in term of α.

(iii) The Henry's constant (K_H, i.e. the slope of the isotherm at low P/P_0).

(iv) The heat of adsorption (Q_{st}, kJ mol^{-1}).

Metal-Organic Framework Composites - Volume I   Materials Research Forum LLC
Materials Research Foundations **53** (2019) 170-176   https://doi.org/10.21741/9781644900291-8

For mesoporous MOFs, the water sorption isotherms are typically of Type IV or V, and show a hysteresis loop due to capillary condensation. For microporous hydrophilic MOFs, the water adsorption isotherms can be Type I, II or IV depending on the chemical nature of the MOFs, and there is no hysteresis unless the adsorption of water results in a structural distortion, degradation, or open metal sites are present.

3. Water adsorption mechanisms

Water adsorption in MOFs follows three mechanisms[4]:

(i) Chemisorption on open metal sites. The strong interaction between water and the open metal sites results in large K_H, Type I isotherm, and high regeneration temperature, e.g. 195°C for the regeneration of (Zn)MOF-74 [6]. The adsorption and desorption of water on open metal sites is often accompanied by local or global structural deformation, leading to gradual degradation of MOFs upon repeated cycling [7].

(ii) Physisorption in the form of layers or clusters. Physisorption of water in microporous, hydrophilic MOFs is initiated by nucleation on primary adsorption sites(typically close to polar, hydrophilic centers), formation and growth of water clusters (evidenced by the step shape of the isotherm).

(iii) Capillary condensation. Capillary condensation is an irreversible process, and can be observed for MOFs with effective pore diameters larger than a critical diameter (e.g. 20.76 Å for water at 25 °C). Therefore, capillary condensation is commonly observed in mesoporous MOFs, displaying a hysteresis loop between the adsorption and the desorption branch.

Generally, the chemisorbed water molecules is reflected by small thermal ellipsoids, and physisorbed water molecules close to the SBUs exhibit larger thermal ellipsoids, and those located loosely in the center of the pore show the largest thermal ellipsoids.

4. Water harvesting processes

In temperature-triggered capture and release of atmospheric water, water is adsorbed at night and released during the day (e.g. by using the solar power). The process contains three steps:

(i) Adsorption of water from air at low temperatures and high RH at night.

(ii) Desorption at elevated temperatures during daytime by the solar thermal heating in an enclosed unit. The energy absorbed by the MOFs is spent on three processes: overcoming the MOF-water interactions, increasing the temperature of MOFs, and dissipating due to heat loss[8]. For a thermally insulated adsorbent container, the majority energy is spent on breaking the MOF-water interactions, and MOFs with low isosteric heat of adsorption is desirable. The capture cycle can be evaluated by the release efficiency (η_R),

$$\eta_R = m_{released}/m_{captured} \tag{1}$$

where $m_{released}$ and $m_{captured}$ are the mass of released and captured water, respectively. High release efficiency can be achieved by using MOFs with high solar absorptivity, high thermal conductivity, small heat capacity, along with adsorbent container with large surface-to-volume ratio.

(iii) Condensation at low temperatures in the condenser. Cooling of the humid air leads to increased relative humidity (RH), and water condensation occurs when the dew point is reached (100% RH). The cooling energy is spent on three processes: decreasing the temperature of the humid air, dewing, and dissipating due to heat loss. Considering the large specific heat of phase change for water, most of the cooling energy is spent on condensation. The main challenge for water production in desert is to maintain the condenser temperature below the dew point. This can be achieved by three methods: using an efficient condenser with large surface area and enhanced convective heat and mass transfer, employing a water sorption unit with thermal insulation structure (maintaining a low condenser temperature while heating the MOF), and applying an IR reflective coating to all exposed surfaces (minimizing the solar thermal incalescence).

5. Criteria for MOFs

The criteria for MOFs for water adsorption from air mainly include the following factors:

(i) High water stability and cycling stability.

(ii) High working capacity with temperature and/or pressure swing. MOFs need to show a high reversible water uptake between 10% and 30% RH. That means that MOFs interact relatively strongly with water while not too strong to allow for temperature swing adsorption (TSA) using low-grade and renewable energy sources (e.g. solar-thermal heating, waste heat, geothermal energy) or pressure swing adsorption (PSA). Table 1 gives some typical MOFs for water adsorption.

(iii) High thermal conductivity and low heat capacity. Low heat capacity allows for short response times with respect to temperature changes[9].Due to the open structures and low atomic number density, the thermal conductivity of MOFs is comparatively low. To enhance the thermal conductivity and absorptive properties, composites of (Fe)MIL-

100/multiwalled carbon nanotubes (MWCNTs)[10], and MOF-801/nonporous graphite (33 wt %) were reported.

Table 1 Typical MOFs for water adsorption.

	$q_{max}(g\ g^{-1})$	α	Pore size(nm)	Topology	Ref.
MOF-801	0.28-0.36	0.08	0.48,0.56,0.74	fcu	[4]
UiO-66	0.44	0.36	0.74, 0.84	fcu	[4]
MOF-841	0.51	0.26	0.92	flu	[4]
(Al)MIL-100	0.5	0.32	2.5,3.9	mtn	[11]
(Cr)MIL-101	1.4	0.46	2.934	mtn	[12]
(Al)MIL-53	0.09	0.24	0.7-1.3	sra	[12]
CAU-10	0.36	0.15-0.25	0.7	yfm	[13]
(Mg)MOF-74	0.54	0.05	1.11	etb	[14]

6.   Factors affecting adsorption of MOFs

The factors affect the water adsorption include the linkers, defects of MOFs, etc.

(i) Linkers. Polar linkers capable of hydrogen bonding are hydrophilic, showing low values of α in water adsorption isotherms. Linkers with large aromatic system typically result in high α. Introduction of heteroatoms capable of hydrogen bonding interactions allows for hydrophilic pore environments and low α. For all substituents of linkers, the maximum capacity decline due to the decreased free pore volume (evidenced by nitrogen adsorption isotherms). The hydrophilicity of a substituent is correlated to its hydration number, pKa, and hydrogen-bond donor/acceptor ability, rather than its dipole moment[15]. Hydrophilic substituent attached directly to the hydrophobic linker shift the inflection point to lower relative pressures. Nitro groups, esters, and ketones are considered hydroneutral rather than hydrophilic.

(ii) Defects. It was reported that the isotherms of single crystal and powder samples of MOF-801 and UiO-66 show different inflection points and adsorption capacities[4]. These were ascribed to the presence of defects in the powder sample rendering the pores more hydrophilic and increasing the pore volume[16]. It is worth noting that BET surface areas are relatively insensitive toward the defects, whereas water adsorption isotherms show significant changes depending on the concentration of defects.

Conclusions and outlooks

The high porosity of MOFs combined with their wide structural diversity renders them promising candidates for water adsorption. Understanding of the water adsorption isotherm, kinetics and mechanism in MOFs provides the basis for the design of next generation MOFs with promising performance.

On the other hand, considering the practical applications in water adsorption, the following aspects should be paid more attention in future studies: (i) Fabrication of MOFs films, monolithic and composites. (ii) Toxicology. As partial dissolution of the MOFs can result in contamination, the use of linkers and metals with a low LD_{50}(e.g. H_2BDC or H_3BTC, and Al, Ti, or Fe) is preferable for the preparation of MOFs [17].(iii) Device-engineering for applications[18].

References

[1] P. H. Gleick, 1993.Water in Crisis: A Guide to the Worlds Fresh Water Resources, Oxford University Press, New York.

[2] K.-C. Park, S. S. Chhatre, S. Srinivasan, R. E. Cohen, G. H. McKinley, Optimal design of permeable fiber network structures for fog harvesting, Langmuir, 29(2013) 13269-13277.

[3] Paul A. Kallenberger, Michael Froba, Water harvesting from air with a hygroscope salt in a hydrogel-derived matrix, Comm. Chem. 1(2018)28/DOI:10.1038

[4] H. Furukawa, F. Gándara, Y.-B. Zhang, J. Jiang, W. L. Queen, M. R. Hudson, O. M. Yaghi, Water adsorption in porous metal frameworks and related materials, J. Am. Chem. Soc. 136(2014) 4369-4381.

[5] Markus J. Kalmutzki, Christian S. Diercks, Omar M. Yaghi, Metal–Organic Frameworks for Water Harvesting from Air, Adv. Mater. 30(2018)1704304.

[6] W. S. Drisdell, R. Poloni, T. M. McDonald, J. R. Long, B. Smit, J. B. Neaton, D. Prendergast, J. B. Kortright, Probing Adsorption Interactions in Metal Organic Frameworks using X-ray Spectroscopy, J. Am. Chem. Soc. 135(2013) 18183.

[7] P. D. Dietzel, R. E. Johnsen, R. Blom, H. Fjellvåg, Structural changes and coordinatively unsaturated metal atoms on dehydration of honeycomb analogous microporous metal-organic frameworks, Chem. Eur. J. 14(2008)2389-2397.

Metal-Organic Framework Composites - Volume I   Materials Research Forum LLC
Materials Research Foundations **53** (2019) 170-176   https://doi.org/10.21741/9781644900291-8

[8] Farhad Fathieh, Markus J. Kalmutzki, Eugene A. Kapustin, Peter J. Waller, Jingjing Yang, Omar M. Yaghi, Practical water production from desert air, Sci. Adv. 4(2018) eaat3198

[9] B. L. Huang, A. J. H. McGaughey, M. Kaviany, Thermal conductivity of a metal-organic framework (MOF-5), Part I: molecular dynamic simulations, Int. J. Heat Mass Trans. 50(2007) 393.

[10] N. U. Qadir, S. A. Said, R. B. Mansour, K. Mezghani, A. Ul-Hamid, Synthesis, characterization, and water adsorption properties of a novel multi-walled carbon nanotube/MIL-100(Fe) composite, Dalton Trans. 45(2016) 15621.

[11] F. Jeremias, A. Khutia, S. K. Henninger, C. Janiak, J. Mater. Chem. 22(2012) 10148.

[12] J. Canivet, J. Bonnefoy, C. Daniel, A. Legrand, B. Coasne, D. Farrusseng, Structure-property relationships of water adsorption in metal–organic frameworks, New J. Chem. 38(2014) 3102.

[13] H. Reinsch, M. A. van der Veen, B. Gil, B. Marszalek, T. Verbiest, D. De Vos, N. Stock, Structures, sorption characteristics, and nonlinear optical properties of a new series of highly stable aluminum MOFs, Chem. Mater. 25(2013) 17-26.

[14] P. M. Schoenecker, C. G. Carson, H. Jasuja, C. J. Flemming, K. S. Walton, Effect of water adsorption on retention of structure and surface area of metal-organic-frameworks, Ind. Eng. Chem. Res. 51(2012) 6513-6519.

[15] N. Sagawa, T. Shikata, Are all polar molecules hydrophilic? Hydration numbers of nitro compounds and nitriles in aqueous solution, Phys. Chem. Chem. Phys. 16(2014) 13262.

[16] P. Ghosh, Y. J. Colón, R. Q. Snurr, Water adsorption in UiO-66: The importance of defects, Chem. Commun. 50(2014) 11329-11331.

[17] P. Horcajada, T. Chalati, C. Serre, B. Gillet, C. Sebrie, T. Baati, J. F. Eubank, D. Heurtaux, P. Clayette, C. Kreuz, Porous metal-organic-framework nanoscale carriers as a potential platform for drug delivery and imaging, Nat. Mater. 9(2010) 172-178.

[18] H. Kim , S. Yang , S.R. Rao, S. Narayanan , E.A. Kapustin , H. Furukawa , A.S. Umans , O.M. Yaghi , E.N. Wang , Water harvesting from air with metal-organic frameworks powered by natural sunlight, Science 365 (2017)430–434.

Metal-Organic Framework Composites - Volume I
Materials Research Foundations **53** (2019) 177-214

Materials Research Forum LLC
https://doi.org/10.21741/9781644900291-9

Chapter 9

Flexibility in Metal-Organic Frameworks: A Fundamental Understanding

Christia Jabbour[2 †], Noor Aljammal[2, †], Tatjána Juzsakova[4], Francis Verpoort[1,2,3,*]

[1] College of Arts and Sciences, Khalifa University of Science and Technology, PO Box 127788, Abu Dhabi, UAE.

[2] Center for Environmental and Energy Research (CEER), Ghent University Global Campus, 119 Songdomunhwa-Ro, Yeonsu-Gu, Incheon, 406-840 South Korea

[3] Laboratory of Organometallics, Catalysis and Ordered Materials, State Key Laboratory of Advanced Technology for Materials Synthesis and Processing, Wuhan University of Technology, Wuhan, China.

[4] Institute of Environmental Engineering, University of Pannonia, Veszprem, 10 Egyetem St., 8200 Veszprém, Hungary.

[†] Christia Jabbour and Noor Aljammal both authors have contributed equally to this work.

[*] francis.verpoort@ghent.ac.kr

Abstract

Metal-organic frameworks (MOFs) had until recently the reputation of being one of the most critical porous materials. Their flexibility, however, has gained a lot of attention due to the wide selection of possible combinations between metal nods and/or ligands. Nonetheless, it is not always easy to identify the source of flexibility. This chapter focuses on the origin of flexibility, and the substantial geometrical changes that can occur due to external stimuli, such as temperature, pressure, light, gas or solvent adsorption. Flexibility control methods have also been discussed along with possible characterization techniques to help to identify the source of flexibility. Practical applications of flexible MOFs in gas separation and other processes are also discussed. In this respect, several prized examples covered by the literature are present to help in a comprehensive understanding in terms of design and structure tunability of flexible MOFs.

Keywords

Metal-Organic Frameworks, Flexibility, Mechanical Properties, Secondary Building Unit, Characterization

Metal-Organic Framework Composites - Volume I   Materials Research Forum LLC
Materials Research Foundations **53** (2019) 177-214   https://doi.org/10.21741/9781644900291-9

Contents

Metal-Organic Framework Composites - Volume I
Materials Research Foundations **53** (2019) 177-214

Materials Research Forum LLC
https://doi.org/10.21741/9781644900291-9

1. Introduction about MOFs and its structure

In this chapter, we seek to discover insights to the following question: which state of organic linkers, metal nodes, and properties of metal-organic framework determines their flexibility?

Several investigations focused on developing un-rigid catalysts and adsorbents. In this respect, some metal-organic frameworks (MOFs) featured with unique flexibility that is not observed in other porous solids.

The literature reported more than 20,000 wide variety of MOFs, constructed from the assembly of a large diversity of both organic linker and inorganic nodes. Metal-ligand coordination is used to organize scaffolding into sophisticated, multi-dimensional architectures. Theoretically and according to an infinite number of combinations and permutations between various metal centers and organic linkers, thousands of MOFs have been reported. However, the frameworks containing a flexible nature and dynamic properties are rather limited.

A thorough understanding of the flexible properties of MOFs requires some knowledge on the connection between the ligands and the metal ions, and this, in turn, means that we must have insight in the whole structure and bonding capabilities of the framework.

The highly ordered networks have structural transformability and seem inherent to the MOFs synthesis and investigations. The outstanding progress in this topic has been reviewed during the last few years. Both, Kitagawa's and Ferey's group took the early lead through a variety of studies on flexible MOFs [1,2] and they named it as the third-generation of porous coordination polymers (PCPs). The pioneering work of these scientists and their outstanding discoveries in MOFs flexibility is highly appreciated. Kitagawa investigated the inclusion/evacuation of guests within the coordination polymer framework. He classified six possible flexible structures according to the dimensionality of the network [1]. Flexible MOFs show diverse structures depending on the different metal centers/clusters and ligands involved. Their flexibility and dynamical behaviors can be simplified in several ways since general mechanisms can be found behind the various apparent phenomena.

After briefly recalling the history of this discovery, this chapter will analyze the different parameters influencing this phenomenon, as well as advances in the fundamental understanding of numerous structural features of the framework, in both its inorganic and organic parts. Open questions and avenues of recent advances in the characterization of flexibility based on recently reported examples will be pointed out. Recent applications of this family of solids are mentioned in the last section of this chapter.

2.   Origin of MOFs flexibility

The design and synthesis of MOFs always begin with the selection of the scaffolding elements as bridging organic spacers and metal nodes. The structure and functionality depend strongly on either the two components or the nature and type of connection linking them. The rational and intelligent design of the coordinating ligand and metal nodes becomes of paramount importance when attempting to create flexible scaffold-like material.

Virtually, most of the researches have been adequately focusing on the synthesis of the scaffold-like compounds using the supramolecular building blocks (SBBs) or the secondary building units (SBUs) [3–6]. Particularly, rigid MOFs with certain topologies are generated using fixed organic building blocks (based on phenyl rings or multiple bonds, rigid benzene di-, tri-, and tetra-carboxylic acids, terephthalic acid, azolate-based ligands, as well as their derivatives). Such frameworks usually show relatively high thermal and mechanical stability and are capable of retention the porosity upon guest solvents removal.

Generally, the key to classifying flexible elements includes determining which parts of the crystalline framework are not flexible. The functionality of an organic ligand, as a connector, is vital for the creation of network structures and therefore the multi bonding capacity regarding coordination, hydrogen bonding, and attractive π-stacking is necessary to design and tune crystal structures [1]. However, the nature of the resulting frameworks cannot be determined by the linkers' flexibility or rigidity, since in some cases, flexible ligands can be used to build both robust and flexible MOFs. In this particular case, the choice of the source of flexibility exerts a profound influence on the nature of the resulting MOFs. In general, MOFs with a flexible ligand (FL-MOF) are capable of responding to external stimuli, meaning that the entire framework is dynamic [7]. The behavior of the flexible MOFs toward external stimuli such as temperature, pressure, light, gas or solvent adsorption may depend on the metal compound choice, or on whether the linkers can rotate, twist or bend. The nature of the MOFs' response to external stimuli can significantly improve its performances in several applications such as storage, separation, sensing, and others.

In contrast with the prolific production of rigid MOFs and by the vast diversity of metal ions and organic ligands, yet, at present, the design, synthesis, and applications of flexible MOFs are somewhat overlooked. Given few highlights concerning the flexible MOFs, herein, we provide a review covering their design, structures, and characterization.

2.1. Flexibility of functionalized linkers

Flexible linkers are defined as linkers that can adapt themselves to different conformations and consequently producing diverse symmetries during the self-assembly process. Aside from the effect of some synthetic variables, such as temperature, pressure and solvents, the nature of the organic linker(s) plays a major role in the overall topology of the frameworks [8,9].

Applying specific organic synthetic strategies with a particular selection of the electron withdrawing or donating nature of the "contributing" functional linkers, and the incorporation of extended conjugated aromatic systems is a key element in modifying the backbone flexibility of the framework. Furthermore, the shape, the size and the shared angle between the functional group added onto the main linker, play a role in identifying the structure of the MOF and its overall network topology [8,10,11]. Asymmetric linkers also contribute to tuning the topological flexibility of the network [12,13]. Smart linker functionalization is the introduction of functional groups on the organic linkers of the MOFs. Hence, the inclusion of such functional groups can tailor the framework flexibility by the substituent effects at the linker [14]. This topic is further explained in this upcoming section of this chapter.

There are different types of flexible ligands that can affect the conformation of the linker and the resulting framework dynamic behavior: (i) aliphatic carbon chain; this type of linker is the main source of flexible ligand as the long backbone of these chains can bend, rotate around bonds, and reorient themselves, giving rise to more flexible MOFs [15,16]. (ii) aromatic rings with the capability of rotation or movement of dangling side chains of the organic ligands, and (iii) ligands with an open structure. The last type depends on the formation of coordination bonds with the central metal atom M. The decrease of the dentation gives multiple degrees of rotational freedom of the ligand around the inorganic moiety. Additionally, the ratio of M/L and linker/linker is a major-league factor that plays a vital role in determining the degree of flexibility [17,18].

The response of these linkers to external stimuli can occur through a variety of mechanisms including a rotation of the organic linkers, a hinge motion of the carboxylate groups that usually attach ligands to the metal nodes or displacement of sub-networks relative to each other. Benchmark linkers such as carboxylate linkers and functionalized 1,4-benzenedicarboxylate (BDC) linkers are the most common linkers explored so far. Recently these types of linkers have been influential in the field of flexibility recognition of MOFs.

Metal-Organic Framework Composites - Volume I   Materials Research Forum LLC
Materials Research Foundations **53** (2019) 177-214   https://doi.org/10.21741/9781644900291-9

2.1.1 Carboxylate linkers

Empirical rules have been proposed regarding the impact of the carboxylate linkers on the flexibility of MOFs. The first rule stated that ditopic carboxylate ligands, which are linked to two metal clusters or SBUs, are favorable for the design of flexible MOFs. The second rule states that the use of the tri- or tetra-topic carboxylate linkers prohibit the breathing in MOFs [2]. Presumably, utilizing these principles can alert the flexibility of the MOFs.

The use of BDC or terephthalate is vital to construct a family of flexible MOFs. The linker rotation of the BDC-type linker has been also investigated by Kitagawa and his group. They examined the layers of $[Cd_2(PDC)_2]_n$ which contains the cadmium metal centers connected with the pyridinedicarboxylic acid linker (PDC) [19]. Upon the adsorption of polar guest molecules, such as water, a rotational behavior in the linker has led to the opening of the pores [19].

Particular functionalization of BDC linkers can modify the conformational flexibility [20]. A library of functionalized BDC-type linkers, with minimal additional dangling side groups and diverse functionalities by varying chain length at different positions of the benzene core, were generated by Fischer's group [20]. The extra decoration on the pores, by using functionalized linker molecules, can add some complexity to their properties [21,22]. For example, adding a group which can interact through their hydrogen bonding such as $-NH_2$,-COOH and OH, will lead to a higher degree of deprotonation and higher density of open metal sites, and hence the ratio of M/L will increase [18,23].

The fascinating properties of MIL-53 and MIL-88 attracted attention as a prototypical example for experimental and theoretical flexibility investigations [24,25]. Ahnfeldt et al. investigated the insertion of amine-bearing organic linkers in MIL-53. The structure contains linkages of $AlO_4(OH)_2$-octahedra with 1,4-benzenedicarboxylic acid, such inclusion resulted in a slight expansion of the unit cell volume [26].

Yaghi and coworkers have borne out the significance of mixing various organic linkers, as another chief factor that can tune the MOFs flexibility. They used 8 types of functionalized 1,4-BDC organic linkers to synthesize 18 one-phase multivariates (MTV) MOF-5. The resulting series of MOFs contains distinct functionalities [27]. Aromatic carboxylate linkers enable synthesis of many topologically fascinating structures [28], though its functional property can be improved by using a mixed ligand system, i.e., using a semi-flexible tricarboxylic acid ligand H_3L and pyridyl-based co-ligand under solvothermal/hydrothermal conditions to synthesize Cd(II) coordination polymers. The aromatic polycarboxylate co-ligands have an important effect on the features and the overall structures of Cdbmb polymers, concerning their constructions, flexibility, coordination modes, etc. [29,30].

Based on some basic examples with confirmed structures under various type of ligands, the flexibility of MOFs has mainly been attributed to a conformation change of the flexible organic linkers [7] [31].

2.2. Flexibility of the metal nodes

Structural flexibility has been observed and reported in inorganic frameworks, indicating the vast potential and important role of the inorganic components [32–35]. Structural flexibility due to metal nodes, can be caused by (i) the changed coordination environment of metal ions [36,37] and (ii) the deformed configuration/connectivity of secondary building units (SBUs) containing metal ions in response to the removal/binding of coordinative molecules [38]. For example, the presence of some unpaired electrons in the chains of the metal centers might block the breathing nature of the structure by puckering and slightly modifying the distances between adjacent metal centers [39].

For MIL-53, several structures can be composed by varying the metal M as such [M(OH) $(BDC)_2$]n. (M = Cr [40], Sc [41], In [42], Ga [43], Al [44], Fe [45]). Different flexibility behaviors were observed by simply changing the choice of the metal center. In the manner depicted in Figure 1 [46], MIL-53(Cr) showed a better ability for pore opening upon dehydration [40]. A transition from large pores (LP) to narrow pores (NP) was detected upon applying a mercury intrusion-extrusion test using a mechanical pressure below 500 MPa on MIL-53 (Cr) and MIL-53 (Al). A penetration of mercury within pores larger than 3 nm takes place. The result of this test revealed that MIL-53(Al) solid was shown to exhibit an irreversible contraction, while MIL-53 (Cr) undergoes a reversible structure change under similar conditions. These result attributed to the presence of corner-sharing chains of $AlO_4(OH)_2$ octahedra which lead to stiffness of structure upon inducing a pressure [47].

In summary, in this section, it has been proven that the flexibility will not only depend on the nature of the metal ions and/ or the choice of organic likers, but also on their connectivity. The upcoming section will focus on the effect of flexibility on the overall structural behavior during different applications, such as adsorption and desorption.

Materials Research Forum LLC
https://doi.org/10.21741/9781644900291-9

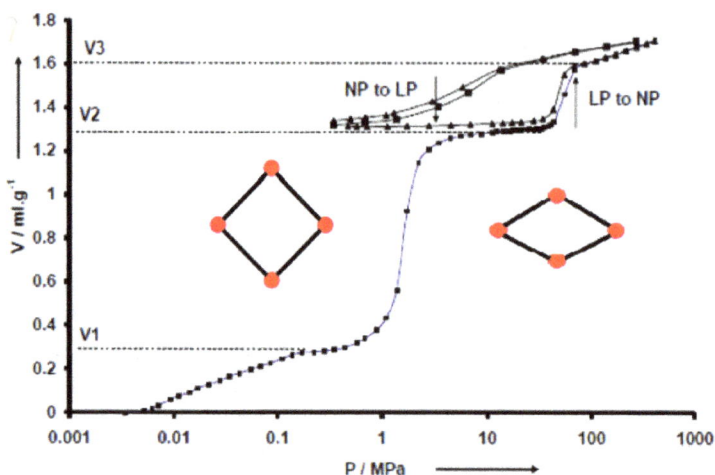

Fig. 1. *LP to NP transition in MIL-53 (Cr) sample upon intrusion-extrusion of mercury: cumulative intruded volume versus applied pressure. The arrows show the steps that correspond to the LP to NP transition. Squares: first intrusion-extrusion cycle. Triangles: second intrusion-extrusion cycle. Adapted from [46].*

3. General aspects of framework flexibility

Flexible MOFs behave in a remarkable stimuli-responsive fashion. The challenging mission is the design of site-specific, stimuli-responsive controlled MOFs. The form of the stimuli has a bearing on the significance of the results obtained, especially in the measurement of different induced responses. In this respect, physical and chemical stimuli are needed to stimulate the dynamic behavior of the flexible MOFs. Moreover, each response for each type of stimuli has a great impact on the architectural framework, such as phase change, gate opening, and change in the cell parameters.

Natural stimuli used to provoke the structural transition can be derived from:

(i) Insertion/removal of guest molecules [48,49], expansion during the incorporation or removal of guest molecules, dynamic frameworks can expand or contract which subsequently allow a variety of guest exchange.

(ii) Exposure to external pressure. The effect of pressure on a material is directly related to its mechanical stability, critical for many types of commercial applications [50–52].

(iii) Exposure to thermal induction. When MOFs are exposed to temperature as external stimuli, thermal responsive action can be detected such as changing the linker's conformation (e.g. rotation of aromatic rings or the movement of dangling side chains of the organic ligands) [25,52–54].

(iv) Light can also be used as a stimulus for the dynamic movement of MOFs [55,56]. Switching between cis-trans / LP-NP transition upon interaction with light [55–57].

3.1   Breathing, swelling and linker rotation

Substantial geometrical changes can occur as a result of external stimuli exposure of the "weak points" in the framework. The structural changes occur mainly through defined transitions between different crystalline states. The weak points can be classified into three groups. MIL-53 has provided the foundation for extensive research for the breathing phenomena. Figure 2 shows the three principle breathing cases of MIL-53.

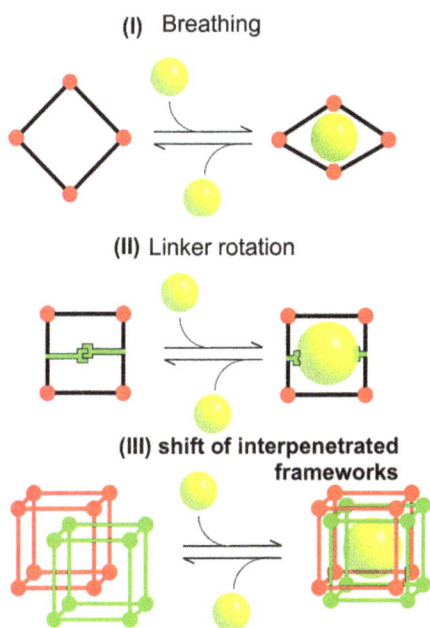

(I) Breathing

(II) Linker rotation

(III) shift of interpenetrated frameworks

Fig. 2. Principal types of breathing: (I) flexibility of the IBUs, (II) structural changes of the linker molecules, and (III) shift of interpenetrated or interwoven frameworks. Adapted from [58].

- The inorganic building units (IBU), **Figure 2**(II) [2]. The dynamic behavior of these materials is subjected to a variety of mechanisms such as: (i) ordered rotation of the ligand molecules; (ii) a displacement of sub-networks which are connected by van der Waals interactions [59,60]; (iii) a hinge motion of the linkers [2,61,62]. These dynamic behaviors promote the opening or closing of pores and consequently affect the loss/uptake of guest molecules. The best-investigated example for weak IBU is carboxylate ligands that are capable of switching their binding mode, such as the rotation of the ligand around the O–O-axis of the carboxylate groups, this process called as knee-cap mechanism

- Low energy and reversible structural changes of the organic linkers can also act as a weak point that leads to different linker isomers, see **Figure 2**(II). The reversible photoisomerization of azobenzene molecules with alternating UV/visible light irradiation is a good example of such type of flexible transformation [63].

- The free pore volume of interlocked linkers and interpenetrated networks can also act as a weak point. In this case, breathing arises by shifting the individual frameworks against each other (**Figure 2**(III)).

It has been pointed out that despite the variety of MOFs; only several classes exhibit the unique breathing phenomena. In the early stage of research, most dynamic behaviors were related to the presence or removal of guest molecules, known as the host-guest interaction [1,64]. Kitagawa et al., classifies three types of MOFs exhibiting flexibility under the inclusion/evacuation of guests within the coordination framework, depending on the dimensionality of the polymer network (**Figure 3**).

The first is considered the most well investigated flexible MOFs. Pillared MOFs, are 1D channels with diamond or square cross-section. The pillaring linkers of the frameworks are flexible as they can expand or shrink (elongated or shortened) upon incorporation or removal of the guest molecule [65,66].

In a rigid 2D layers MOF, the layers are covalently connected with flexible pillars [67]. Generally, upon a strong framework–guest interaction with the so-called sponge-like MOFs, a change in the framework cell volume can occur. Hence, elongation of pillars leads to the expansion of frameworks during absorption of guest molecules while the frameworks shrink upon the removal of guest molecules, the rotation of organic moieties, resulted from strong guest-host interactions, persuades the volume change [67,68]. It is worthy to mention that the framework topology maintains their single crystallinity upon the adsorption of guest molecules. A good example of this case can be presented by a bilayer framework synthesized from BTC^{3-} (BTC = 1,3,5-benzenetricarboxylate) ligand and bismacrocyclic Ni(II). This network showed a sponge-like behavior, shrinking and swelling according to the number of guest solvent molecules contained within the networks [69].

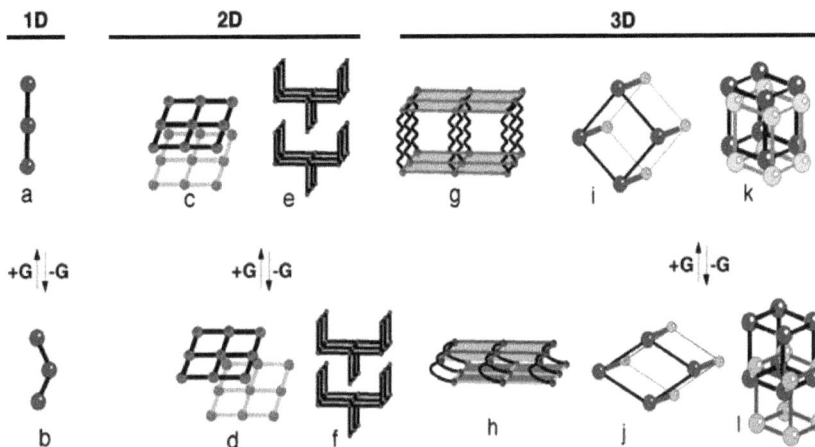

Fig. 3. *The illustration of dynamic behaviors of the metal-organic frameworks' (MOFs) structure under the incorporation/removal of guest molecule. Adapted from [64].*

The last case is presented by interpenetrated 3D grids that contain layers connected covalently with flexible pillars. The introduction of guest molecules causes the sliding of the adjacent interpenetrated net and hence causing the pores opening or closing upon the adsorption of guest molecules [65,70].

MIL-53 (Al, Cr) is a 3D metal (III) terephthalates having 1D pore channels. This MOF containing disordered BDC linkers, switched from NP to LP upon hydration and dehydration respectively, with a 50% volume increase without any change in the topology (Figure 4) [71]. Such a large breathing effect can be explained by the absence of the hydrogen-bond interactions between the hydrogen atoms of the water molecules and the oxygen atoms of the carboxylic group and the $\mu 2$-hydroxo group.

MOFs that exhibit large flexibility, without defined transitions, are rare and are not well understood. The prototypical name of such continuous breathing behavior is often referred to as "swelling behavior." The swelling mode is characterized by gradual enlargement of the MOF unit cell volume without a change in the shape of the unit cell shape or in the space group. MIL-88 undergoes very large changes in the pore size during solvation and desolvation [66]. It could swell upon immersion in liquids with variations in unit cell volume from 85% to 230% depending on the nature/length of the organic spacer (see Figure 5) [25,72].

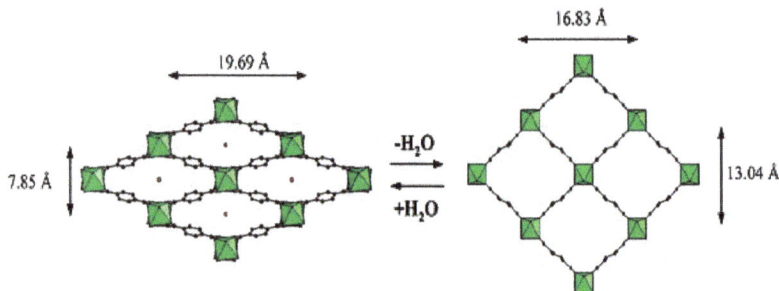

Fig. 4. Breathing behavior of MIL-53 (Al,Cr). Adapted from [68].

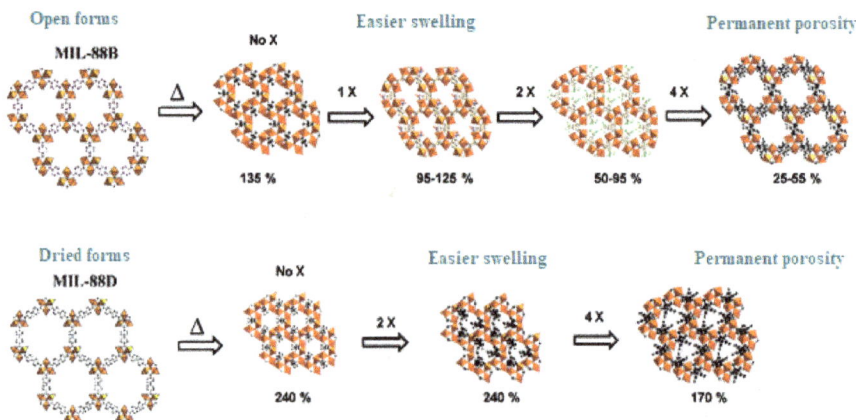

Fig. 5. Illustration for swelling phenomena of the structure and the flexibility of the MIL-88B and MIL-88 D modified solids as a function of the number of functional groups (X) per spacer (on the top). Adapted from [25].

The amplitudes of swelling, corresponding to the structural switching from the open to the closed forms, are given under each structure, and it is calculated by the following formula:

$$\text{Swelling } \% = \frac{V_{OP} - V_{dry}}{V_{drt}} \times 100 \qquad (1)$$

Ramsahye and his group studied the impact of n-alkanes adsorption on MIL-88 (**Figure 6).** During adsorption, the external surface of the structure opens the pore sufficiently for the first molecules to be able to diffuse into the framework and thus initiating a pore swelling [73]. Note that this solid exhibits only a small cell volume contraction of about 20% upon drying and thus bears a pore size in its dried state over 5 Å.

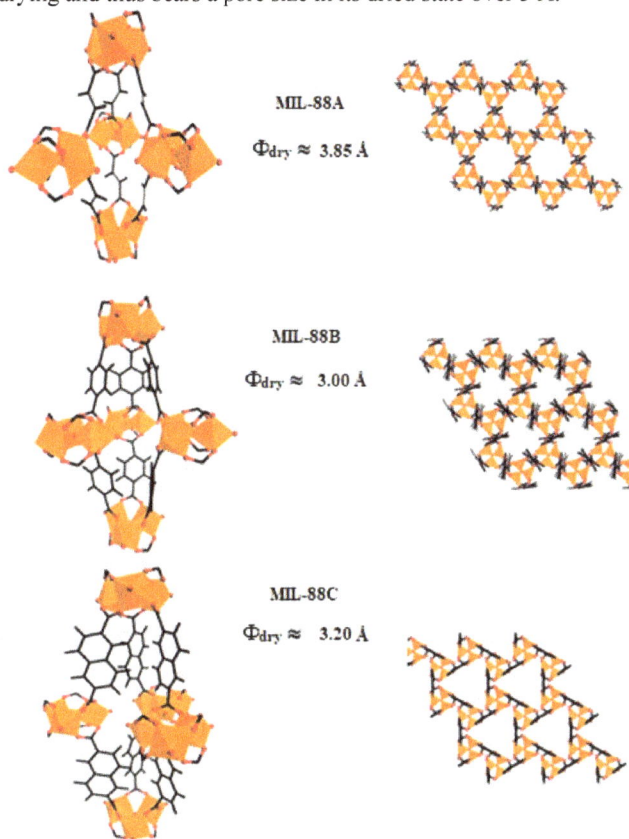

MIL-88A

$\Phi_{dry} \approx 3.85$ Å

MIL-88B

$\Phi_{dry} \approx 3.00$ Å

MIL-88C

$\Phi_{dry} \approx 3.20$ Å

Fig. 6. Structures of MIL-88 (A,B,C) with the corresponding linkers (fumarate, terephthalate, naphthalenedicarboxylate) with different pore swelling. Adapted from [73].

Linker rotation or gate opening behavior is described as a (continuous) transition where the fragments of a linker can rotate around a certain axis. It is important to note that linker rotation is not essentially accompanied by crystallographic phase transitions

[39]. **Figure 7** shows the rotational behavior of the aromatic ring within the linker structure around a certain axis within the void. The MOF structure flexibility of two pyrazolate (pz) rings of H_2Me_4bpz can rotate around the central C–C single bond during the intercalation and deintercalation of Li-ions [75]. This kind of dynamic behavior resulted in significant change in the inner geometry of the pore, as well as the positions of possible gas uptake [74].

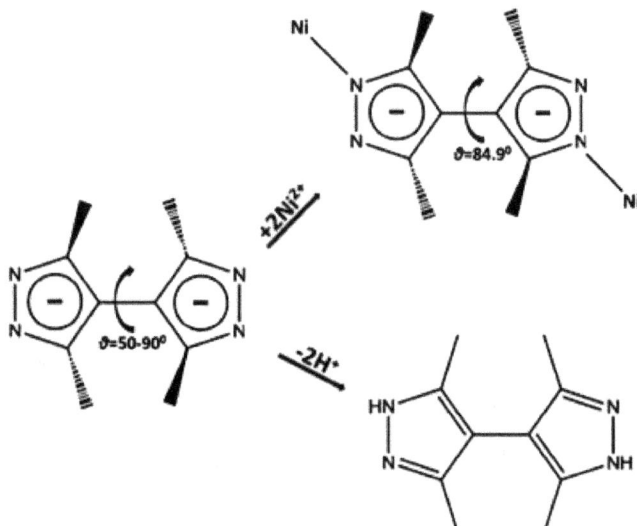

Fig. 7. *The rotated structure of ligand* H_2Me_4bpz *and its coordination mode. Adapted from [75]*

The distinguishing feature between **Linker rotation /gate opening behavior** and breathing phenomena is that a change in the unit cell of the crystal occurs while in case of linker rotation the rotational movement results in the expansion of the pore windows.

Prized example of a reversible linker reorientation can be observed on the linker of ZIF-8 (methylimidazole linker) during the application of very high pressure (14–700 bar). Hence the size of the pore window increased as the implemented pressure increased [76], **Figure 8**.

Materials Research Forum LLC
https://doi.org/10.21741/9781644900291-9

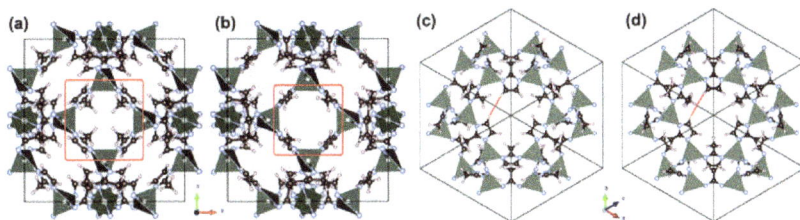

Fig. 8. *Linker reorientation can be observed on the linker of ZIF-8 viewing from the axis direction (a) and the diagonal direction (c), corresponding structure of ZIF-8 loaded with N_2, (b) and (d). Adapted from [77].*

3.2 Thermoresponsivity

Some of the flexible MOFs show a transition state from narrow pore form at low temperatures, to large pore structures when the threshold temperature of the phase transition is reached. Increasing the side chain length tends to cause a decrease in the transition temperature. In contrary, the more polar the linker, the higher the transition temperature, as pillars rotation eventually ends up in the opening of the pore space [78,79].

Going back to our prominent example, increasing the temperature in MIL-53(Sc) favors an increase in the rotational motion of the phenyl rings and thus a transition to the open form. Therefore, changing the nature of the aromatic spacer has a strong impact on the flexible character of these solids [41]. For the comparison of the framework opening in nitro functionalized MIL-53(Sc) MIL-53(Sc)-cp (closed pores) (a) and MIL-53(Sc)- np (b), See Figure 9.

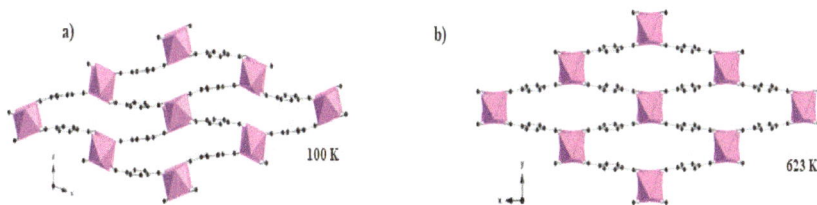

Fig. 9. *Comparison of the framework opening in nitro functionalized MIL-53(Sc) MIL-53(Sc)-cp (a) and MIL-53(Sc)- np (b). Adapted from [41].*

3.3 Mechanical properties, elasticity

To date, some pressure-induced behaviors of flexible MOFs have been investigated [80–83]. Frameworks with metal polyhedra, bridged by the organic linkers, is a key signature of highly flexible and high elastic MOFs, as it can relax or rotate when external pressure is induced [84]. In many cases, compression is considered fully reversible after the pressure has changed back to the initial conditions [51,85,86]. However, some structural transformations resulting from pressure stimuli are not necessarily reversible.

There are four possible unit cell transformations of MOFs upon applying a mechanical pressure; the first is amorphization, e.g., pressure-induced amorphization of ZIF-4 [87]. The second is compression, e.g., the linear compressibility of MIL-53(Al) [87]. The third is known as the reversible phase transition, e.g., ZIF-8 [88], and the fourth possibility is the phase formation (bond rearrangement), e.g., pressure-induced bond rearrangement in [tmenH$_2$][Er(HCOO)$_4$]$_2$ [50]. However, a truly flexible material should show reversible behavior under external parameters such as applied pressure.

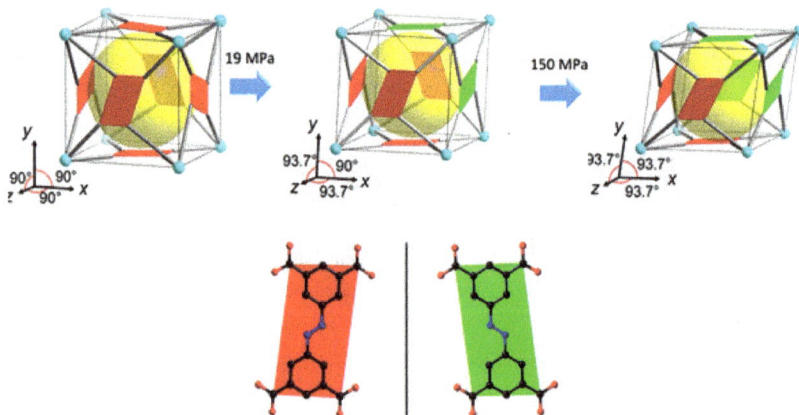

Fig. 10. *Illustration of possible transformations that MIL-127 can undergo upon applying mechanical pressure such as inclination of the unit cell and flipping of linkers. Adapted from [89].*

ZIF-8 is one of the hallmarks of MOFs in this context as its cell volume can be decreased rapidly by 5% when a pressure of 0.34 GPa is applied. However, ZIF-8 featured with crystal-to-amorphous transitions upon pressure induction [88]. MIL-127 is another interesting example to study the nature of stimuli-responsive, as the findings concerning

MIL-127 elasticity are provoking on some levels. Figure 10 demonstrates the structural transformation of MIL-127 under different pressure, increasing the pressure from 19 MPa to 150 MPa buttoned up a unit cell symmetry, distorted off and a flipping of a fraction of the non-planner ABTC (3,3′,5,5′-azobenzenetetracarboxylate) organic linkers framing the unit cell; this phase change resulted in a cell compression [89].

3.4 Photoresponsive

MOFs demonstrate a unique behavior upon interaction with light "photoswitches phenomena." A structural transformation in the framework usually is used as a stimulus for the dynamic movement of flexible MOFs.

Coudert classified the photoresponsive frameworks into three main categories [90].

The first generation contains MOFs with thermally or optically functional group attached to the overall framework. Generally, this kind of MOFs are synthesized by using azobenzene and its derivatives owing to their photochromic molecules that can undergo clean, efficient, and extrinsic reversible photoisomerization [91]. Modrow and coworker have pioneered this fact as they examined photoluminescent properties of 3-azo-phenyl-4,4'-bipyridine in ethanol by successfully recording the switch using UV/Vis spectra. Exposure of the resulting MOF to UV-light of a wavelength of 365 nm triggers the cis-trans transformation of the linker (See Figure 11) [92] .

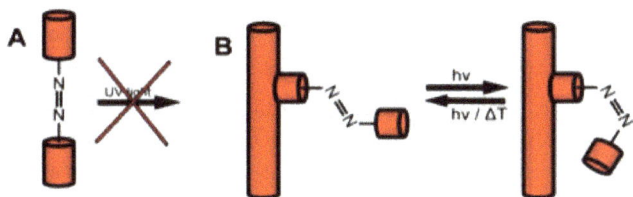

Fig. 11*. Photoswitchable azobenzene side groups. (A) Azo-linker as part of the backbone of the MOF, (B) azo-group covalently attached to the inner pore wall and extended into the pore. Adapted from [92].*

Brown and co-workers succeeded to develop a photo-switched azobenzene functionalized linker hosted by IRMOF-74-III framework $[Mg_2(C_{26}H_{16}O_6N_2)]n$. When this MOF was exposed to light, switching from the trans-conformer to the cis-conformer could be noticed. Moreover, a distance increment between the para-carbon atoms in azobenzene molecule from 8.3 Å to 10.3 Å was also noticed [92].

It was noted that this reversibility (cis-trans switch-ability) and guest's photo isomerization could lead to drastic changes in the gas adsorption properties. Park and his group examined the reversible alteration of CO_2 uptake upon light exposure or thermal treatment, particularly, when the azobenzene is in the cis form, the framework has a significantly lower CO_2 uptake in contrast to the material framework when the linker is in the trans-form [57]. A reversible cis-trans switching of azobenzene was performed by irradiation light with a wavelength of 440 nm.

This concept was pursued by other groups and further exploited with a different host framework, e.g., it was noted that some host frameworks forbid and block the *cis-trans* transformation as in case of azobenzene in MIL-53(Al) [93].

The second generation was synthesized by incorporation of a photoresponsive unit directly in the "backbone" of the linker as a side chain. It was verified that the azobenzene side-chain fully retained its photoisomerization ability, with a change in pore dimensions upon light exposure [92,94–96]. In analogy to the previous examples, Modrow and coworkers [94] prepared a Cr-MIL-101-amide and Cr-MIL-101-urea, by introducing the azo functional group to the Cr-MIL-101-NH$_2$ mesoporous cages. *Cis/trans* isomerization was tested by methane adsorption experiments. Upon interaction of the prepared MOFs with alternating irradiation of UV-Vis light, a variation in the CH$_4$ sorption behavior was observed, e.g., an increase of the cis-isomer concentration. Hence, a rise in the CH$_4$ adsorption capacity due to the change of the polarity, dipole moment and free space for each conform was detected [57,94].

The third generation focuses on the change in the framework itself, upon exposure to light. This kind of dynamic photo-switching MOFs is difficult to synthesize, and it is considered as the most affected light-induced transformations. Lyndon et al. succeeded to develop a triple interpenetrated framework (Zn(AzDC)(4,4'- (BPE)$_{0.5}$) with multiple photoisomerizable linkers. The switching experiments conducted by exposure of the material to UV-light, to examine the uptake and release of CO_2, revealed that 64 % adsorbed gas released upon broadband radiation. Moreover, a fully reversible response have been noticed [91]. In summary, different modes of attachment lead to different photoswitching behavior.

4.   How to control MOFs flexibility

Even though flexible MOFs show a distinct architectural framework, their structure can be tuned and designed by controlling the different metal clusters and ligands involved during synthesis. It is important to note that flexibility can be introduced into the backbone of MOFs through different reactions, such as Friedel–Craft acylation of an aromatic ring or Lewis acid catalyzed alkylation, reactions involving the amide formation and nucleophilic substitution [97,98].

4.1 Metal ion

The framework dynamics and the flexibility of any MOF highly depend on the type of metal-node and its connectivity with the linker molecule in the solid framework. For MIL-53, the choice of metal shows different response towards guest molecules as simple as water. Even though Cr^{3+} and Fe^{3+} possess similar electronegativities and ionic radii, MIL-53(Fe) and MIL-53(Cr) showed different behavior towards the sorption of CO_2 [45].

A cobalt-based MOF, also known as Co(BDP) (BDP=1,4- benzene dipyrazolate), can undergo five structural transitions, during the adsorption of nitrogen at 77 K. Breathing occurs in the square void channels of the MOF that is composed of tetrahedral coordinated Co II centers linked in two orthogonal directions by bridging BDP^{2-} coordinating benzenedipyrozolate ligands (Figure 12). [52,99].

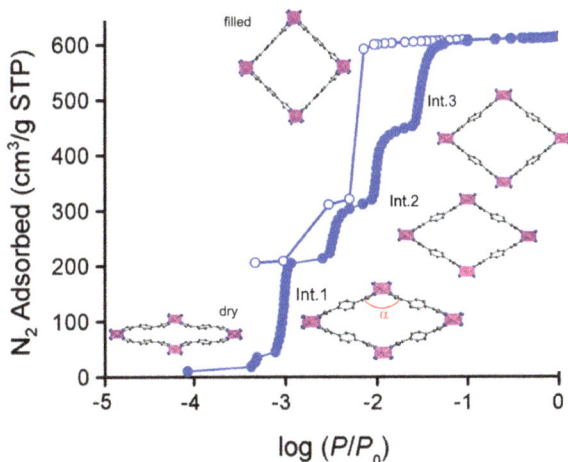

Fig. 12. N_2 adsorption isotherm at 77 K, indicating the five pressure-dependent phases of Co(BDP). The dry and filled symbols correspond to adsorption and desorption, respectively, and intermediate states (Int) correspond to adsorption at different pressures. Co, N, and C atoms denoted in pink, blue, and gray spheres. Adapted from [99].

Metal-Organic Framework Composites - Volume I
Materials Research Forum LLC
Materials Research Foundations **53** (2019) 177-214
https://doi.org/10.21741/9781644900291-9

Co(II) adopts a square planar coordination geometry in all phases (from dry to Int.3), showing the same topology and symmetry, except when the molecule is filled. When saturation is reached, the framework adapts a tetrahedral form. This transformation is caused by the long 0.26 nm Co-N bond [99].

Solvent-assisted ligand incorporation (SALI)

When active species fail to be incorporated into the framework of a catalyst during synthesis, alternative techniques are adopted, such as the solvent assisted ligand incorporation (SALI), which relies on acid-base chemistry. The SALI technique is used to efficiently incorporate various active species into Zirconium based MOFs [100]. Such catalytic functions are commonly inserted either at the linker, or at the zirconium-oxide node, or encapsulated in the pores [101]. Perfluoro alkane carboxylates of various chain lengths (C_1–a_9) were successfully incorporated on the Zr_6 nodes of NU-1000 through the SALI technique by Deria and co-workers. Catalytic functionality is introduced through the ionic bonding of the OH groups on the NU-1000 node and the carboxylate group of the perfluorinated chain (Figure 13) [102–104]

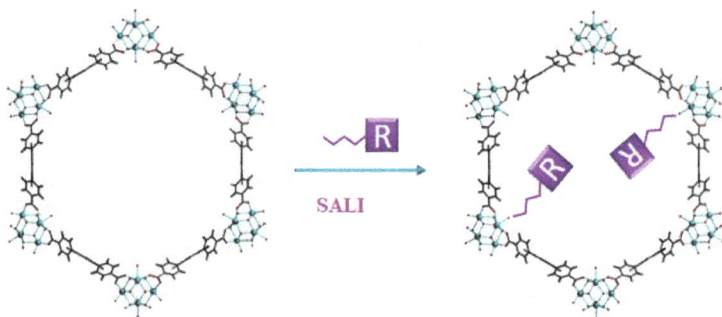

Fig. 13. *Solvent-assisted ligand incorporation onto Zr-nodes of NU-1000. Adapted from [101].*

4.2   Ligand (functional group)

Choosing the proper organic linker, or adding side chains to the ligands, extra species, during synthesis will determine the dynamics of the framework. The nature of the extra-framework species can highly affect the breathing phenomena.

Metal-Organic Framework Composites - Volume I   Materials Research Forum LLC
Materials Research Foundations 53 (2019) 177-214   https://doi.org/10.21741/9781644900291-9

4.2.1 Secondary building block unit (SBU)

The adsorption behavior of JLU-Liu4 (JLU – JiLin University), containing two formic acid anions directly coordinated to a zinc metal ion, was compared after the systematic substitution of monocarboxylates in the SBU of the MOF. The utilization of monocarboxylic acids with different backbones during the synthesis resulted in the formation of isomorphous frameworks. The incorporation of small carboxylic acids such as formic acid and acetic acid resulted in a gate opening behavior during adsorption. However, increasing the number of carbon atoms in the backbone will lead to a decrease in the gate pressure. Bulkier carboxylates such as benzoic acid and cinnamic acid show similar adsorption properties, but different structural transformation when incorporated within the backbone. Classical breathing behavior was observed for the MOF containing benzoic acid, while the MOF containing cinnamic acid encountered a step-wise increase of the unit cell volume [105].

4.2.2 Linker substitution

Flexibility in MOFs can be controlled by adjusting the linker substitution. Handke et al. studied the influence of linker substitution on network porosity in isostructural Ag-MOFs through 15 different substituted 3-(1,2,4-triazole-4-yl)-5 benzamidobenzoate ligands. Sorption studies show the effect caused by linker substitution on their appearances, causing structural transformation, thus affecting the gate-opening behavior [106].

Taddei and co-workers, were the first to synthesis water stable isoreticular phosphonate based Cu-MOFs, containing both phosphonate and N-heterocyclic linkers. They studied the flexibility of the MOF by varying the choice of linkers. A reversible breathing effect during hydration and dehydration was noticed when N,N,N',N'-tetrakis (phosphonomethyl) hexamethylenediamine (denoted as L) and 1,2-bis-(4-pyridyl)ethane (denoted as etbipy) were used as organic ligands for the construction of $Cu_3(L)(etbipy)_2$ framework. However, when replacing either of the two ligands by a more rigid linker, (etbipy with 4,4'-bipyridyl), the transition from large pore to narrow pores, during dehydration was irreversible [107].

Serre and co-workers studied the adsorption behavior of a series of isoreticular Cr (III) or Fe(III) dicarboxylates labeled MIL-88A-D {dicarbox = fumarate (88A); terephthalate (1,4-BDC) (88B); 2,6-naphthalene dicarboxylate 2,6-NDC) (88C); and 4-4'-biphenyl dicarboxylate (4-4'-BPDC) (88D)}. The results show breathing motion, due to the interaction of guest molecules with the MIL-88 framework, without apparent bond breaking in the network. However, the nature of the organic linker played a big role in the flexibility of each MOF. By calculating the ratio of the cell volume of the open form to that of the dried solid, they deduced that the longer the length of the linker, the higher the unit cell parameters (from 1.85 for MIL-88A to 3.3 for MIL-88D [24]. The presence of bulky functional groups (such as $4CH_3$, 4F, and $2CF_3$) will force the framework to swell in order to be able to accommodate guest molecules.

4.2.3 Introduction of bulky side chains at the flexible edge of the MOF

The steric hindrance highly depends on the functional groups within the structure of the MOF. Horcajada et al. studied the flexibility of MIL-88 after few modifications on the ligands. MIL-88 is known for its high flexibility, as they can be synthesized using the controlled secondary building units (SBU) approach [24,108]. This approach is applicable to trimeric inorganic species (three metallic octahedral shared by μ_3-O), predetermined by the SBU to be used in the synthesis of the MIL [109]. MIL-88 was modified by introducing functional groups into the aromatic rings. Iron(III) terephthalate MIL-88 and iron(III) 4,40 -biphenyl dicarboxylate MIL-88 exhibit different flexibilities due to the steric and intermolecular interactions between the phenyl rings. A proportional decrease in the breathing amplitude was observed as a function of the size and number of functional group introduced [25]

4.2.4 Linker rotation

The flexibility of MOFs is strongly affected by the addition of functional groups since they induce steric hindrance and affect the host-guest interaction. Changes in the pore opening are observed as well as a change in the host-guest interaction of the MOF when chemical groups of various properties are introduced through the aromatic linker of MIL-53 [110]. The introduction of 4 CH_3 groups in MIL-53(Cr), causes the phenyl rings to rotate by 90°, which prevents the adsorption of guest molecules [111].

The flexibility of some MOFs can be controlled upon adsorption/desorption of specific gases and liquids by controlling the breathing phenomena [62]. Depending on the adsorbed species, the size and the structure of the pores can be adjusted from large rectangular to narrow trapezoids. In principle, for MIL-53, such molecules are CO_2, water, and hydrocarbons [87]. Hamon et al., studied the effect of flexibility of MIL-53(Cr) on the breakthrough curves, during the adsorption of a mixture of CO_2 and CH_4 [112]. The structure flexibility plays an important role in the separation of CO_2 and CH_4. They examined the NP/LP fraction through Raman spectroscopy, by varying the ratio of CO_2 to CH_4. As the concentration of CO_2 increases, the narrow pores open up and CH_4 molecules are pushed out of the NP to accommodate CO_2 molecules (see Figure 14).

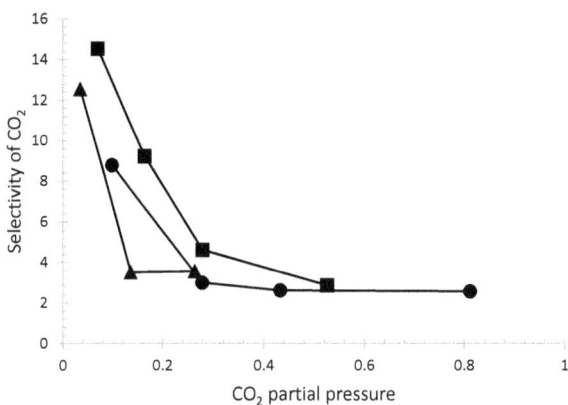

Fig. 14. *Breakthrough measurements of CO_2 selectivity on MIL-53(Cr). CO_2-CH_4 gas feed, 75-25 (■), 50-50 (●) and 25-75 (▲). Adapted from [112]*

The flexible nature of MIL-88, exhibits large swelling when hydrated, due to the absence in the bipyramid of any linkage between its "equatorial" trimers, allowing the adsorption of a wide range of molecules. The swelling is caused by two degrees of freedom, the first coming from the free rotation of the COO groups around the single C-C bond, and the second is caused by the change in the dihedral angle between the COO plane and the trimers plane (Fe-Fe-Fe) [113].

4.3 Post-synthetic modification (crystal size)

Functional groups such as rigid phenyl rings or heterocycles tend to reduce the flexibility of the ligands. Linkers can be rotationally predefined either during synthesis or through post-synthetic modifications (PSM). Chiral functions can be incorporated into MOFs by introducing chiral building blocks [114]. Chemical reactions that involve covalent bond formation with the MOF framework can be considered for such modification.

The flexibility of a catalyst has a major effect on its catalytic applications [115,116]. Fairen-Jimenez et al. compared the adsorption of N_2 on ZIF-8 at ambient pressure and after its modification at high pressure (1.47 GPa), denoted as ZIF-8AP and ZIF-8HP, respectively. Results show that by increasing the pressure, a reversible structural change was observed through a reorientation of the imidazolate linkers, which led to the increase in the accessible pore volume and the size of the window. A 30% increase of N_2 adsorption was denoted for ZIF-8HP compared to that of ZIF-8AP [76].

Heneke and co-workers observed a relation between the gate opening and the substituted linker. Flexibility was observed when alkyl ether groups were introduced onto the ligands of the functionalized $[Zn_2(BDC)_2(dabco)]_n$ material. Such groups tend to control the sorption behavior by acting as molecular gates for guest molecules [20,117].

5. Characterization of Flexibility in MOFs

The combination of spectroscopy and diffraction in the development of *in-situ* monitoring techniques gives a deep understanding of the dynamic structural changes in flexible MOFs. Highlighting the importance of such a phenomenon is a must, in order to achieve the ideal intended design of MOFs [39].

Characterization techniques such as NMR (Nuclear Magnetic Resonance), allows the determination of flexibility through the detection of structural changes of the host during adsorption. ^{129}Xe NMR spectroscopy relies on the adsorption of xenon to study the xenon chemical shift as an indicator of the extended transition from LP to NP. In-situ ^{129}Xe adsorption NMR experiments on the flexible MIL-53(Al) material have been performed by Springuel-Huet et al. at different pressures and temperatures [118]. They relied on the chemical shift of Xe to study the local Xe amount and the transformation from LP to NP form.

Another way to characterize the mechanical properties of flexible microporous MOFs is by studying its porosity through mercury intrusion, which often is non-wetting and requires a specific pressure to enter the pores of the solid. Mercury porosimetry is a technique that allows studying the transition of LP to NP just by applying mechanical pressure [46].

Infrared spectroscopy is a vibrational method often used to determine chemical bonding in addition to structural changes of the host framework and the interaction of the skeleton with the guest molecule. IR spectroscopy is used to study the pyrazolate ring stretch vibrations in Co(BDP), during N_2 adsorption at 100 K. Results show that upon increasing the N_2 pressures, the C-N stretching vibrations of the pyrazolate rings undergo different transition states. *In-situ* infrared spectra were recorded to show the transition from NP to intermediate phase ip1 and ip2 (at 0.01 and 0.023 bar, respectively) [99].

Powder X-ray diffraction (PXRD) is another analytical technique that can be used to determine structural transformation in flexible MOFs [47]. Devic and co-workers used a combination of techniques in order to evaluate the effect of the functional group X, of a series of functionalized flexible MIL-53(Fe)–X during adsorption. Different diffraction patterns were obtained, indicating the effect of the functional group of the pore opening during adsorption [119].

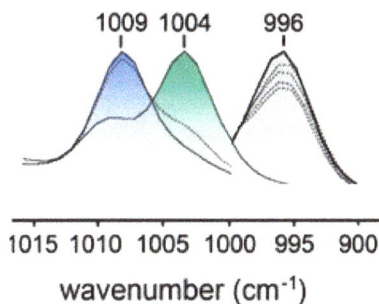

Fig. 15. *In situ infrared spectra of Co(BDP), during N_2 adsorption at 100 K. Phase transitions from the dry form (gray) to ip1 (green) and ip2 (blue). Adapted from [99].*

6. Applications

Much thought has been given to the rational design of flexible MOFs, and this has led to the belief that it holds distinct potential in a plethora of applications from that of the rigid MOF. Nevertheless, the industrial application of flexible MOFs did not yet reach a major breakthrough in high-tech applications. Only a few flexible MOFs are offered on a commercial scale, to mention just a few, e.g., ZIF-8 and MIL-53(Al) (corresponding to Basolites A100 and Basolites Z1200 respectively by Sigma Aldrich). Recently, the potentiality of flexible MOFs have triggered an interest towards limited industrial applications such as selective gas separation [120–122], molecular recognition [123], catalytic process [124], sensing [125] , and biomedical application (i.e., controlled drug release) [126]. The employment of flexible MOFs in catalytic reactions is not extensively developed yet, due to their instability under different operating conditions. Therefore, a more optimistic statement cannot be concluded, because the breadth of these studies is still not large.

Hence, at least for the near future, further development in the scaling up process of flexible MOFs is needed, for its industrial application. Which should be closely associated with the development of synthesis and characterizing strategies.

Conclusion

The fascinating topic in MOFs flexibility, specifically how ligands, metal nodes, both rigid and flexible, are organized and assembled into various structures have attracted much attention due to the thousands of possibilities in designing, tailoring, or controlling framework for developing MOF materials with excellent performances. This Chapter shed light on the linking of metal nods and ligands of flexible frameworks and tries to identify the causes of this remarkable feature. Manipulation of metal-ligand coordination along with their general responsive stimuli has led to a full understanding of the flexible structure-property of metal-organic frameworks. Flexible MOF materials showing responsiveness to general responsive stimuli, host-guest systems were implemented in several types of research to demonstrate the type and potential of flexible MOFs. The work described above is set to revolutionize our thinking about flexible MOFs and may have repercussions far beyond the confines of the original discovery. In conclusion, based on the inspiring teaching of the research of flexible MOFs in the current state, a brighter future can be expected.

Abbreviation

ABTC	Azobenzenetetra carboxylate
BDC	Benzene dicarboxylate
BDP	1,4-benzene dipyrazolate
BHE-bpb	2,5-bis(2-hydroxyethoxy)-1,4-bis(4-pyridyl)benzene
BPE	Bis-pyridyl ethylene
BTC	Benzene tricarboxylate
H_3L	5-(2-carboxybenzyloxy)
HKUST-1	Hong Kong University of Science and Technology
IBU	Inorganic building unit
IR	Infrared spectroscopy
JLU-Liu4	JiLin University
LP	Large pores
MIL	Materials of Institute Lavoisier
MTV	multivariable
NMR	Nuclear Magnetic Resonance
NP	Narrow pores
PCP	Porous coordination polymers
PDC	Pyridine dicarboxylic acid
PSM	Post-synthetic modifications

Metal-Organic Framework Composites - Volume I
Materials Research Foundations **53** (2019) 177-214

Materials Research Forum LLC
https://doi.org/10.21741/9781644900291-9

PXRD	Powder X-ray diffraction
Pz	pyrazolate
Pzdc	2,3-pyrazinedicarboxylate
SALI	Solvent assisted ligand incorporation
SBBs	Supramolecular building blocks
SBUs	Secondary building units
ZIF	Zeolitic Imidazolate Framework
ZIF-8	$[Zn(mIm)_2]n$ (mIm, also Im) = 2- methylimidazole,

References

[1]   S. Kitagawa, M. Kondo, Functional micropore chemistry of Crystalline metal complex-assembled compounds, Bull. Chem. Soc. Jpn. 71 (1998) 1739.

[2]   G. Férey, C. Serre, Large breathing effects in three-dimensional porous hybrid matter: Facts, analyses, rules and consequences, Chem. Soc. Rev. 38 (2009) 1380–1399.

[3]   M. O'keeffe, M.A. Peskov, S.J. Ramsden, O.M. Yaghi, The reticular chemistry structure resource (RCSR) database of, and symbols for, crystal nets, Acc. Chem. Res. 41 (2008) 1782–1789.

[4]   D.J. Tranchemontagne, Z. Ni, M. O'Keeffe, O.M. Yaghi, Reticular chemistry of metal-organic polyhedra, Angew. Chemie - Int. Ed. 47 (2008) 5136–5147.

[5]   J.J. Perry VI, J.A. Perman, M.J. Zaworotko, Design and synthesis of metal-organic frameworks using metal-organic polyhedra as supermolecular building blocks, Chem. Soc. Rev. 38 (2009) 1400–1417.

[6]   D.J. Tranchemontagne, J.L. Mendoza-Cortés, M. O'Keeffe, O.M. Yaghi, Secondary building units, nets and bonding in the chemistry of metal-organic frameworks, Chem. Soc. Rev. 38 (2009) 1257–1283.

[7]   Z.-J. Lin, Jian Lü, M.H. And, R. Cao, Metal-organic frameworks based on flexible ligands (FL- MOFs): Structures and applications, Int. J. Pharma Bio Sci. 3 (2012) P59–P65.

[8]   M. O'Keeffe, O.M. Yaghi, Deconstructing the crystal structures of metal organic frameworks, Chem. Rev. 112 (2012) 675–702.

[9]   N. Stock, S. Biswas, Synthesis of metal-organic frameworks (MOFs): Routes to various MOF topologies, morphologies, and composites, Chem. Rev. (2012) 933–969.

[10]   D. Banerjee, J.B. Parise, Recent advances in s-block metal carboxylate networks, Cryst. Growth Des. 11 (2011) 4704–4720.

[11]   A.M. Plonka, D. Banerjee, J.B. Parise, Effect of ligand structural isomerism in formation of calcium coordination networks, Cryst. Growth Des. 12 (2012) 2460–2467.

[12]   M. Ahmad, M.K. Sharma, R. Das, P. Poddar, P.K. Bharadwaj, Syntheses, crystal structures, and magnetic properties of metal-organic hybrid materials of Co(II) using flexible and rigid nitrogen-based ditopic ligands as spacers, Cryst. Growth Des. 12 (2012) 1571–1578.

[13]   O.M. Yaghi, H. Li, C. Davis, D. Richardson, T.L. Groy, Synthetic strategies, structure patterns, and emerging properties in the chemistry of modular porous solids, Acc. Chem. Res. 31 (1998) 474–484.

[14]   M. Eddaoudi, J. Kim, N. Rosi, D. Vodak, J. Wachter, M. O'Keeffe, O.M. Yaghi, Systematic design of pore size and functionality in isoreticular MOFs and their application in methane storage, Science 80. 295 (2002) 469–472.

[15]   G. Wang, K. Leus, S. Couck, P. Tack, H. Depauw, Y.Y. Liu, L. Vincze, J.F.M. Denayer, P. Van Der Voort, Enhanced gas sorption and breathing properties of the new sulfone functionalized COMOC-2 metal organic framework, Dalt. Trans. 45 (2016) 9485–9491.

[16]   J. Wieme, L. Vanduyfhuys, S.M.J. Rogge, M. Waroquier, V. Van Speybroeck, Exploring the flexibility of MIL-47(V)-type materials using force field molecular dynamics simulations, J. Phys. Chem. C. 120 (2016) 14934–14947.

[17]   T. Lescouet, E. Kockrick, M. Pera-titus, S. Aguado, D. Farrusseng, Homogeneity of flexible metal – organic frameworks containing mixed linkers, J. Mater. Chem. (2012) 10287–10293.

[18]   P.D.C. Dietzel, R. Blom, H. Fjellvåg, Base-induced formation of two magnesium metal-organic framework compounds with a bifunctional tetratopic ligand, Eur. J. Inorg. Chem. (2008) 3624–3632.

[19]   J. Seo, R. Matsuda, H. Sakamoto, C. Bonneau, S. Kitagawa, A pillared-layer coordination polymer with a rotatable pillar acting as a molecular gate for guest molecules, J. Am. Chem. Soc. 131 (2009) 12792–12800.

Metal-Organic Framework Composites - Volume I   Materials Research Forum LLC
Materials Research Foundations **53** (2019) 177-214   https://doi.org/10.21741/9781644900291-9

[20]   S. Henke, A. Schneemann, A. Wu, R.A. Fischer, Directing the breathing behavior of pillared-layered metal−organic frameworks via a systematic library of functionalized linkers bearing flexible substituents, J. Am. Chem. Soc. 134 (2012) 9464–9474.

[21]   S. Biswas, T. Ahnfeldt, N. Stock, New functionalized flexible Al-MIL-53-X (X= - Cl, -Br, -CH$_3$,-NO$_2$,-(OH)$_2$) solids: syntheses, characterization, sorption, and breathing behavior, Inorg. Chem. 50 (2011) 9518–9526.

[22]   P. Müller, F. Wisser, V. V Bon, R. Grünker, I. Senkovska, S. Kaskel, Post-synthetic paddle-wheel crosslinking and functionalization of 1,3-phenylenebis(azanetriyl)tetrabenzoate based MOFs, Chem. Mater. 27 (2015) 2460–2467.

[23]   L. Pan, T. Frydel, M.B. Sander, X. Huang, J. Li, The effect of pH on the dimensionality of coordination polymers, Inorg. Chem. 40 (2001) 1271–1283.

[24]   C. Serre, C. Mellot-Draznieks, S. Surblé, N. Audebrand, Y. Filinchuk, G. Férey, Role of solvent-host interactions that lead to very large swelling of hybrid frameworks, Science 80. 315 (2007) 1828–1831.

[25]   P. Horcajada, F. Salles, S. Wuttke, T. Devic, D. Heurtaux, G. Maurin, A. Vimont, M. Daturi, O. David, E. Magnier, N. Stock, Y. Filinchuk, D. Popov, C. Riekel, G. Férey, C. Serre, How linker's modification controls swelling properties of highly flexible iron(III) dicarboxylates MIL-88, J. Am. Chem. Soc. 133 (2011) 17839–17847.

[26]   T. Ahnfeldt, D. Gunzelmann, T. Loiseau, D. Hirsemann, J. Senker, G. Férey, N. Stock, Synthesis and modification of a functionalized 3D open-framework structure with MIL-53 topology, Inorg. Chem. 48 (2009) 3057–3064.

[27]   H. Deng, C.J. Doonan, H. Furukawa, R.B. Ferreira, J. Towne, C.B. Knobler, B. Wang, O.M. Yaghi, Multiple functional groups of varying ratios in metal-organic frameworks, Science 80. 327 (2010) 846–850.

[28]   S. Sen, S. Neogi, A. Aijaz, Q. Xu, P.K. Bharadwaj, Structural variation in Zn(ii) coordination polymers built with a semi-rigid tetracarboxylate and different pyridine linkers: Synthesis and selective CO$_2$ adsorption studies, Dalt. Trans. 43 (2014) 6100–6107.

[29]   C. Xu, L. Li, Y. Wang, Q. Guo, X. Wang, H. Hou, Y. Fan, Three-dimensional Cd(II) coordination polymers based on semirigid bis(methylbenzimidazole) and aromatic polycarboxylates: Syntheses, topological structures and photoluminescent properties, Cryst. Growth Des. 11 (2011) 4667–4675.

[30]   A.K. Gupta, K. Tomar, P.K. Bharadwaj, Structural diversity of Zn(II) based coordination polymers constructed from a flexible carboxylate linker and pyridyl co-linkers: Fluorescence sensing of nitroaromatics, New J. Chem. 41 (2017) 14505–14515.

[31]   X.N. Hua, L. Qin, X.Z. Yan, L. Yu, Y.X. Xie, L. Han, Conformational diversity of flexible ligand in metal-organic frameworks controlled by size-matching mixed ligands, J. Solid State Chem. 232 (2015) 91–95.

[32]   J. Ple´vert, T.M. Gentz, A. Laine, H. Li, V.G. Young, O.M. Yaghi, M. O'Keeffe, A Flexible germanate structure containing 24-ring channels and with very low framework density, J. Am. Chem. Soc. 123 (2001) 12706–12707.

[33]   D.C.S. Souza, V. Pralong, A.J. Jacobson, L.F. Nazar, A reversible solid-state crystalline transformation in a metal phosphide induced by redox chemistry, Science 80. 296 (2002) 2012–2015.

[34]   T.G. Amos, A.W. Sleight, Negative thermal expansion in orthorhombic $NbOPO_4$, J. Solid State Chem. 160 (2001) 230–238.

[35]   T. Takaishi, K. Tsutsumi, K. Chubachi, A. Matsumoto, Adsorption induced phase transition of ZSM-5 by p-xylene, J. Chem. Soc. - Faraday Trans. 94 (1998) 601–608.

[36]   Q. Chen, Z. Chang, W.C. Song, H. Song, H. Bin Song, T.L. Hu, X.H. Bu, A controllable gate effect in cobalt(II) organic frameworks by reversible structure transformations, Angew. Chemie - Int. Ed. 52 (2013) 11550–11553.

[37]   J. Tian, L. V. Saraf, B. Schwenzer, S.M. Taylor, E.K. Brechin, J. Liu, S.J. Dalgarno, P.K. Thallapally, Selective metal cation capture by soft anionic metal-organic frameworks via drastic single-crystal-to-single-crystal transformations, J. Am. Chem. Soc. 134 (2012) 9581–9584.

[38]   J. Seo, C. Bonneau, R. Matsuda, M. Takata, S. Kitagawa, Soft secondary building unit: Dynamic bond rearrangement on multinuclear core of porous coordination polymers in gas media, J. Am. Chem. Soc. 133 (2011) 9005–9013.

[39]   A. Schneemann, V. Bon, I. Schwedler, I. Senkovska, S. Kaskel, R.A. Fischer, Flexible metal-organic frameworks, Chem. Soc. Rev. 43 (2014) 6062–6096.

[40]   F. Millange, C. Serre, G. Férey, Synthesis, structure determination and properties of MIL-53as and MIL-53ht: the first Cr^{III} hybrid inorganic–organic microporous solids: $Cr^{III}(OH)\cdot\{O_2C–C_6H_4–CO_2\}\cdot\{HO_2C–C_6H_4–CO_2H\}_x$, Chem. Commun. (2002) 822–823.

[41]   J.P.S. Mowat, V.R. Seymour, J.M. Griffin, S.P. Thompson, A.M.Z. Slawin, D. Fairen-Jimenez, T. Düren, S.E. Ashbrook, P.A. Wright, A novel structural form of MIL-53 observed for the scandium analogue and its response to temperature variation and CO_2 adsorption, Dalt. Trans. 41 (2012) 3937–3941.

[42]   E. V. Anokhina, M. Vougo-Zanda, X. Wang, A.J. Jacobson, In(OH)BDC·0.75BDCH$_2$ (BDC = benzenedicarboxylate), a hybrid inorganic-organic vernier structure, J. Am. Chem. Soc. 127 (2005) 15000–15001.

[43]   A. Boutin, M.A. Springuel-Huet, A. Nossov, A. Gédéon, T. Loiseau, C. Volkringer, G. Férey, F.X. Coudert, A.H. Fuchs, Breathing transitions in MIL-53(Al) metal-organic framework upon Xenon adsorption, Angew. Chemie - Int. Ed. 48 (2009) 8314–8317.

[44]   X. Qian, B. Yadian, R. Wu, Y. Long, K. Zhou, B. Zhu, Y. Huang, Structure stability of metal-organic framework MIL-53 (Al) in aqueous solutions, Int. J. Hydrogen Energy. 38 (2013) 16710–16715.

[45]   F. Millange, N. Guillou, R.I. Walton, J.M. Grenèche, I. Margiolaki, G. Férey, Effect of the nature of the metal on the breathing steps in MOFs with dynamic frameworks, Chem. Commun. (2008) 4732–4734.

[46]   I. Beurroies, M. Boulhout, P.L. Llewellyn, B. Kuchta, G. Férey, C. Serre, R. Denoyel, Using pressure to provoke the structural transition of metal-organic frameworks, Angew. Chemie - Int. Ed. 49 (2010) 7526–7529.

[47]   P.G. Yot, Q. Ma, J. Haines, Q. Yang, A. Ghoufi, T. Devic, C. Serre, V. Dmitriev, G. Férey, C. Zhong, G. Maurin, Large breathing of the MOF MIL-47(V^{IV}) under mechanical pressure: A joint experimental-modelling exploration, Chem. Sci. 3 (2012) 1100–1104.

[48]   S. Kitagawa, R. Kitaura, S. Noro, Functional hybrid porous coordination polymers, Angew. Chem. Int. Ed. 43 (2004) 2334–2375.

[49]   M.D. Allendorf, R. Medishetty, R.A. Fischer, Guest molecules as a design element for metal-organic frameworks, MRS Bull. 41 (2016) 865–869.

[50]   E.C. Spencer, M.S.R.N. Kiran, W. Li, U. Ramamurty, N.L. Ross, A.K. Cheetham, Pressure-induced bond rearrangement and reversible phase transformation in a metal-organic framework, Angew. Chemie - Int. Ed. 53 (2014) 5583–5586.

[51]   K.J. Gagnon, C.M. Beavers, A. Clearfield, MOFs under pressure: The reversible compression of a single crystal, J. Am. Chem. Soc. 135 (2013) 1252–1255.

[52]   A. Clearfield, Flexible MOFs under stress: Pressure and temperature, Dalt. Trans. 45 (2016) 4100–4112.

[53]   C.A. Fernandez, P.K. Thallapally, B.P. McGrail, Insights into the temperature-dependent "breathing" of a flexible fluorinated metal-organic framework, ChemPhysChem. 13 (2012) 3275–3281.

[54]   J.P. Zhang, X.M. Chen, Optimized acetylene/carbon dioxide sorption in a dynamic porous crystal, J. Am. Chem. Soc. 131 (2009) 5516–5521.

[55]   N. Yanai, T. Uemura, M. Inoue, R. Matsuda, T. Fukushima, M. Tsujimoto, S. Isoda, S. Kitagawa, Guest-to-host transmission of structural changes for stimuli-responsive adsorption property, J. Am. Chem. Soc. 134 (2012) 4501–4504.

[56]   F. Luo, C. Bin Fan, M.B. Luo, X.L. Wu, Y. Zhu, S.Z. Pu, W.Y. Xu, G.C. Guo, Photoswitching CO_2 capture and release in a photochromic diarylethene metal-organic framework, Angew. Chemie - Int. Ed. 53 (2014) 9298–9301.

[57]   J. Park, D. Yuan, K.T. Pham, J.-R. Li, and H.-C.Z. Andrey Yakovenko, Reversible alteration of CO_2 adsorption upon photochemical or thermal treatment in a metal−organic framework, J. Am. Chem.Soc. 134 (2012) 99–10212.

[58]   S. Kaskel, The Chemistry of Metal–Organic Frameworks, 2007.

[59]   P. Kanoo, R. Matsuda, M. Higuchi, S. Kitagawa, T.K. Maji, New interpenetrated copper coordination polymer frameworks having porous properties, Chem. Mater. 21 (2009) 5860–5866.

[60]   T.K. Maji, R. Matsuda, S. Kitagawa, A flexible interpenetrating coordination framework with a bimodal porous functionality, Nat. Mater. 6 (2007) 142–148.

[61]   F.X. Coudert, A. Boutin, M. Jeffroy, C. Mellot-Draznieks, A.H. Fuchs, Thermodynamic methods and models to study flexible metal-organic frameworks, ChemPhysChem. 12 (2011) 247–258.

[62]   C.R. Murdock, B.C. Hughes, Z. Lu, D.M. Jenkins, Approaches for synthesizing breathing MOFs by exploiting dimensional rigidity, Coord. Chem. Rev. 258–259 (2014) 119–136.

[63]   Y.-T.L. Heng Song, Chao Jing, Wei Ma, Tao Xie, Reversible photoisomerization of azobenzene molecules on single gold nanoparticle surface, Chem. Commun. 52 (2016) 2984–2987.

[64]   S. Kitagawa, K. Uemura, Dynamic porous properties of coordination polymers inspired by hydrogen bonds, Chem. Soc. Rev. 34 (2005) 109–119.

[65]   R. Kitaura, S. Kitagawa, Y. Kubota, T.C. Kobayashi, K. Kindo, Y. Mita, A. Matsuo, M. Kobayashi, H.C. Chang, T.C. Ozawa, M. Suzuki, M. Sakata, M. Takata, Formation of a one-dimensional array of oxygen in a microporous metal-organic solid, Science 80. 298 (2002) 2358–2361.

[66]   E.J. Carrington, C.A. McAnally, A.J. Fletcher, S.P. Thompson, M. Warren, L. Brammer, Solvent-switchable continuous-breathing behaviour in a diamondoid metal-organic framework and its influence on CO_2 versus CH_4 selectivity, Nat. Chem. 9 (2017) 882–889.

[67]   D. Maspoch, D. Ruiz-Molina, K. Wurst, N. Domingo, M. Cavallini, F. Biscarini, J. Tejada, C. Rovira, J. Veciana, A nanoporous molecular magnet with reversible solvent-induced mechanical and magnetic properties, Nat. Mater. 2 (2003) 190–195.

[68]   C. Serre, F. Millange, C. Thouvenot, M.N. S, G. Marsolier, D. Loue"r, G. Fe´rey, Very large breathing effect in the first nanoporous chromium(III)-based solids: MIL-53 or $Cr^{III}(OH), \{O_2C-C_6H_4-CO_2\}, \{HO_2C-C_6H_4-CO_2H\}_x, H_2O_y$, J. Am. Chem. Soc. 124 (2002) 13519–13526.

[69]   M.P. Suh, Metal-Organic frameworks and porous coordination polymers: properties and applications, Bull. Jpn. Soc. Coord. Chem. 65 (2015).

[70]   L. Carlucci, G. Ciani, M. Moret, D.M. Proserpio, S. Rizzato, Polymeric layers catenated by ribbons of rings in a three-dimensional self-assembled architecture: A nanoporous network with spongelike behavior, Angew. Chemie - Int. Ed. 39 (2000) 1506–1510.

[71]   P.L. Llewellyn, S. Bourrelly, C. Serre, Y. Filinchuk, G. Férey, How hydration drastically improves adsorption selectivity for CO_2 over CH4 in the flexible chromium terephthalate MIL-53, Angew. Chemie - Int. Ed. 45 (2006) 7751–7754.

[72]   C. Mellot-Draznieks, C. Serre, S. Surblé, N. Audebrand, G. Férey, Very large swelling in hybrid frameworks: A combined computational and powder diffraction study, J. Am. Chem. Soc. 127 (2005) 16273–16278.

[73]   N.A. Ramsahye, T.K. Trung, L. Scott, F. Nouar, T. Devic, P. Horcajada, E. Magnier, O. David, C. Serre, P. Trens, Impact of the flexible character of MIL-88 iron(III) dicarboxylates on the adsorption of n-alkanes, Chem. Mater. 25 (2013) 479–488.

[74]   Y. Yan, D.I. Kolokolov, I. Da Silva, A.G. Stepanov, A.J. Blake, A. Dailly, P. Manuel, C.C. Tang, S. Yang, M. Schröder, Porous Metal-Organic Polyhedral

Frameworks with Optimal Molecular Dynamics and Pore Geometry for Methane Storage, J. Am. Chem. Soc. 139 (2017) 13349–13360.

[75]   T. An, Y. Wang, J. Tang, Y. Wang, L. Zhang, G. Zheng, A flexible ligand-based wavy layered metal-organic framework for lithium-ion storage, J. Colloid Interface Sci. 445 (2015) 320–325.

[76]   D. Fairen-Jimenez, S.A. Moggach, M.T. Wharmby, P.A. Wright, S. Parsons, T. Düren, Opening the gate: Framework flexibility in ZIF-8 explored by experiments and simulations, J. Am. Chem. Soc. 133 (2011) 8900–8902.

[77]   M.E. Casco, Y.Q. Cheng, L.L. Daemen, D. Fairen-Jimenez, E. V. Ramos-Fernández, A.J. Ramirez-Cuesta, J. Silvestre-Albero, Gate-opening effect in ZIF-8: The first experimental proof using inelastic neutron scattering, Chem. Commun. 52 (2016) 3639–3642.

[78]   M. Alaghemandi, R. Schmid, Model study of thermoresponsive behavior of metal-organic frameworks modulated by linker functionalization, J. Phys. Chem. C. 120 (2016) 6835–6841.

[79]   S. Henke, A. Schneemann, R.A. Fischer, Massive anisotropic thermal expansion and thermo-responsive breathing in metal-organic frameworks modulated by linker functionalization, Adv. Funct. Mater. 23 (2013) 5990–5996.

[80]   S.C. McKellar, S.A. Moggach, Structural studies of metal-organic frameworks under high pressure, Acta Crystallogr. Sect. B Struct. Sci. Cryst. Eng. Mater. 71 (2015) 587–607.

[81]   A.J. Graham, A.M. Banu, T. Düren, A. Greenaway, S.C. McKellar, J.P.S. Mowat, K. Ward, P.A. Wright, S.A. Moggach, Stabilization of scandium terephthalate MOFs against reversible amorphization and structural phase transition by guest uptake at extreme pressure, J. Am. Chem. Soc. 136 (2014) 8606–8613.

[82]   P. Zhao, T.D. Bennett, N.P.M. Casati, G.I. Lampronti, S.A. Moggach, S.A.T. Redfern, Pressure-induced oversaturation and phase transition in zeolitic imidazolate frameworks with remarkable mechanical stability, Dalt. Trans. 44 (2015) 4498–4503.

[83]   S.C. McKellar, J. Sotelo, A. Greenaway, J.P.S. Mowat, O. Kvam, C.A. Morrison, P.A. Wright, S.A. Moggach, Pore Shape Modification of a Microporous Metal-Organic Framework Using High Pressure: Accessing a New Phase with Oversized Guest Molecules, Chem. Mater. 28 (2016) 466–473.

[84] S.K. Elsaidi, M.H. Mohamed, D. Banerjee, P.K. Thallapally, Flexibility in metal – organic frameworks : A fundamental understanding, Coord. Chem. Rev. 358 (2018) 125–152.

[85] K.W. Chapman, G.J. Halder, P.J. Chupas, Guest-dependent high pressure phenomena in a nanoporous metal-organic framework material, J. Am. Chem. Soc. 130 (2008) 10524–10526.

[86] W. Li, M.R. Probert, M. Kosa, T.D. Bennett, A. Thirumurugan, R.P. Burwood, M. Parinello, J.A.K. Howard, A.K. Cheetham, Negative linear compressibility of a metal-organic framework, J. Am. Chem. Soc. 134 (2012) 11940–11943.

[87] P. Serra-Crespo, A. Dikhtiarenko, E. Stavitski, J. Juan-Alcañiz, F. Kapteijn, F.X. Coudert, J. Gascon, Experimental evidence of negative linear compressibility in the MIL-53 metal-organic framework family, CrystEngComm. 17 (2015) 276–280.

[88] K.W. Chapman, G.J. Halder, G.J. Halder, P.J. Chupas, Pressure-Induced Amorphization and Porosity Modification in a Metal-Organic Framework, J. Am. Chem. Soc. 131 (2009) 17546–17547.

[89] S. Yuan, X. Sun, J. Pang, C. Lollar, J.S. Qin, Z. Perry, E. Joseph, X. Wang, Y. Fang, M. Bosch, D. Sun, D. Liu, H.C. Zhou, PCN-250 under pressure: sequential phase transformation and the implications for MOF densification, Joule. 1 (2017) 806–815.

[90] F.X. Coudert, Responsive metal-organic frameworks and framework materials: Under pressure, taking the heat, in the spotlight, with friends, Chem. Mater. 27 (2015) 1905–1916.

[91] R. Lyndon, K. Konstas, B.P. Ladewig, P.D. Southon, P.C.J. Kepert, M.R. Hill, Dynamic photo-switching in metal-organic frameworks as a route to low-energy carbon dioxide capture and release, Angew. Chemie - Int. Ed. 52 (2013) 3695–3698.

[92] A. Modrow, D. Zargarani, R. Herges, N. Stock, The first porous MOF with photoswitchable linker molecules, Dalt. Trans. 40 (2011) 4217–4222.

[93] D. Hermann, H. Emerich, R. Lepski, D. Schaniel, U. Ruschewitz, Metal-organic frameworks as hosts for photochromic guest molecules, Inorg. Chem. 52 (2013) 2744–2749.

[94] A. Modrow, D. Zargarani, R. Herges, N. Stock, Introducing a photo-switchable azo-functionality inside Cr-MIL-101-NH_2 by covalent post-synthetic modification, Dalt. Trans. 41 (2012) 8690–8696.

[95]   S. Bernt, M. Feyand, A. Modrow, J. Wack, J. Senker, N. Stock, A mixed-linker ZIF containing a photoswitchable phenylazo group, Eur. J. Inorg. Chem. (2011) 5378–5383.

[96]   R.D. Mukhopadhyay, V.K. Praveen, A. Ajayaghosh, Photoresponsive metal-organic materials: Exploiting the azobenzene switch, Mater. Horizons. 1 (2014) 572–576.

[97]   Y. Zou, M. Park, S. Hong, M.S. Lah, A designed metal-organic framework based on a metal-organic polyhedron, Chem. Commun. (2008) 2340–2342.

[98]   Y.M. Lu, Y.Q. Lan, Y.H. Xu, Z.M. Su, S.L. Li, H.Y. Zang, G.J. Xu, Interpenetrating metal-organic frameworks formed by self-assembly of tetrahedral and octahedral building blocks, J. Solid State Chem. 182 (2009) 3105–3112.

[99]   F. Salles, G. Maurin, C. Serre, P.L. Llewellyn, C. Knöfel, H.J. Choi, Y. Filinchuk, L. Oliviero, A. Vimont, J.R. Long, G. Férey, Multistep N_2 breathing in the metal-organic framework Co(1,4-benzenedipyrazolate), J. Am. Chem. Soc. 132 (2010) 13782–13788.

[100] P. Deria, W. Bury, J.T. Hupp, O.K. Farha, Versatile functionalization of the NU-1000 platform by solvent-assisted ligand incorporation, Chem. Commun. 50 (2014) 1965–1968.

[101] M. Rimoldi, A.J. Howarth, M.R. Destefano, L. Lin, S. Goswami, P. Li, J.T. Hupp, O.K. Farha, Catalytic Zirconium/Hafnium-Based Metal-Organic Frameworks, ACS Catal. 7 (2017) 997–1014.

[102] I. Hod, W. Bury, D.M. Gardner, P. Deria, V. Roznyatovskiy, M.R. Wasielewski, O.K. Farha, J.T. Hupp, Bias-switchable permselectivity and redox catalytic activity of a ferrocene-functionalized, thin-film metal−organic framework compound, J. Phys. Chem. Lett. 6 (2015) 586–591.

[103] P. Deria, Y.G. Chung, R.Q. Snurr, J.T. Hupp, O.K. Farha, Water stabilization of Zr6-based metal−organic frameworks via solvent-assisted ligand incorporation, Chem. Sci. 6 (2015) 5172–5176.

[104] P. Deria, W. Bury, I. Hod, C. Kung, O. Karagiaridi, J.T. Hupp, O.K. Farha, MOF functionalization via solvent-assisted ligand incorporation: phosphonates vs carboxylates, Inorg. Chem. 54 (2015) 2185–2198.

[105] V. Bon, N. Kavoosi, I. Senkovska, P. Müller, J. Schaber, D. Wallacher, D.M. Többens, U. Mueller, S. Kaskel, Tuning the fl exibility in MOFs by SBU, Dalt. Trans. 45 (2016) 4407–4415.

[106] M. Handke, H. Weber, M. Lange, J. Mo, J. Lincke, R. Gla, R. Staudt, H. Krautscheid, Network flexibility: Control of gate opening in an isostructural series of Ag-MOFs by linker substitution, Inorg. Chem. 53 (2014) 7599–7607.

[107] M. Taddei, F. Costantino, A. Ienco, A. Comotti, P. V Dau, S.M. Cohen, Synthesis, breathing, and gas sorption study of the first isoreticular mixed-linker phosphonate based metal–organic frameworks, Chem. Commun. 49 (2013) 1315–1317.

[108] S. Surble, C. Serre, C. Mellot-draznieks, A new isoreticular class of metal-organic-frameworks with the MIL-88 topology, Chem. Commun. (2006) 284–286.

[109] C. Serre, F. Millange, S. Surble, G. Ferey, A Route to the synthesis of trivalent transition-metal porous carboxylates with trimeric secondary building units, Communication. 43 (2004) 6285–6289.

[110] T. Devic, P. Horcajada, C. Serre, F. Salles, G. Maurin, D. Heurtaux, G. Clet, A. Vimont, J. Grene, B. Le Ouay, F. Moreau, E. Magnier, Y. Filinchuk, J. Lavalley, M. Daturi, I. Charles, G. Montpellier, Functionalization in flexible porous Solids : Effects on the Pore Opening and the Host - Guest Interactions, J. Am. Chem. Soc. (2010) 1127–1136.

[111] C. Serre, F. Millange, T. Devic, N. Audebrand, W. Van Beek, Synthesis and structure determination of new open-framework chromium carboxylate MIL-105 or Cr (OH).{$O_2C-C_6(CH_3)_4-CO_2$}.nH_2O, Mater. Res. Bull. 41 (2006) 1550–1557.

[112] L. Hamon, P.L. Llewellyn, T. Devic, A. Ghoufi, G. Clet, V. Guillerm, G.D. Pirngruber, G. Maurin, C. Serre, G. Driver, W. Van Beek, E. Jolimaı, A. Vimont, M. Daturi, S. Je, A.A. V Orientale, Co-adsorption and separation of CO_2 - CH_4 mixtures in the highly flexible MIL-53(Cr) MOF, JACS. 53 (2009) 17490–17499.

[113] C. Mellot-draznieks, C. Serre, S. Surble, N. Audebrand, D.V.S.Y. V, A. V Etats-unis, R. Cedex, U.V. De France, B. V Saint-michel, Very large lwelling in hybrid frameworks : A combined computational and powder diffraction study, JACS. 127 (2005) 16273–16278.

[114] Z. Wang, S.M. Cohen, Postsynthetic modification of metal-organic frameworks, Chem. Soc. Rev. 38 (2009) 1315–1329.

[115] E. Haldoupis, T. Watanabe, S. Nair, D.S. Sholl, Quantifying large effects of framework flexibility on diffusion in MOFs : CH_4 and CO_2 in ZIF-8, ChemPhysChem. 13 (2012) 3449–3452.

[116] N.A. Ramsahye, G. Maurin, S. Bourrelly, P.L. Llewellyn, T. Loiseau, On the breathing effect of a metal – organic framework upon CO_2 adsorption : Monte Carlo compared to microcalorimetry experiments, Chem. Commun. (2007) 3261–3263.

[117] S. Henke, R. Schmid, J. Grunwaldt, R.A. Fischer, Flexibility and sorption selectivity in rigid metal–organic frameworks, Chem. Eur. J. (2010) 14296–14306.

[118] M.-A. Springuel-Huet, A. Nossov, Z. Adem, F. Guenneau, C. Volkringer, T. Loiseau, A. Ge, P. Marie, Xe NMR study of the framework flexibility of the porous hybrid MIL-53(Al), JACS. 53 (2010) 11599–11607.

[119] T. Devic, F. Salles, S. Bourrelly, B. Moulin, G. Maurin, P. Horcajada, C. Serre, A. Vimont, J.-C. Lavalley, H. Leclerc, G. Clet, M. Daturi, P.L. Llewellyn, Y. Filinchuk, G. Ferey, Effect of the organic functionalization of flexible MOFs on the adsorption of, J. Mater. Chem. 22 (2012) 10266–10273.

[120] E. Quartapelle Procopio, T. Fukushima, E. Barea, J.A.R. Navarro, S. Horike, S. Kitagawa, A soft copper(II) porous coordination polymer with unprecedented aqua bridge and selective adsorption properties, Chem. -A Eur. J. 18 (2012) 13117–13125.

[121] S. Horike, Y. Inubushi, T. Hori, T. Fukushima, S. Kitagawa, A solid solution approach to 2D coordination polymers for CH_4/CO_2 and CH_4/C_2H_6 gas separation: Equilibrium and kinetic studies, Chem. Sci. 3 (2012) 116–120.

[122] J. Kim, W.Y. Kim, W.S. Ahn, Amine-functionalized MIL-53(Al) for CO_2/N_2 separation: Effect of textural properties, Fuel. 102 (2012) 574–579.

[123] B. Chen, S. Xiang, G. Qian, Metal-organic frameworks with functional pores for recognition of small molecules, Acc. Chem. Res. 43 (2010) 1115–1124.

[124] R.K. Das, A. Aijaz, M.K. Sharma, P. Lama, P.K. Bharadwaj, Direct crystallographic observation of catalytic reactions inside the pores of a flexible coordination polymer, Chem. - A Eur. J. 18 (2012) 6866–6872.

[125] N. Klein, C. Herzog, M. Sabo, I. Senkovska, J. Getzschmann, S. Paasch, M.R. Lohe, E. Brunner, S. Kaskel, Monitoring adsorption-induced switching by 129Xe NMR spectroscopy in a new metal-organic framework $Ni_2(2,6\text{-ndc})_2(\text{dabco})$, Phys. Chem. Chem. Phys. 12 (2010) 11778–11784.

[126] P. Horcajada, C. Serre, G. Maurin, N.A. Ramsahye, F. Balas, M. Vallet-Regí, M. Sebban, F. Taulelle, G. Férey, Flexible porous metal-organic frameworks for a controlled drug delivery, J. Am. Chem. Soc. 130 (2008) 6774–6780.

Metal-Organic Framework Composites - Volume I
Materials Research Foundations **53** (2019) 215-234

Materials Research Forum LLC
https://doi.org/10.21741/9781644900291-10

Chapter 10

Metal-Organic Frameworks as Host for Encapsulation of Enzymes

Bhagwati Sharma[1,2], Tridib K. Sarma*[1]

[1]Discipline of Chemistry, Indian Institute of Technology Indore, Simrol, Khandwa Road, Indore, Madhya Pradesh 486002, India

[2]Materials Research Centre, Malaviya National Institute of Technology, Jaipur-302017, India

* tridib@iiti.ac.in

Abstract

Enzymes are a class of highly selective and efficient natural catalysts. The use of enzymes for catalysis often requires them to be immobilized onto solid supports so that they can be efficiently recycled, the contamination due to presence of enzymes in the product can be minimized and their application in biomedicine can be explored. Various strategies for the use of metal-organic frameworks (MOFs) as host material for the encapsulation of enzymes have been discussed. Recent developments on the preparative strategies and applications of MOF encapsulated enzyme with special focus on catalysis are summarized in this chapter. The enhancement/retention of enzymatic activity of the composite material compared to free enzymes in denaturation conditions and advantages of encapsulation of the enzymes has been reviewed.

Keywords

Enzymes, Metal-Organic Frameworks, Encapsulation, Catalyst, Reusability

Contents

1.   Introduction

Nature has provided us a variety of biomolecules with complex structures and specific functions. Among the variety of biomolecules, enzymes form a special group as they are known to catalyze specific reactions under specific conditions [1,2]. Undoubtedly, enzymes are one of the most important biological molecules for life and function way superior to artificial catalysts in catalyzing several biological transformations. Enzymes consist of amino acids arranged in a linear sequence, which fold to give a complex structure with specific catalytically active sites. As a result, the enzymatically catalyzed reactions are highly selective and proceed with substantially high turnover number and lower reaction time [3]. Therefore, employing enzymes as industrial catalysts has been a focal point of research in recent years [4-6]. Despite the enormous advantages, there are several limitations associated with the use of enzymes as catalysts in industry. Natural enzymes have a narrow pH range and cannot withstand highly acidic or alkaline conditions, have lower thermal stabilities and salt tolerance and mostly function under aqueous condition [6]. Any change in these conditions, which are a prime requisite for industrial purposes (high temperature and organic solvents) leads to unfolding of the enzymes, resulting in their denaturation leading to deactivation of their catalytic properties. Furthermore, natural enzymes are homogeneous catalysts and thus are a source of contamination in the final product, making the purification steps a cumbersome effort. Therefore, the use of enzymes in a heterogeneous manner is an alternative strategy to improve their utility for practical industrial applications [7,8]. One such heterogeneous approach involves the immobilization of the enzymes onto solid supports [6,9]. Immobilization of the enzymes on solid surfaces offers fine control over the reaction processes and increases the stability of the enzyme under the operational pH and temperature conditions [10]. Further the handling of the enzymes gets easier and they can be recovered and reused, thus reducing their cost [11]. Although immobilization has several advantages, it is imperative to understand that the physico-chemical properties of the enzymes might undergo changes upon immobilization depending on the type of methods used for immobilization and the structural parameters of the matrix [12,13]. Therefore, it is vital to identify the right solid matrix and method for the immobilization of enzymes, so that there is minimal loss in the enzymatic activity after immobilization. Several materials such as hydrogels, sol-gel matrices, mesoporous silica, nanoparticles and carbon materials have been employed as host for the immobilization of enzymes [14-19]. Nevertheless, all these materials suffer from disadvantages such as small pore sizes, leaching of enzymes, lower surface area, rational structural design and denaturation of the enzymes during loading [20,21].

Metal-organic frameworks are a new class of crystalline and highly porous hybrid organic-inorganic materials. MOFs can be built from a variety of combinations of inorganic nodes (metal ions) and organic linkers. The versatility in the geometry and connecting ability of metal ions and ligands allow for the fabrication of MOFs for specific functions [22-26]. The ease with which the pore size and surface area in a MOF can be tuned and the mild synthetic conditions used in fabrication of MOFs suggest that they can serve as excellent host materials for immobilization/encapsulation of enzymes. MOFs can be designed to sustain harsh conditions such as high temperature and presence of chemicals [27], which is important for the use of enzymes as catalysts in industries. Furthermore, uniform loading and minimal leaching of the enzymes can be attained owing to the crystalline and ordered structure of the MOFs [28]. Moreover, the enzyme encapsulated MOF serves as heterogeneous catalysts, and can effortlessly be recovered from the reaction mixture [29]. The recovered catalyst can be employed for further cycles, thus reducing the cost of the catalyst and efforts required for the separation of natural enzyme. This chapter deals with the progress on the encapsulation of enzymes within MOFs using different strategies and the advantages of such encapsulation processes.

2. Encapsulation of enzymes in MOFs

There are two general techniques employed for the encapsulation of enzymes in MOFs: the first technique is based on *in situ* encapsulation of the enzyme within the MOF, which involves the formation of MOFs around the enzyme (*de novo* encapsulation), while the second strategy which is also known as post synthetic modification involves the formation of MOF followed by the incorporation of the enzyme within the pores of the MOF [30]. Both methods have advantages with respect to the improved stability, activity and efficiency. However, it is also important to note that conditions such as the environment used for the synthesis of the MOF as well as the type of MOF also play a crucial role in efficient encapsulation of the enzymes. The crystallite size of MOFs must also be considered during the selection of supported catalytic systems consisting of enzymes for specific reactions.

The *in situ* encapsulation strategy can be achieved by the well-known co-precipitation method, wherein enzymes are encapsulated during the synthesis of MOFs. On the other hand, post synthetic modification strategy can be employed by using cage and channel type MOFs. Examples of enzyme encapsulation using both these strategies are discussed in this chapter.

2.1   In situ enzyme encapsulation within the MOFs

The *in situ* encapsulation of enzyme within a MOF network involves the formation of MOF in a solution containing the enzyme to be encapsulated. A careful selection of the conditions that favor the formation of MOF, but do not bring about conformational changes in the structure of the enzyme is highly desired in this method [31]. Most of the MOFs have pore sizes smaller than the size of the folded enzyme, hence the *in situ* approach can be effective in encapsulating larger enzymes in the MOFs. Due to their mild synthetic conditions, Zeolitic imidazolate frameworks (ZIFs) consisting of Zinc ions as the metal constituent connected by imidazolate linkers are the most common MOFs used for the encapsulation of the enzymes in the *in situ* approach [29].

In their seminal work, Lyu *et al.* successfully incorporated cytochrome c (Cyt *c*) into the ZIF-8 [32]. In their method, Lyu *et al.* first mixed Cyt *c* with polyvinylpyrollidone (PVP) followed by addition of the mixture to a methanolic solution of zinc nitrate hexahydrate and 2-methylimidazole. PVP was used to allow proper dispersion and stabilization of Cyt *c* in methanol. They also confirmed that Cyt *c* molecules were embedded within the ZIF-8 and not merely adsorbed on its surface by calcining as prepared composite at 325 $^{\circ}$C. Electron microscopic studies after the calcination process revealed the presence of small pores with diameters ranging from 5 to 20 nm, resulting from the removal of Cyt *c* and its aggregates. Such cavities were absent in pure ZIF-8 crystals after calcination, clearly indicating that Cyt *c* was embedded within ZIF-8. The embedded Cyt *c* exhibited a 10-fold increase in the peroxidase activity when compared to pure native Cyt *c*. Control experiments revealed that mere presence of Zn^{2+} ions in the solution also led to an increase in activity. The enhancement in the activity of Cyt *c* after encapsulation in ZIF-8 was attributed to the interaction of Cyt *c* with Zn^{2+} ions in the framework.

In another excellent work, Liao *et al.* embedded the enzyme catalase (CAT) into ZIF-90 and ZIF-8 and found that the stability of the enzyme is drastically enhanced even when exposed to denaturation conditions [33]. CAT is known to denature in the presence of urea and at temperatures above 62 $^{\circ}$C. The nonpolar groups in the core of the proteins get solvated by water molecules in the presence of urea and the hydrophobic interactions in

Metal-Organic Framework Composites - Volume I   Materials Research Forum LLC
Materials Research Foundations **53** (2019) 215-234   https://doi.org/10.21741/9781644900291-10

the enzyme are disrupted resulting in unfolding of the enzymes and a loss in their activity. Liao *et al.* found that the exposure of native CAT to urea or heating the enzyme at 80 °C, led to a complete loss in activity of the enzyme. However, the CAT embedded in ZIF-90 or ZIF-8 retained substantial activity even when exposed to these denaturation conditions. This result clearly indicates that the ZIF provided a protective coating to the enzyme even under harsh conditions. The stability under such extreme denaturation conditions was attributed to the confinement of the enzyme molecules in the cavities of the MOF. Due to the confinement, the enzyme molecules are restricted towards a change in their structural conditions, hence retaining the activity. To prove this hypothesis, the authors carried out the catalytic reaction in the presence of 3-amino-1,2,4-triazole (3-AT), which can covalently bind to the CAT-H_2O_2 complex and inhibit the activity of the enzyme without unfolding the enzyme. It was found that the activity of CAT@ZIF-90 composite ceased upon introduction of 3-AT, as the confinement in MOF is unable to restrict the binding of 3-AT to the CAT in the MOF.

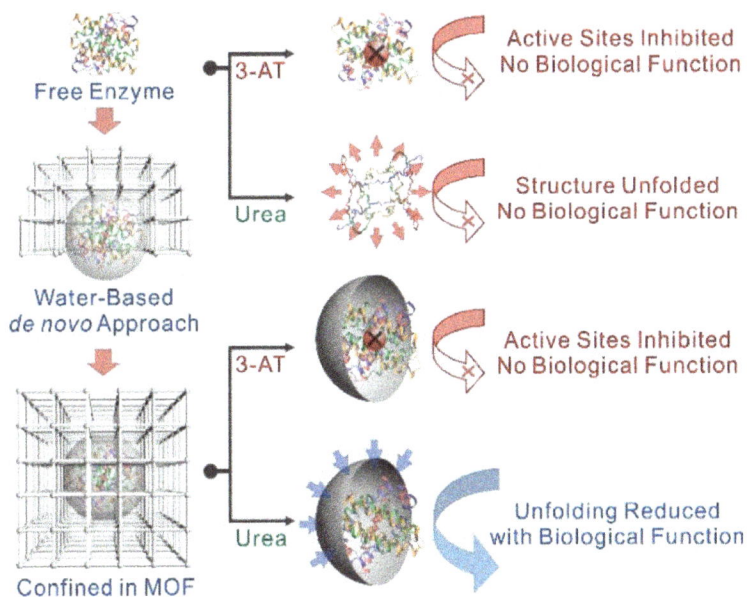

Fig. 1. Schematic representation of the confinement of CAT in the cavities of ZIF-8 or ZIF-90, which prevents a change in their structural conformation and provides protection against heat or urea. Reproduced with permission from ref. 33. Copyright 2017 American Chemical Society.

Metal-Organic Framework Composites - Volume I   Materials Research Forum LLC
Materials Research Foundations **53** (2019) 215-234   https://doi.org/10.21741/9781644900291-10

Liang *et al.* employed a similar strategy to encapsulate biomolecules within ZIF-8 [34]. This system and methodology could be used to encapsulate DNA as well as oligonucleotides in addition to enzymes. The method involved the synthesis of MOF in a solution compatible with the desired biomolecule to be encapsulated. Aqueous solutions of zinc acetate and 2-methylimidazole were mixed with the biomolecule of choice, which resulted in the formation of a protective layer of ZIF-8 around the biomolecule. The system loaded with urease and horse radish peroxidase (HRP) were stable up to 80 °C in water and 153 °C in dimethylformamide. Further, the loaded biomolecules could be released by decreasing the pH, as ZIF-8 decomposes at pH below 6. A similar biomineralized MOF has recently been reported by Jùnior *et al.* where they have embedded bacterial collagenase into ZIF-8, which not only provided immobilization to the enzyme but also provided a fine control on their activity [35]. Mesoporous flower shaped CAT@ZIF-8 instead of the normal rhombic dodecahedral shape were synthesized by Cui *et al.* employing a similar biomineralization technique [36]. The prepared flower shaped CAT@ZIF-8 showed high loading of CAT and exhibited catalytic activity 400% higher than the rhombic dodecahedral CAT@ZIF-8 crystals.

Ge and co-workers used polydopamine (PDA) as a bioadhesive to crosslink enzyme-MOF nanocrystals to form micrometer sized particles [37]. They first formed a glucose oxidase-ZIF 8 composite (GOx/ZIF-8) and incubated it with dopamine for 24 hours. Under these conditions, dopamine polymerized to a form PDA on the outside surface of ZIF-8. The initially formed GOx/ZIF-8 nanoparticles were crosslinked by PDA resulting in microparticles (PDA@GOx/ZIF-8). The increase in size of the particles led to a poorer dispersion of the particles in solvents and hence better recyclability of the composite. The PDA@GOx/ZIF-8 showed better temperature stability and solvent stability as compared to native GOx. While the native GOx lost its activity completely in 50% methanol, ethanol or acetone, the PDA@GOx/ZIF-8 retained 90% of the catalytic activity, suggesting that the MOF framework provided excellent protection against denaturation. They ascribed the stability of the composite to the interactions between enzymes and carriers which led to improvements in the ability of enzyme molecules under extreme conditions.

Wu *et al.* developed facile strategy for simultaneously embedding two enzymes in a MOF, to be used as a nanoreactor for tandem reactions [38]. GOx and HRP were co-immobilized on ZIF-8 nanocrystals by mixing aqueous solution of zinc nitrate, enzymes and 2-methylimidazole, followed by stirring for 30 minutes at room temperature and centrifugation. GOx was used to oxidize glucose into gluconic acid and H_2O_2, which is a substrate for the oxidation of 2,2'-azino-bis(3-ethylbenzothiazoline-6-sulphonic acid) (ABTS) by HRP. Using this method, glucose could be determined up to a limit of 0.5

µM. Furthermore, the system showed good tolerance against protease and chelating agents, which could be attributed to the protective layer of ZIF-8 over the enzymes. The small pore window in ZIF-8 prevented the chelating agent from binding to the enzymes at appropriate sites, which subsequently prohibited the conformational changes in the enzyme that normally lead to the modulation in the activity of the enzyme.

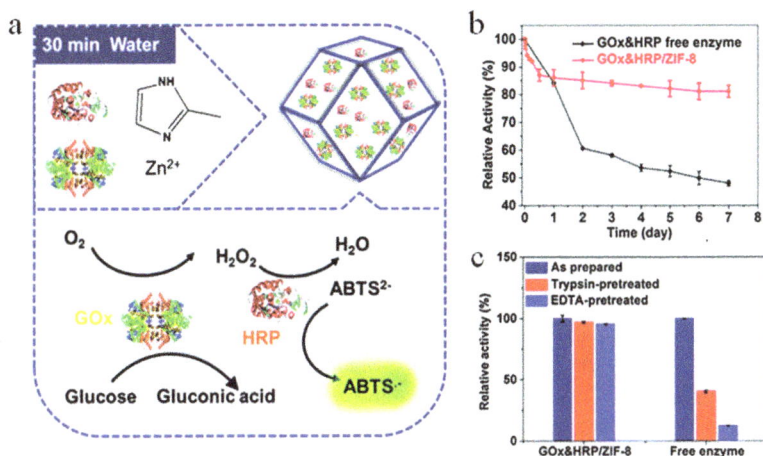

Fig. 2. (a) Schematic representation for the synthesis of multi enzyme incorporated ZIF-8. (b) Stability of the GOx&HRP/ZIF-8 composite compared to free enzymes at pH 7.4 in PBS at room temperature. (c) Stability of the GOx&HRP/ZIF-8 composite compared with free enzymes incubated in pH 7.4 PBS containing protease (trypsin 1 mg mL^{-1}) or chemicals (1 wt% of EDTA). Reproduced with permission from ref. [38]. Copyright 2015 The Royal Society of Chemistry.

2.2 Post-synthetic enzyme encapsulation within the MOFs

As the term suggests, in the post-synthetic method, MOF is first synthesized and then the enzymes are encapsulated in the MOF matrix by simple soaking of the MOF crystals in a solution of the enzyme, usually from a few minutes to several hours under physiological conditions. Unlike the *in situ* synthesis, this method is advantageous owing to the fact that the synthesis of the MOF can be performed under harsh conditions. However, the synthesized MOF must be stable at the operating conditions of the enzyme. For instance, most enzymes function at a certain pH in aqueous buffer solutions, and hence the MOF used for the encapsulation must be stable to the salts present in the buffer solution. Thus in addition to the pH of the solution, the chemical composition of the solution also plays a crucial role in determining the encapsulation of the enzyme into the MOF and their stability. The post-synthetic enzyme immobilization can be divided into three main categories, *viz.* surface attachment, covalent linkage and pore encapsulation. While weak intermolecular forces, such as H-bonding, π-π stacking and van der Waals forces are the driving forces in case of surface attachment and pore encapsulation, the covalent attachment technique involves the formation of covalent bonds between the enzyme and MOF [29]. The presence of amine groups on the surface of enzyme facilitates bonding to the MOF surface through the carboxylic acid groups on the surface of the MOF and *vice versa*. However, surface attachment and covalent linkage methods primarily leads to the adsorption of the enzymes on the surface of the MOF and they are not encapsulated within the MOF, therefore, these two techniques are not discussed here.

2.2.1 Pore encapsulation

The pore encapsulation method involves the diffusion of the enzyme into the pores of the MOF. The pore size of the MOF plays a deciding role in encapsulation of the enzymes through this approach. In order to achieve the successful diffusion of enzymes into the MOFs, it is essential that the size of the pore and that of the enzymes are similar. Since, most of the MOFs are mesoporous, this strategy can be used to incorporate small enzymes into the MOFs. In addition to the small pore size, the enzyme must be physically adsorbed on the pores of the MOFs, which is not only important to check the leaching of the enzyme from the MOF, but also to prevent the denaturing agents from diffusing through the cavities of the MOF to access the enzyme [29].

Ma and coworkers, immobilized microperoxidase-11 (MP-11) into a terbium based mesoporous MOF (Tb-mesoMOF) constructed using triazine-1,3,5-tribenzoate as ligand [39]. MP-11 is an enzyme, frequently employed for the oxidation of organic molecules in the presence of H_2O_2 and has dimensions of about 3.3 × 1.7 × 1.1 nm. On the other hand, Tb-mesoMOF has internal cavities with sizes of 4.1, 3.0 and 0.9 nm. The sizes of these cavities (4.1 and 3.0 nm) are appropriate for the encapsulation of MP-11. The incorporation of MP-11 in Tb-mesoMOF involved the immersion of fresh crystals of the MOF into a buffer solution of MP-11 followed by incubation at 37 °C and subsequent washing with buffer to remove the enzyme adsorbed on the surface of MOF (MP-11@Tb-mesoMOF). The activity of MP-11@Tb-mesoMOF was then compared with the activity of free MP-11 and MP-11@MCM-41by monitoring the oxidation of 3,5-di-*tert*-butyl-catechol (DTBC) to *o*-quinone by H_2O_2. MP-11@Tb-mesoMOF showed superior conversion ability as the aggregation of MP-11 in solution could be prevented due to encapsulation within the Tb-mesoMOF. Furthermore, although free MP-11 is active only

Metal-Organic Framework Composites - Volume I   Materials Research Forum LLC
Materials Research Foundations **53** (2019) 215-234   https://doi.org/10.21741/9781644900291-10

in HEPES buffer and loses its activity in methanol, the encapsulation of MP-11 in Tb-mesoMOF significantly improved its solvent adaptability as well as reusability.

In a similar manner, the same group immobilized myoglobin (Mb), which has dimensions about $2.1 \times 3.5 \times 4.4$ nm, within the Tb-mesoMOF [40]. The analysis of the sizes of the cavities after the immobilization of Mb (Mb@Tb-mesoMOF) revealed the presence of cavity of size around 0.8 nm and the cavities with sizes 4.1 and 3.0 nm were not available after the incorporation of Mb. This clearly suggested that Mb molecules were encapsulated into the MOF through the mesopores with sizes 4.1 and 3.0 nm, while the micropores (0.8 nm) were still vacant and could be used for the diffusion of the substrate. Mb@Tb-mesoMOF catalyst was employed for the oxidation of two substrates ABTS and 1,2,3-trihydroxybenzene (THB) in the presence of H_2O_2. When ABTS was used as a substrate, a significantly lower reaction rate (almost nil) was observed with Mb@Tb-mesoMOF compared to free Mb and Mb@SBA-15. However, for the oxidation of THB, initial reaction rate similar to that for free Mb and Mb@SBA-15 was observed. The difference in the rate of reaction using two different substrates was attributed to the sizes of the substrates. While THB, a small molecule having dimensions 0.57×0.58 nm^2 could easily approach the enzyme through the small cavity (0.8 nm) in the MOF, the approach of the relatively larger ABTS (1.01×1.73 nm^2) was not possible through such tiny micropore.

The mechanism of entry of Cyt c into the Tb-mesoMOF was also studied by Ma *et al.* using fluorescence spectroscopy [41]. Cyt c is a structurally robust, heme containing electron transfer protein and has a dimension of $2.6 \times 3.2 \times 3.3$ nm. It consists of a four tyrosine residues and one tryptophan residue, which are responsible for the fluorescence in Cyt c. In the native folded state, the tryptophan residue is located $\sim 5\text{Å}$ from the heme edge and the emission due to the tryptophan is quenched by the heme group. Upon unfolding by using guanidine HCl, the tryptophan residue in Cyt c gets exposed to solvent. This leads to the tryptophan residue moving farther away from the heme group and an emission centered at around 353 nm is observed. The window openings of the mesoporous cavities of the MOF are small than the actual size of Cyt c, but the diameter of the cavity of the cages is larger than the size of Cyt c. Therefore, the migration of Cyt c into the cavities of Tb-mesoMOF would require the protein to undergo some unfolding. The steady state fluorescence studies of the Cyt c@Tb-mesoMOF showed a peak with maximum at 353 nm, similar to that of the protein unfolded using guanidine HCl. This result demonstrated that Cyt c underwent some conformational changes leading to unfolding of the protein during the process of entry into the cavities of the MOF, proving that the conformation of the proteins is also an important factor, when considering their incorporation into a MOF.

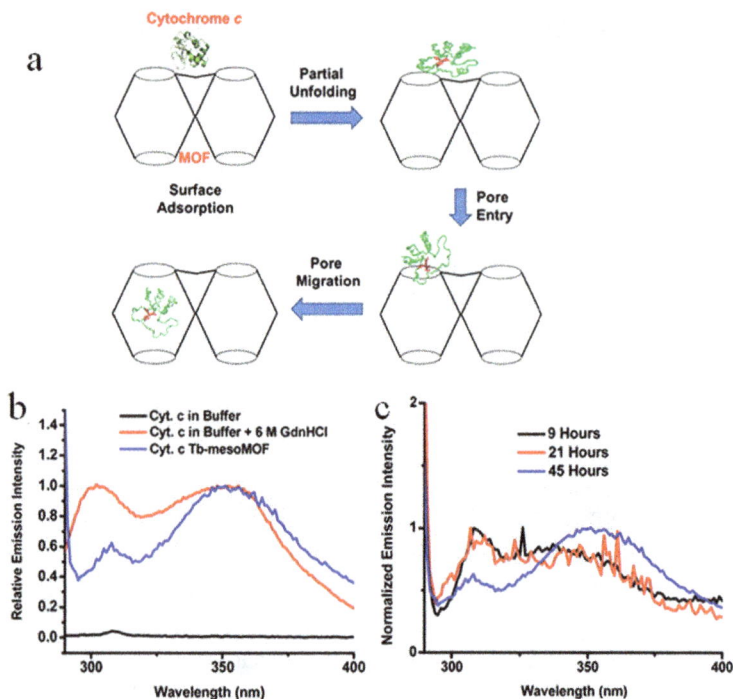

Fig. 3. (a) Plausible mechanism for the translocation of Cyt c into the cavities of Tb-
mesoMOF. The mechanism comprises of four steps: The protein is adsorbed on the
surface of MOF (step1) and then undergoes partial unfolding (step 2). In step 3, the
unfolded protein is partitioned between the surface and exterior pores and finally
migration of protein in the interior cavities of MOF take place (step 4). (b) Fluorescence
spectra of Cyt c, Cyt c denatured by guanidine HCl, and Cyt c@Tb-mesoMOF. (c)
Fluorescence spectra of Cyt c@Tb-mesoMOF at different incubation time of incubation.
Reproduced with permission from ref. 41. Copyright 2017American Chemical Society.

Zhou *et al.* designed and prepared mesoporous MOFs with even larger hierarchical
cavities for immobilization of several enzymes [42]. They prepared several highly stable,
single molecule trap containing MOFs using trivalent metal ions, namely PCN-333(M)
(M= Al(III), Fe(III), Sc(III)). PCN-333(Al) exhibited cages with three different pore sizes
of diameter 1.1, 3.4 and 5.5 nm. It also showed an excellent tolerance of pH and was
therefore used for the immobilization of three enzymes of different dimensions, namely,
HRP, Cyt c and MP-11. It was observed that depending on the dimensions of the enzyme

and cavities, only one enzyme per cage for larger enzymes (HRP and Cyt c) and multiple enzymes per cage for smaller enzyme (MP-11) could be encapsulated in PCN-333(Al). The favorable interaction between the cage and the enzyme led to an extremely high enzyme loading in PCN-333(Al). In all the three cases, enzyme after immobilization showed higher stability compared to the free enzyme, but a lower K_m value for the oxidation of o-phenylenediamine or ABTS than the free enzyme, suggesting that maximum conversion rate could be achieved using a lower substrate concentration in case of immobilized enzymes. Similarly, Steunou et al. also successfully embedded MP-11 into nanoparticles of MIL-101 (Cr) [43]. The enzyme retained its activity, could be recycled and showed long term stability as well as good resistance to acidic conditions.

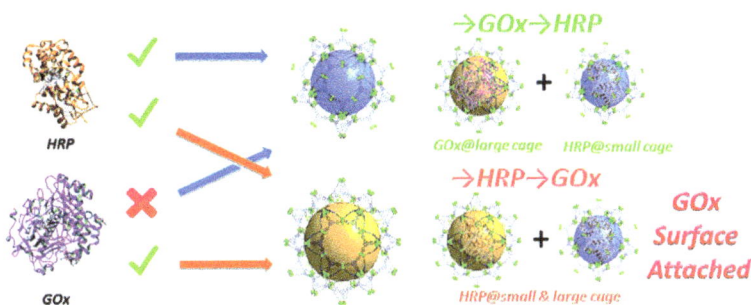

Fig. 4. Graphic representation of the results from the stepwise encapsulation of GOx and HRP in PCN-888. Reproduced with permission from ref. 44. Copyright 2016 The Royal Society of Chemistry.

Zhou et al. also fabricated PCN-888, another Al-MOF and used it for coupling two enzymes in a tandem reactor [44]. PCN-888 has three different cages with sizes 6.0 nm, 5.0 nm and 2.0 nm. GOx, which has a dimension of $6.0 \times 5.2 \times 7.7$ nm can be accommodated in the larger cage, while HRP with dimensions $4.0 \times 4.4 \times 6.8$ nm can be accommodated in both the larger and the medium cage. Therefore, it is necessary to follow a stepwise encapsulation order with GOx being allowed to be encapsulated first, followed by HRP, such that GOx and HRP are exclusively distributed in the larger (6.2 nm) and medium (5.0 nm) cages respectively, giving a tandem bioreactor. The smaller cages with size 2.0 nm are vacant and allow for substrate and product transfer. The as prepared bio-nanoreactor showed initial reaction rates similar to the free enzymes, however, the reaction rate was still slower when compared to the free enzyme. Although the reaction rate is slower than the free enzymes, the method is advantageous, as it can be recycled and shows negligible enzyme leaching. Further, the enzymes encapsulated in PCN-888 show a greater stability than the free enzymes in the presence of trypsin at 37 °C.

In addition to the mesoporous cage type MOF, channel type MOFs have also been employed for the encapsulation of enzymes. Similar to the cage type MOFs, channel type MOFs can have several channels with different cavity sizes. Pisklak et al. for the first time employed a Cu based channel type MOF for the encapsulation of MP-11 [45]. They observed that the encapsulated enzyme retained its peroxidase activity for the oxidation of amplex ultrared to resorufin by H_2O_2. They also performed experiments to study the leaching of the enzyme from the MOF, and found that the MOF showed less leaching compared to other support systems such as mesoporous benzene silica. Yaghi and co-workers synthesized a series of MOFs, named IRMOF-74 series by expanding the MOF systematically [46]. The pore sizes could be tuned from 1.4 nm (IRMOF-74-I) to 9.8 nm (IRMOF-74-XI). Several biomolecules such as Vitamin B12 (VB12), Mb and green fluorescent protein (GFP) were encapsulated in IRMOF-74-IV, IRMOF-74-VII and IRMOF-74-IX respectively. They found that GFP, which has a length of 4.5 nm and diameter of 3.4 nm could be encapsulated in the IRMOF-74-IX (apertures slightly smaller than IRMOF-74-XI) without a change in conformation of the protein. This was clearly evidenced by confocal microscopy, which showed fluorescence properties typical of GFP. Furthermore, the crystallinity of the MOF was retained even after encapsulation of the biomolecules as indicated by the powder X-ray diffraction pattern.

Hierarchical channel type MOFs which feature windows between the channels are quite attractive for enzyme encapsulation as one of the channels can be employed to accommodate the enzyme based on the size of the channel and enzyme, while the other channel remains empty [30]. Substrate molecules can then easily be incorporated into the MOFs through the other available channel and can easily diffuse to the site of catalytic reaction through the window between the channels. Farha et al. employed a zirconium based MOF, NU-1000 for the encapsulation of an esterase, cutinase [47]. NU-1000 has hexagonal channels with diameter of 3.1 nm and triangular channels with edge length of 1.5 nm and feature windows that connect the two channels. Furthermore, NU-1000 is stable to both elevated temperatures (450 °C) as well as pH (1-11). Thus NU-1000 serves as excellent host for the encapsulation of enzymes. Cutinase, which has a smallest axis of 3.0 nm was thus accommodated into the hexagonal channels and the triangular channels which were unoccupied, were utilized for diffusion of the substrate to the site of activity (enzyme) in the hexagonal channel through the window between the two channels.

3.   Advantages of enzyme encapsulation

Encapsulation of enzymes into MOFs provides physical confinement to the encapsulated enzyme, which in turn provides protection to the enzyme against harsh reaction conditions. Furthermore, the catalyst used for large scale industrial production must be cost-effective. Encapsulation of enzymes into the MOF matrix has helped researchers attain these goals. We discuss here three most important advantages of encapsulation, *viz.* recyclability, catalytic activity, and stability which make the encapsulated enzymes superior compared to free enzymes.

3.1   Recyclability

The most important improvement of encapsulated enzymes over the free enzymes is their recyclability, which significantly contributes in cost reduction of the catalysts. Free enzymes are excellent catalysts, but are homogeneous in nature, *i.e.* their separation from the reaction mixture requires tedious, time-consuming and costly separation processes. Encapsulation of the enzyme in the MOFs makes the enzyme heterogeneous in nature and they can be separated from the reaction mixtures by simple separation techniques such as centrifugation. MOFs that have pore windows smaller than the actual size of the enzymes prevent leaching of the enzyme and contribute towards their recyclability. MP-11@Tb-mesoMOF and Mb@Tb-mesoMOF retained their catalytic activities during six and fifteen successive cycles respectively [39,40]. Similarly, HRP encapsulated within PCN-333 retained about 80% activity when recycled five times. On the contrary, nearly 70% activity was lost during the first cycle when the same enzyme was immobilized onto SBA-15 [42].

3.2   Catalytic activity

In comparison to the free enzymes, several MOF encapsulated enzymes have demonstrated increased activity. It has been observed that enzymes such as MP-11 self-aggregate in water, which results in poor yields of the product [39]. In such a scenario, it is important to prevent the aggregation of enzymes to achieve higher conversion rates. It has been found that when MP-11 is encapsulated within Tb-mesoMOF or Cu-MOF, it shows a four-fold and ten-fold enhancement in the catalytic activity respectively [39,45]. Furthermore, the compatibility of size of the enzyme with the apertures of the MOF also plays a crucial role in determining the activity of the encapsulated enzyme. If the size of the substrate is large than the apertures of the pores of the MOF, a lower activity will be observed. This size selective catalysis was proved by taking Mb@Tb-mesoMOF as an example [40]. When THB was used as a substrate, Mb@Tb-mesoMOF showed higher activity compared to free Mb. However, when ABTS (size greater than the apertures of the pores) was used as catalyst, almost negligible activity was observed, as due to the larger size, ABTS could not enter the pores of the MOF to access Mb.

3.3   Stability

One of the most common disadvantages of enzymes is their environmental instability. Enzymes usually function under aqueous conditions and physiological temperature. This mild reaction conditions prove to be a barrier in the application of enzymes for industrial catalysis. MOFs provide the enzymes protection against such harsh conditions used in industrial catalysis. HRP encapsulated within ZIF-8 showed retention of 85% of its activity in boiling water or DMF [34]. In another report CAT embedded within ZIF-8 showed similar stability against temperature as well as urea [33]. Similarly, lipase@ZIF-8 also showed exceptional stability against denaturation conditions [48]. Digestion by trypsin is another denaturation route for enzymes. It was observed that a bi-enzyme loaded ZIF-8 retained 90% of its activity upon treatment with trypsin for 30 minutes at 37 °C, while the free enzyme lost complete activity under similar conditions [38].

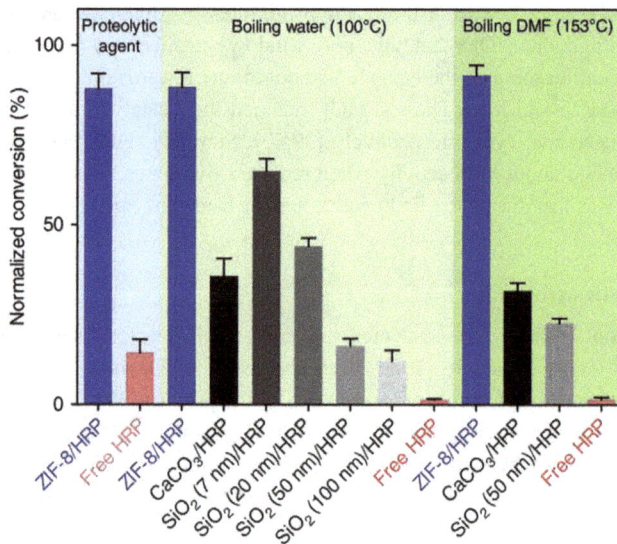

Fig. 5. Comparison of the stability of free HRP, ZIF-8 encapsulated HRP and HRP immobilized on other supports such as CaCO₃ and SiO₂ under different denaturation conditions. Reproduced with permission from ref. 34. Copyright 2015 Nature Publishing Group.

Metal-Organic Framework Composites - Volume I   Materials Research Forum LLC
Materials Research Foundations **53** (2019) 215-234   https://doi.org/10.21741/9781644900291-10

4.   Summary and outlook

In summary, the application of MOFs as effective support system for the stabilization of enzymes against denaturation is discussed. The library of known frameworks along with their known physico-chemical properties and stability coupled with the high surface area and porous nature make MOFs ideal candidates for the encapsulation of enzymes. For effective applications, it is highly desired for the enzymes to preserve their natural activity while maintaining the native three dimensional structure. Most of the examples discussed herein, except a few, show a lower reaction rate when compared to the free enzyme. It is important to understand that the lower initial rates might be due to the size constraints associated with the reactants/products. If the reactant (or product) cannot diffuse through the pores, then the observed initial rates might have contribution only from the enzyme that is adsorbed on the surface of the MOF, even if the enzyme encapsulated within the pores is assumed to be active. Nevertheless, these examples contribute extensively to our understanding of different strategies of enzyme encapsulation. Furthermore, strategies now exist for the preparation of enzyme encapsulated MOF that also promote the diffusion of reactants/products together with the protection they provide against denaturation conditions.

Although the encapsulation of enzymes by MOFs has shown to provide sufficient protection, the area is still in infancy. There is still a huge scope for the development of efficient synthetic strategies. For instance, the pore encapsulation method provides sufficient protection and resistance to enzymes against digestion by proteases, such as trypsin, however, this is true only for those enzymes whose dimensions match with the cavity size of the MOF. While in situ synthesis method has been used to overcome such limitations, the in situ synthesis itself suffers from the requirement of mild, aqueous synthetic conditions. Therefore, developments of general strategies which are applicable to most of the enzymes and circumvent the above limitations are highly desirable. Moreover, the diffusion mechanism of enzymes into the pores of the MOF is still not very clear and requires further investigation. Further, MOFs themselves are known to function as efficient catalysts for various transformations, therefore, the study of synergistic catalysis by MOFs and enzymes might prove to be an excellent area of research and requires further study. Although the MOF-enzyme composites have shown tremendous potential for catalytic applications in the preliminary studies, there is a huge scope for development of synthetic methods for various organic transformations on industrial scale or biomedical applications. Advancement in these areas is very important for utilizing the true potential of MOF encapsulated enzymes for different applications.

Abbreviations

MOF	Metal-organic framework
Cyt *c*	Cytochrome *c*
PVP	Polyvinylpyrollidone
CAT	Catalase
3-AT	3-amino-1,2,4-triazole
HRP	Horseradish peroxidase
PDA	Polydopamine
GOx	Glucose oxidase
MP-11	Microperoxidase-11
ABTS	2,2'-azino-bis(3-ethylbenzothiazoline-6-sulphonic acid)
DTBC	di-tert-butyl-catechol
Mb	Myoglobin
THB	1,2,3-trihydroxybenzene
VB12	Vitamin B12
GFP	Green fluorescent protein

References

[1]   S. J. Benkovic, S. Hammes-Schiffer, A Perspective on Enzyme Catalysis, Science 301 (2003) 1196-1202.

[2]   U. T. Bornscheuer, G. W. Huisman, R. J. Kazlauskas, S. Lutz, J. C. Moore, K. Robins, Engineering the third wave of biocatalysis, Nature 485 (2012) 185-194. https://doi.org/10.1038/nature11117

[3]   S. F. M. van Dongen, J. A. A. W. Elemans, A. E. Rowan, R. J. M. Nolte, Processive catalysis, Angew. Chem. Int. Ed. 53 (2014) 11420-11428. https://doi.org/10.1002/anie.201404848

[4]   G. Hills, Industrial use of lipases to produce fatty acid esters, Eur. J. Lipid Sci. Technol. 105 (2003) 601–607. https://doi.org/10.1002/ejlt.200300853

[5]   H. Griengl, H. Schwab, M. Fechter, The synthesis of chiral cyanohydrins by oxynitrilases, Trends Biotechnol. 18 (2000) 252–256. https://doi.org/10.1016/s0167-7799(00)01452-9

[6]   R. DiCosimo, J. McAuliffe, A. J. Poulose, G. Bohlmann, Chem. Soc. Rev. 42 (2013) 6437–6474.

[7]   Z. Zhou, M. Hartmann, Progress in enzyme immobilization in ordered mesoporous materials and related applications, Chem. Soc. Rev. 42 (2013) 3892–3912.

[8]   Z. Zhou, M. Hartmann, Recent progress in biocatalysis with enzymes Immobilized on mesoporous hosts, Top. Catal. 55 (2012) 1081-1100. https://doi.org/10.1007/s11244-012-9905-0

[9]   M. C. R. Franssen, P. Steunenberg, E. L. Scott, H. Zuilhof, J. P. M. Sanders, Immobilised enzymes in biorenewables production, Chem. Soc. Rev. 42 (2013) 6491–6533. https://doi.org/10.1039/c3cs00004d

[10]   K. Hernandez, R. Fernandez-Lafuente, Control of protein immobilization: Coupling immobilization and site-directed mutagenesis to improve biocatalyst or biosensor performance, Enzyme Microb. Technol. 48 (2011) 107-122. https://doi.org/10.1016/j.enzmictec.2010.10.003

[11]   R. A. Sheldon, Cross-linked enzyme aggregates (CLEAs): stable and recyclable biocatalysts, Biochem. Soc. Trans. 35 (2007) 1583-1587. https://doi.org/10.1042/bst0351583

[12]   N. R. Mohamad, N. H. C. Marzuki, N. A. Buang, F. Huyop, R. A. Wahab, An overview of technologies for immobilization of enzymes and surface analysis techniques for immobilized enzymes, Biotechnol. Biotechnol. Equip. 29 (2015) 29 205-220. https://doi.org/10.1080/13102818.2015.1008192

[13]   S. Datta, L. R. Christena, Y. R. S. Rajaram, Enzyme immobilization: an overview on techniques and support materials, 3 Biotech 3 (2013) 1-9. https://doi.org/10.1007/s13205-012-0071-7

[14]   H. Liang, S. Jiang, Q. Yuan, G. Li, F. Wang, Z. Zhang, J. Liu, Co-immobilization of multiple enzymes by metal coordinated nucleotide hydrogel nanofibers: improved stability and an enzyme cascade for glucose detection, Nanoscale 8 (2016) 6071-6078. https://doi.org/10.1039/c5nr08734a

[15]   A. Küchler, J. N. Bleich, B. Sebastian, P. S. Dittrich, P. Walde, Stable and simple immobilization of proteinase K inside glass tubes and microfluidic channels, ACS Appl. Mater. Interfaces 7 (2015) 25970–25980. https://doi.org/10.1021/acsami.5b09301

[16]   E. Magner, Immobilisation of enzymes on mesoporous silicate materials, Chem. Soc. Rev. 42 (2013) 6213–6222. https://doi.org/10.1039/c2cs35450k

[17]   C. –K. Lee, A. –N. Au–Duong, Enzyme immobilization on nanoparticles: recent applications, in: H. N. Chang (Eds.), Emerging areas in bioengineering, Wiley-VCH, Weinheim, 2018, pp. 67-80. https://doi.org/10.1002/9783527803293.ch4

[18]   I. V. Pavlidis, M. Patila, U. T. Bornscheuer, D. Gournis, H. Stamatis, Graphene-based nanobiocatalytic systems: recent advances and future prospects, Trends Biotechnol. 32 (2014) 312–320. https://doi.org/10.1016/j.tibtech.2014.04.004

[19]   W. Feng, P. Ji, Enzymes immobilized on carbon nanotubes, Biotechnol. Adv. 29 (2011) 889–895.

[20]   R. A. Sheldon, Enzyme immobilization: the quest for optimum performance, Adv. Synth. Catal. 349 (2007) 1289-1307. https://doi.org/10.1002/adsc.200700082

[21]   K. Y. Lee, S. H. Yuk, Polymeric protein delivery systems, Prog. Polym. Sci. 32 (2007) 669-697.

[22]   H.-C. Zhou, S. Kitagawa, Metal–organic frameworks (MOFs), Chem. Soc. Rev. 43 (2014) 5415-5418. https://doi.org/10.1039/c4cs90059f

[23]   L. E. Kreno, K. Leong, O. K. Farha, M. Allendorf, R. P. Van Duyne, J. T. Hupp, Metal–organic framework materials as chemical sensors, Chem. Rev. 112 (2012) 1105-1125. https://doi.org/10.1021/cr200324t

[24]   K. Sumida, D. L. Rogow, J. A. Mason, T. M. McDonald, E. D. Bloch, Z. R. Herm, T.-H. Bae, J. R. Long, Carbon dioxide capture in metal–organic frameworks, Chem. Rev. 112 (2012) 724-781. https://doi.org/10.1021/cr2003272

[25]   P. Horcajada, R. Gref, T. Baati, P. K. Allan, G. Maurin, P. Couvreur, G. Fèrey, R. E. Morris, C. Serre, Metal–organic frameworks in biomedicine, Chem. Rev. 112, (2012) 1232-1268. https://doi.org/10.1021/cr200256v

[26]   T. Zhang, W. Lin, Metal–organic frameworks for artificial photosynthesis and photocatalysis, Chem. Soc. Rev. 43 (2014) 5982-5993. https://doi.org/10.1039/c4cs00103f

[27]   S. Yuan, L. Feng, K.Wang, J. Pang,M. Bosch, C. Lollar, Y. Sun, J. Qin, X. Yang, P. Zhang, Q. Wang, L. Zou, Y. Zhang, L. Zhang, Y. Fang, J. Li, H. –C. Zhou, Adv. Mater. 30 (2018) 1704303. https://doi.org/10.1002/adma.201704303

[28]   J. Mehta, N. Bhardwaj, S. K. Bhardwaj, K.-H. Kim, A. Deep, Recent advances in enzyme immobilization techniques: metal-organic frameworks as novel substrates, Coord. Chem. Rev. 322 (2016) 30–40. https://doi.org/10.1016/j.ccr.2016.05.007

[29]   X. Lian, Y. Fang, E. Joseph, Q. Wang, J. Li, S. Banerjee, C. Lollar, X. Wang, H. –C. Zhou, Enzyme–MOF (metal–organic framework) composites, Chem. Soc. Rev. 46 (2017) 3386-3401. https://doi.org/10.1039/c7cs00058h

[30]   M. B. Majewski, A. J. Howarth, P. Li, M. R. Wasielewski, J. T. Hupp, O. K. Farha, Enzyme encapsulation in metal–organic frameworks for applications in catalysis, CrystEngComm, 19 (2017) 4082-4091. https://doi.org/10.1039/c7ce00022g

[31]   E. Gkaniatsou, C. Sicard, R. Ricoux, J. –P. Mahy, N. Steunou, C. Serre, Metal–organic frameworks: a novel host platform for enzymatic catalysis and detection, Mater. Horiz. 4 (2017) 55-63. https://doi.org/10.1039/c6mh00312e

[32]   F. Lyu, Y. Zhang, R. N. Zare, J. Ge, Z. Liu, One-pot synthesis of protein-embedded metal–organic frameworks with enhanced biological activities, Nano Lett. 14 (2014) 5761–5765. https://doi.org/10.1021/nl5026419

[33]   F. –S. Liao, W. –S. Lo, Y. –S. Hsu, C. –C. Wu, S. –C. Wang, F. –K. Shieh, J. V. Morabito, L. –Y. Chou, K. C. –W. Wu, C. –K. Tsung, Shielding against unfolding by embedding enzymes in metal–organic Frameworks via a de Novo approach, J. Am. Chem. Soc. 139 (2017) 6530-6533. https://doi.org/10.1021/jacs.7b01794

[34]   K. Liang, R. Ricco, C. M. Doherty, M. J. Styles, S. Bell, N. Kirby, S. Mudie, D. Haylock, A. J. Hill, C. J. Doonan, P. Falcaro, Biomimetic mineralization of metal-organic frameworks as protective coatings for biomacromolecules, Nat. Commun. 6 (2015) 7240. https://doi.org/10.1038/ncomms8240

[35]   O. B. Jùnior, A. Bedran-Russo, J. B. S. Flor, A. F. S. Borges, V. F. Ximenes, R. C. G. Frem, P. N. Lisboa-Filho, Encapsulation of collagenase within biomimetically mineralized metal–organic frameworks: designing biocomposites to prevent collagen degradation, New. J. Chem. 43 (2019), 1017-1024. https://doi.org/10.1039/c8nj05246h

[36]   J. Cui, Y. Feng, T. Lin, Z. Tan, C. Zhong, S. Jia, Mesoporous metal–organic framework with well-defined cruciate flower-like morphology for enzyme immobilization, ACS Appl. Mater. Interfaces, 9 (2017) 10587-10594. https://doi.org/10.1021/acsami.7b00512

[37]   X. Wu, C. Yang, J. Ge, Z. Liu, Polydopamine tethered enzyme/metal-organic framework composites with high stability and reusability, Nanoscale 7 (2015) 18883-18886. https://doi.org/10.1039/c5nr05190h

[38]   X. Wu, J. Ge, C. Yang, M. Hou, Z. Liu, Facile synthesis of multiple enzyme-containing metal–organic frameworks in a biomolecule friendly environment, Chem. Commun. 51 (2015) 13408-13411. https://doi.org/10.1039/c5cc05136c

[39]   V. Lykourinou, Y. Chen, X. S. Wang, L. Meng, T. Hoang, L. J. Ming, R. L. Musselman, S. Ma, Immobilization of MP-11 into a mesoporous metal–organic framework, MP-11@mesoMOF: A new platform for enzymatic catalysis, J. Am. Chem. Soc. 133 (2011) 10382-10385. https://doi.org/10.1021/ja2038003

[40]   Y. Chen, V. Lykourinou, T. Hoang, L. J. Ming, S. Ma, Size-selective biocatalysis of myoglobin immobilized into a mesoporous metal-organic framework with hierarchical pore sizes, Inorg. Chem. 51 (2012) 9156-9158. https://doi.org/10.1021/ic301280n

[41]   Y. Chen, V. Lykourinou, C. Vetromile, T. Hoang, L. J. Ming, R. W. Larsen, S. Ma, How can proteins enter the interior of a MOF? Investigation of cytochrome c translocation into a MOF consisting of mesoporous cages with microporous windows, J. Am. Chem. Soc. 134 (2012) 13188-13191. https://doi.org/10.1021/ja305144x

[42]   D. Feng, T.-F. Liu, J. Su, M. Bosch, Z. Wei, W. Wan, D. Yuan, Y. P. Chen, X. Wang, K. Wang, X. Lian, Z. Y. Gu, J. Park, X. Zou, H.-C. Zhou, Stable metal-organic frameworks containing single-molecule traps for enzyme encapsulation, Nat. Commun. 6 (2015) 5979. https://doi.org/10.1038/ncomms6979

[43]   E. Gkaniatsou, C. Sicard, R. Ricoux, L. Benahmed, F. Bourdreux, Q. Zhang, C. Serre, J. –P. Mahy, N. Steunou, Enzyme encapsulation in mesoporous metal-organic frameworks for selective biodegradation of harmful dye molecules, Angew. Chem. Int. Ed. 57 (2018) 16141-16146. https://doi.org/10.1002/anie.201811327

[44]   X. Lian, Y.-P. Chen, T.-F. Liu and H.-C. Zhou, Coupling two enzymes into a tandem nanoreactor utilizing a hierarchically structured MOF, Chem. Sci. 7 (2016) 6969-6973. https://doi.org/10.1039/c6sc01438k

[45]   T. J. Pisklak, M. Macías, D. H. Coutinho, R. S. Huang, K. J. Balkus, Hybrid materials for immobilization of MP-11 catalyst, Top. Catal. 38 (2006) 269–278. https://doi.org/10.1007/s11244-006-0025-6

[46]   H. Deng, S. Grunder, K. E. Cordova, C. Valente, H. Furukawa, M. Hmadeh, F. Gandara, A. C. Whalley, Z. Liu, S. Asahina, H. Kazumori, M. O'Keeffe, O. Terasaki, J. F. Stoddart, O. M. Yaghi, Large-pore apertures in a series of metal-organic frameworks, Science 336 (2012) 1018–1023. https://doi.org/10.1126/science.1220131

[47]   P. Li, J. A. Modica, A. J. Howarth, E. L. Vargas, P. Z. Moghadam, R. Q. Snurr, M. Mrksich, J. T. Hupp and O. K. Farha, Toward design rules for enzyme immobilization in hierarchical mesoporous metal-organic frameworks, Chem 1 (2016) 154–169. https://doi.org/10.1016/j.chempr.2016.05.001

[48]   H. He, H. Han, H. Shi, Y. Tian, F. Sun, Y. Song, Q. Li, G. Zhu, Construction of thermophilic lipase-embedded metal–organic frameworks via biomimetic mineralization: A biocatalyst for ester hydrolysis and kinetic resolution, ACS Appl. Mater. Interfaces 8 (2016) 24517-24524. https://doi.org/10.1021/acsami.6b05538

Metal-Organic Framework Composites - Volume I
Materials Research Foundations 53 (2019) 235-257

Materials Research Forum LLC
https://doi.org/10.21741/9781644900291-11

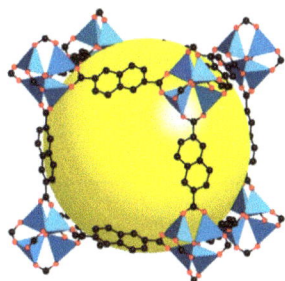

Chapter 11

MOF-Derived Nanocarbons: Synthesis, Properties, and Applications

Boris I. Kharisov*, Cesar Maximo Oliva González, Thelma Serrano Quezada, Oxana V. Kharissova, Yolanda Peña Mendez

Universidad Autónoma de Nuevo León, Monterrey, Mexico

*bkhariss@hotmail.com

Abstract

Nanocarbons, derived from the metal-organic frameworks (MOFs), are reviewed. These nanomaterials represent different carbon allotropes with presence of initial metals in elemental or oxidized forms, depending on the calcination conditions. The main route for obtaining MOF-derived nanocarbons is the pyrolysis at temperatures in the range of 700-1000 °C. The formed nanoporous materials possess useful properties, allowing them to be used for catalytic purposes, in adsorption processes, supercapacitors and solution of environmental problems, among other uses.

Keywords

MOFs, Nanocarbons, Bimetallic Nanocarbons, Catalysis, Pyrolysis

Contents

1.   Introduction

The metal-organic frameworks also known as MOFs are multidimensional structures with large surface area and porosity. These materials are composed of two parts, the first part is a metal ion and the second are organic molecules known as organic linkers (Fig. 1) which are capable of forming coordinate bonds with metal ions. Among the most common linkers are azole molecules, bidentate and tridentate carboxylic acids, due to their facility to form coordination bonds. The metal ion and linkers used to synthesize the MOF, as well as its synthesis process will determine its structure and properties. The great amount of metal ions and binders allows expansive MOF synthesis. This is reflected in about 20,000 reports in the MOFs area for the last 2 decades.

Numerous reports show that MOF composites can be prepared from functionalized materials or assembling MOFs with other materials [1]. Some of MOF composites were prepared from carbon materials 1D and 2D, for example, carbon nanotubes and graphene, respectively. Additionally, the MOFs possessing large pore volumes, high surface areas, and extraordinary tunability of structures and compositions[2], therefore, the MOFs are promising materials promising source for the preparation of self-sacrificing templates and carboneous materials. Carbon-based nanomaterials derived from MOF have great advantages as efficient catalysts or as catalyst supports (Fig. 2), compared to other carbon-based catalysts, the facility to adapt their morphology, high porosity, easy functionalization with a great variety of heteroatoms and metal/metal oxides [3]. Among most current reviews, it is shown that MOF-derived functionalized carbon-based electrocatalysts, including different non-metal heteroatoms (like B, P, and S) and metal-doped (Ni and Zn) doped carbon materials [4]. Furthermore, in addition to the catalytic applications (see below), the uses of MOF-derived carbons include other electrochemical

Metal-Organic Framework Composites - Volume I   Materials Research Forum LLC
Materials Research Foundations **53** (2019) 235-257   https://doi.org/10.21741/9781644900291-11

applications, for example for batteries. In whole, heat treatment of metal–organic frameworks, resulting MOF-derived carbons, belongs to green energy applications [5]. In this chapter, we will analyze most reported Co-, Fe-, and Zn-containing MOFs, and some bimetallics as a source for metallated nanocarbons.

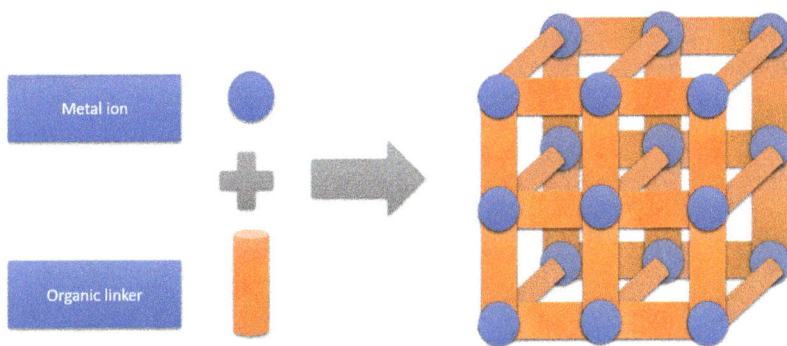

Figure 1. Components and structure of MOFs.

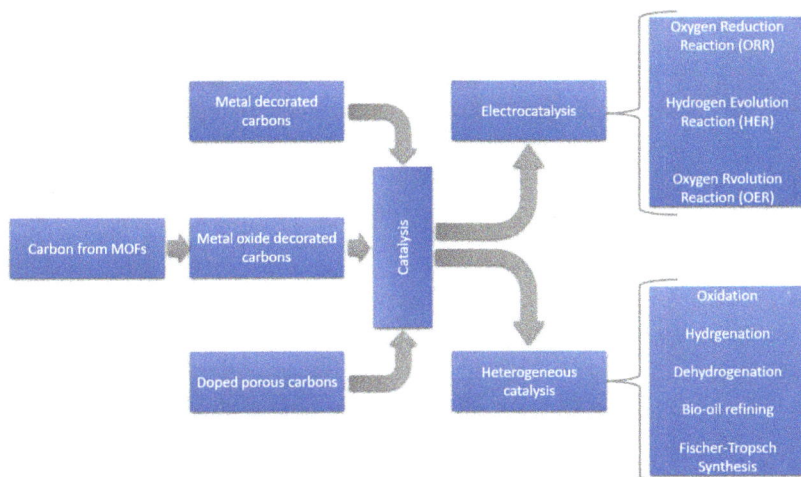

Figure 2. Application of the different types of carbon from MOFs in the catalysis of important reactions.

2. General methods for preparation of MOF-derived nanocarbons

Among the numerous reports to fabricate nanocarbons derived from MOFs, we note that most use is observed for zinc-containing MOF precursors, especially ZIF-8 and MOF-5. As an example, the nanoporous carbons were prepared by the pyrolysis of $C_5H_6O_2$ ((furan-2-yl)methanol, furfuryl alcohol) that was impregnated in the pores of $[Zn_4O(bdc)_3]$ (MOF-5). The porous carbons obtained at temperature of about 1000 °C showed mesoporosity and good electrical conductivity, in contrast to that prepared at 530 °C which showed a poor electrical conductivity. Hence, the samples pyrolyzed at higher temperature showed the best capacitor behaviors.

The second most used MOF for fabricate carbons are the zeolitic imidazolate framework (ZIF), mainly ZIF-8. Since these MOFs contain imidazoles, the obtained carbons were found to be rich in nitrogen, which makes these carbons derived from ZIF-8 promising cathodes of Zn-air and Li–S batteries [6]. For the preparation of these cathodes on silica supports, the templates were coated with MOFs, more specifically ZIF-8, that was synthesized *in situ* on the surface of the silica templates [7] (Fig. 3) and subsequently subjected to thermal treatment. The resulting material presented macro-meso-hierarchical micropores with a superficial area of 2546 $m^2 g^{-1}$ and a total pore volume of 13.42 $cm^{3} g^{-1}$. In terms of its characteristics, suitable to be applied as cathode material, it was observed that it had a capacity of 770 $mAh\,gZn^{-1}$ and a potential density of 197 $mW\,cm^{-2}$. The synthesis of ZIF-8 in situ allows to modulate the spherical morphology of carbon, however, the particles have a great diversity in their average particle size approximately 10.50 nm and bamboo-like carbon nanotubes (abbreviated as B-CNTs) with the average diameter in the range of 40–80 nm appeared on the surface of the modified spherical activate carbon (SAC). A superior catalytic performance toward acetylene hydrochlorination was observed, with the highest acetylene conversion of 81% and the selectivity to VCM above 99%, this can be attributed to the increse in surface area, therefore, more active sites were formed, this allows hidrogen chloride and acetylene to interact more easily, in the process of hydrochlorination.

In the same way the ZIF-8 can be synthesized to produce composites with ZnO nanoparticles embedded in N-doped nanoporous carbons [8] using a carbonization process under steam atmosphere. It was observed that this carbonization process allows obtaining homogeneous ZnO nanoparticles in the carbon matrix. The material obtained by this process has a strong affinity for the CO_2 molecules, besides, it shows a great ability to degrade and adsorb of methylene blue dispersed in water under visible light than the obtained composite [9]. On the other hand, ZIF-8 can be used to generate carbon nanofibers doped ZnO, with high porosity, prepared by the assembly of $[Zn(MeIM)_2]$ (ZIF-8) nanocrystals on tellurium nanowires.

Metal-Organic Framework Composites - Volume I   Materials Research Forum LLC
Materials Research Foundations **53** (2019) 235-257   https://doi.org/10.21741/9781644900291-11

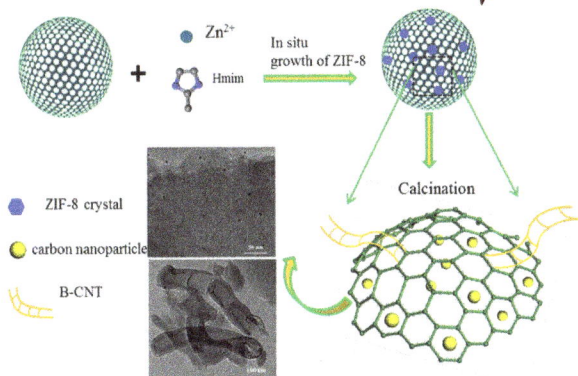

Figure 3. ZIF-8 synthesis on silica templates. Reproduced with permission of Springer.

The carbon nanofibers obtained by this method have a hierarchical pore structure and the material propertis allows them to be electrocatalytically applied for ORR (oxygen reduction reaction) in alkaline media. The material exhibited electrocatalytic activity, good thermal resistance, good capacity to store hydrogen, being able at 77 K to store 2.77 wt.%. All the properties mentioned above are intimately linked to having a high surface area (3405 m^2/g).

3.   Structural peculiarities

The ZIF-8-derived carbon nanosheets are a great example that MOF-derived carbons can have 2D dimensionalities. The carbon nanosheets were synthesized [10] using ZnO sacrificial templates in the form of nanosheets and an agent that allowed controlling the direction of growth.

The carbon nanosheets have a high ion-/electron-transport rates and their high surface area allows them to interact with ions, because of these properties in the material, the carbon nanosheets have potential application in electrochemical energy storage devices, especially as electrodes, however, the graphene sheet-sulfur/carbon composite was prepared from ZIF-8 with a simple process in a single step, also showed to be flexible and have a high electrical conductivity, which gives a potential application as a cathode in lithium and sulfur batteries [11].

In addition, the ZIF-8-derived carbon particles can adopt differents geometrical figures. A clear case of the above mentioned are the microporous carbon polyhedrons (MPCP) with a uniform size and a high porosity, derived from the ZIF-8 polyhedrons from MEIM (2-methylimidazole) as a linker and zinc salts. These composites have sulfur in their pores [12], which allows them to have stable cycles with electrolytes such as DOL/DME and EC/DEC. These characteristics make it applicable in lithium sulfur batteries [13]. Other geometrical figures whit 3D dimensions that can be formed are the carbon cuboids, exhibiting a unique crumpled-sheet assembled porous morphology with a large disparity in the size of their pores. These materials were synthesized from a MOF-5, which has zinc atoms integrated in its structure. In addition, the MOF-DC based Li-HEC shows to have the capacity to support 10000 cycles and only lost 18% of its initial value.

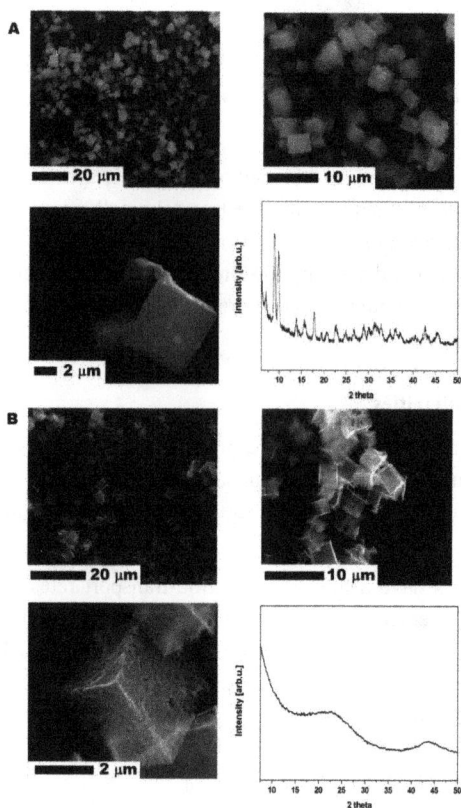

Figure 4. SEM micrographs of MOF-5 (A) and carbon derived from MOF-5 (B) with DRX. Reproduced with permission of the MDPI.

It should be noted that there are patented works where electrodes were made using MOF-5 by solvent evaporation with a binder of PVDF (fluorinated poly(vinylidene)) [14]. Such a method for the preparation of electrodes provides an easy, cost-effective and non-void distribution of nanocomposites. Highly porous carbonized nanoparticles of MOF-5 (Fig. 4), containing carbon, adhere to the membrane based on PVDF, resulting in an increase in the active surface area of the electrode to 847 m^2g. The specific capacity of these samples is approximately 218 F/g. Moreover, an important advantage of the carbon electrode prepared from carbonized MOF-5 is the ability to synthesize MOFs from some wastes, making the material less expensive long-term and longer life have.

4.   Influence of pyrolysis conditions and initial MOF structure on the final structure of nanocarbons

The morphologies and pore sizes of the MOF-derived carbon can be controlled, through the changes in the calcination temperature (600°C-1000°C) [15]. The surface area control and the pore size of the materials allows us to apply in such areas as, for example, hydrogen storage. In case of pyrolyzed MOF-5, whose surface area increased to 2393 $m^2 \cdot g^{-1}$ from 835 $m^2 g^{-1}$, this nanocarbon revealed a capacity for H_2 storage enhanced from 1.9 to 2.7 wt.%. It is necessary to mention that not every time this lead to an improvement of these characteristics: such as the case of the carbonization of the MOF with chromium that is presented below, the decrease of its hydrogen adsorption capacity on its surface, is due to the presence of chromium oxide in the pores of the material. Also, there are other factors during the carbonization process, such as the temperature and the atmosphere used. An example of the importance of the atmosphere is the pure carbon fabrication with multifractal structures, This was achieved by vacuum carbonization of a zinc-based MOF [16], the vacuum pyrolysis of this material gives it numerous macro- and meso- pores, which in turn allow this material to have super-high Li storage capacity (2458 mAh g^{-1}).

Porous carbon materials of MOFs containing zinc (MOF-177, MOF-5), and bioMOF-100), prepared by a simple pyrolysis of their precursors also have a promising capacity for CO_2 capture and were compared to the original MOFs before the thermal treatment. This effect is due to the electrostatic interactions on the surface of the material [17]; its high surface area make it ideal for this application. Some reports mentioned lithium ion batteries, doped with carbonized material from Cu-MOF [18]. The carbonized material has a *shuttle-like structure* and can improve cycle stability and speed capability of the lithium-ion battery when applied as a lithium-ion battery material. The lithium ion battery has a high Coulombic efficiency, high specific load / discharge capacity, excellent speed capability and excellent cycle stability, etc.

Metal-Organic Framework Composites - Volume I   Materials Research Forum LLC
Materials Research Foundations **53** (2019) 235-257   https://doi.org/10.21741/9781644900291-11

The MOF known as ZIF-8C has been shown to be able to generate *graphene quantum dots* doped with nitrogen and soluble in water, using an acid vapor cutting strategy, this material turned out to be a better fluorescence yield and a wider range of applications in photocatalysis, sensors, bioimaging, with respect to the material synthesized by other methods. This set of applications is possible due to the presence of O-functional groups on the surface [19]. This material can serve as a fluorescent detection probe for the selective detection of Fe^{3+} ions with a detection limit of 0.08 μM (with a signal-to-noise ratio of 3). In addition, although it has been shown that MnO is a promising material for anodes in Li ion batteries, taking into account of its high calculated capacity (755 mA Hg^{-1}), low cost and relatively decreased voltage hysteresis (<0.8 V). MnO application as a electrode material for practical purposesd is still hampered by its poor cycle stability and large volume expansion in the process of the loading / unloading. Therefore, it has been proposed that the use of organic-metal frames for the *in situ* manufacture of ultra-fine MnO nanocrystals encapsulated in a porous carbon matrix, this is done by pyrolysis of the MOFs, where the nanopores increase the active sites for store redox ions and improve ion diffusivity to encapsulated MnO nanocrystals. In addition, the highly reversible specific capacity material of 1221 mA Hg^{-1} after 100 cycles [20]. The excellent electrochemical performance can be attributed to its unique structure with MnO nanocrystals dispersed evenly within a porous carbon matrix, which can greatly improve the electrical conductivity and effectively prevent the aggregation of nanocrystals of MnO, and relieve the stress caused by the volumetric change during charging.

MOFs uses as electrodes in batteries have been extensively researched, but these materials can also be applied to hybrid supercapacitors (HSC), to meet the growing market of electric vehicles (EV) and hybrid electric vehicles (HEV). Among the materials studied for this application are the hybrid materials of quantum dots on Nb_2O_5 basis, incrusted into N-doped porous carbon, produced from dodecahedron ZIF-8. These materials have been shown to have excellent high-speed capabilities and long term cyclic stability. It is important to note that assembling an HSC device using this material shows super-high energy electrochemical performance (76.9 W h kg^{-1}) and power density (11 250 W kg^{-1}). Its great capacity retention (cyclical stability) after 4500 cycles in a voltage range of 0.5-3.0 V is from 85% to 5 A g^{-1} [21]. The good performance of these materials indicates the combination of the advantages over current supercapacitors and are promising energy storage materials.

The pyrolysis of MOF with magnetic metals such as Fe or Co, compared to those mentioned earlier in this chapter. However, carbon materials with these metals have many applications, such as Co@C with a surface area of 195.2 m^2/g, the prismatic crystals of MOF-74 can be used to make carbon matrix with magnetic properties and cobalt nanoparticles embedded in their surface by a pyrolysis process in situ. [22], the magnetic carbon derived from MOF-74 showed to have great catalytic properties to reduce 4-nitrophenol to 4-aminophenol and is a catalyst that can be re-used. Also, *microtubular* structures with nano-spheres of cobalt can be obtained from cobalt oxalate microtubes and *in situ* conversion reaction to MOF (ZIF-67) and self-templates for the tubular structure [23]. There are two main reasons to use cobalt oxalate microtubes: first, the morphology of the oxalate allows the ZIF-67 to obtain the *microtubule* structure and second, the cobalt oxalate also serves as a precursor to ZIF-67 by reacting with 2-methylimidazole. This type of materials derived from ZIF-67 present a high porosity, high concentration of nitrogen and CoO nanoparticles in its surface, this last one gives this material to obtain great catalytic and magnetic properties for the reactions of oxidation of alcoholes to esters with O_2 [24]. It is important to note that this material also has on its surface with Co^0, Co^{2+} and Co–OH species.

Additionally, the cobalt ZIF-67 MOFs have been used to coat materials and improve their properties, as an example we have manufacture of magnetic carbon sponges [25], these sponges have micro- and nanopores, magnetic and catalytic properties that characterize the ZIF-67. Among the potential applications of this carbon magnetic sponge are, oil remediation in water, degradation of dyes in water (methylene blue), catalysis and H_2 production, besides, its magnetic properties make it easy to recover [26].

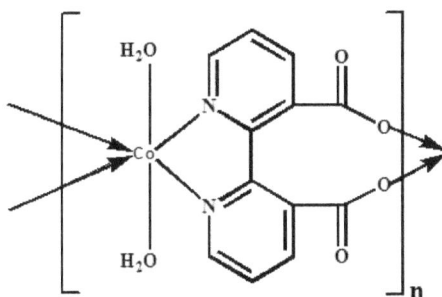

Figure 5. Diagram of interactions between cobalt (ions) and H₂bpdc (linker). Reproduced with permission of MDPI.

The pyrolyzed MOF can be used as source for catalysts fabrication for the ORR (oxygen reduction reactions). These materials can form composites with multi-walled carbon nanotubes (MWCNTs) to increase the active sites. An example is the composition Co-bpdc/MWCNTs (Fig. 5) [27], which undergo a carbonization process generating Co nanoparticles, distributed in the graphitic carbon layers and in the MWCNTs. These nanoparticles have catalytic activity ORR with an initial reduction potential 0.98V and a half wave potential 0.91V.

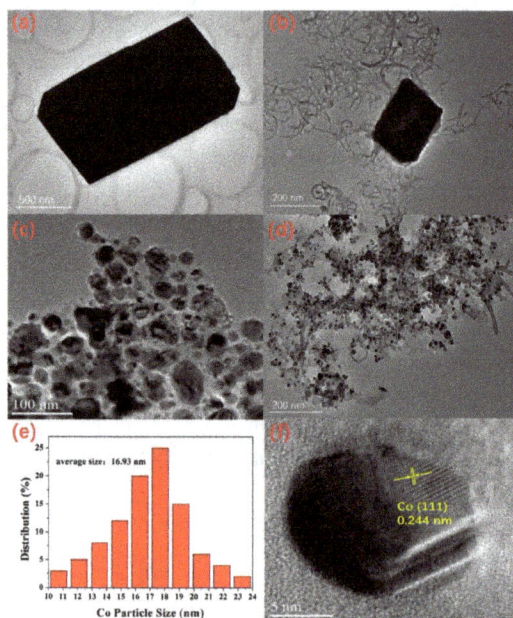

Figure 6. TEM micrographs of Co-bpdc (a); Composite Co-bpdc/MWCNTs with treatment at 100 C (b); Calcined Co-bpdc (c); Composite Co-bpdc/C calcined MWCNTs (d) and finally Co-bpdc HRTEM/MWCNTs (f). Reproduced with permission of MDPI.

On the other hand, carbon composites derived from iron MOFs are usually prepared at high temperatures such as MIL-100 (Fe). Among other applications of this compound, it was found its possible use as material for the manufacture of cathodes in an electro-Fenton system, with the purpose of degrading compounds such as pentadecafluorooctanoic acid [28]. It is also worth noting that the synthesis of this composite, unlike others [29], can be achieved at low temperatures (110°C) by heating in reflux. Apparent formation of Fe nanoparticles or Fe_3O_4 is being studied.

$$Fe(COT)_2 + DME + DMSO \Rightarrow Fe_3O_4 + graphite \quad (1)$$

Materials Research Forum LLC
https://doi.org/10.21741/9781644900291-11

In addition to the mono-metal MOFs of Cu, Co and Fe, roasted MOFs based on Zr, such as the pyrolyzed UiO-66 (Fig. 7), have also been studied, using them for the adsorption elimination of dyes (such as violet glass, named as CV) and some pharmaceuticals (e.g., SA, salicylic acid). The pyrolysis of these MOFs at 900°C generates the formation of a graphite oxide phase and also provides to the material a greater adsorption capacity of the contaminating compounds. The maximum CV absorption (243 mg/g) overcomes the activated carbon that are commercially available. This composition also has an absorption of SA of 102 mg/g, in addition, after doping the material with nitrogen, its adsorption capacity improved to reach 109 mg/g [30]. It should be mentioned, that even though the pH weakly influences these values, the surface area and pore volume were the main factors that influenced the adsorption of the pyrolyzed MOF. Meanwhile, the SA adsorption depends on the electrostatic interactions, for which the pH can play an important role in the adsorption process.

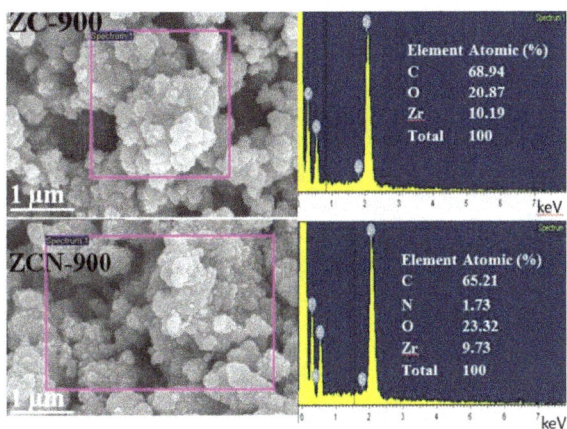

Figure 7. Images of ZC and ZCN treated at 900° C using FE-SEM / EDS. Reproduced with permission of MDPI.

Likewise the MOFs UIO-66 can also be used in flexile supercapacitors if polyaniline is combined, achieving a high specific capacity of 1015 F/g. The specific capacity is only attenuated by 10% when the prepared supercapacitor doubles 800 times, and the specific capacity can be maintained up to 91% by stability testing of the DC load and discharge cycle for 5,000 rounds [31]. The flexible supercapacitor prepared by the invention has good flexibility and electrochemical performance, making this type of materials applicable in flexible electronic devices and energy storage.

5.   Bimetallic MOF-derived carbons

The large number of metals that can be used to generate mono- and bimetalic MOFs, in addition, the ease of the MOF to generate carbon composites has allowed to manufacture bimetallic carbon nanocomposite, due to the aforementioned composites such as CuNi-DABCO-n (Fig. 8, DABCO = 1,4-diazabicyclo[2.2.2]octane) [32]. This composition, like the others mentioned above, has a high catalytic activity and a large surface area, but it should be noted that it has a high stability in alkaline media which is superior to that of the commercially available Pt-based materials. In addition, porous magnetic carbon adsorbent (MPCS) with core-shell structure was prepared by the thermal treatment MOF-5, which was Fe(III)-modified. This MPCS has potential application in the sorption of organic pollutants that we present below in increasing order atrazine< carbamazepine<bisphenol A<norfloxacio<4-nitrophenol.

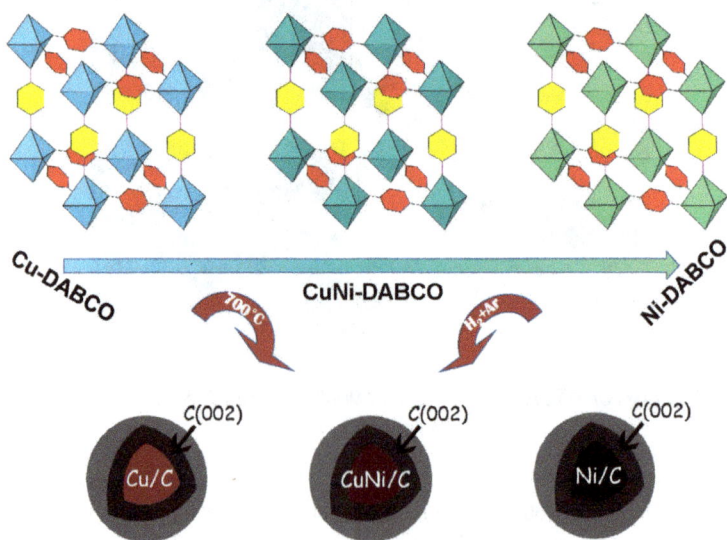

Figure 8. Diagram showing the CuNi/C-n synthesis procedure. Reproduced with permission of the Springer.

Metal-Organic Framework Composites - Volume I   Materials Research Forum LLC
Materials Research Foundations **53** (2019) 235-257   https://doi.org/10.21741/9781644900291-11

Previously, we observed that the process for obtaining ZIF nantubes uses tellurium nanotubes, this technology has also been applied to Zn / Co-based ZIFs the resulting material had high surface area, hierarchical pore structure, abundant Co–N_x active sites[33]. The CNTs doped with N and Co, have high porosity, which makes them ideal for the reduction reactions of oxigene and tri-iodide. Another example of nanocarbon doped with Zn/Co is the C-$ZnCo_2O_4$-ZnO nanorod arrays (NRA) (Fig. 9) and prepared by a template-based solution route on Ti foil substrate and used as high-performance anode for lithium-ion batteries [34]. The carbon derived from the MOF (Fig. 10) on $ZnCo_2O_4$–ZnO nanorods allow the material to be used as a conductor, besides, it can prevent the increase in the volume of the material during the loading-unloading process. This material was used as anode in Li-ion batteries. The $ZnCo_2O_4$–ZnO NRAs revealed a significant increase in the number of cycles that can support, in addition, a great electrochemical performance.

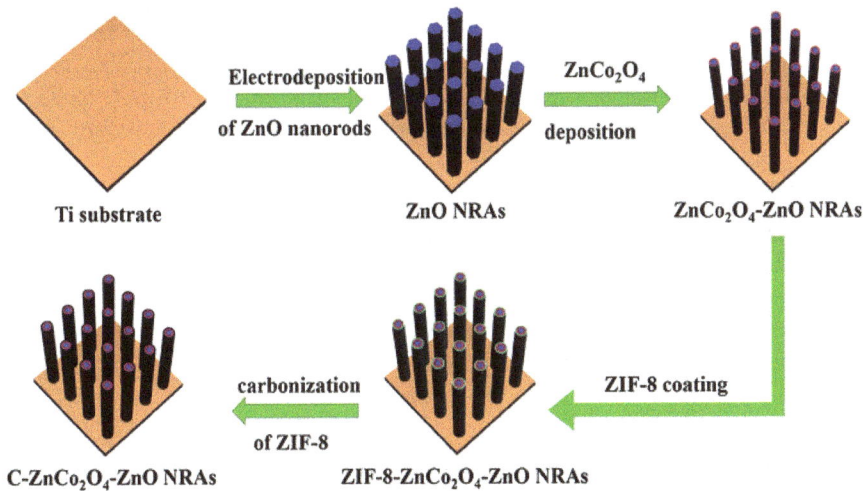

Figure 9. Scheme showing the procedure for the manufacture of the electrodes of C–$ZnCo_2O_4$–ZnO NRAs. Reproduced with permission of Springer.

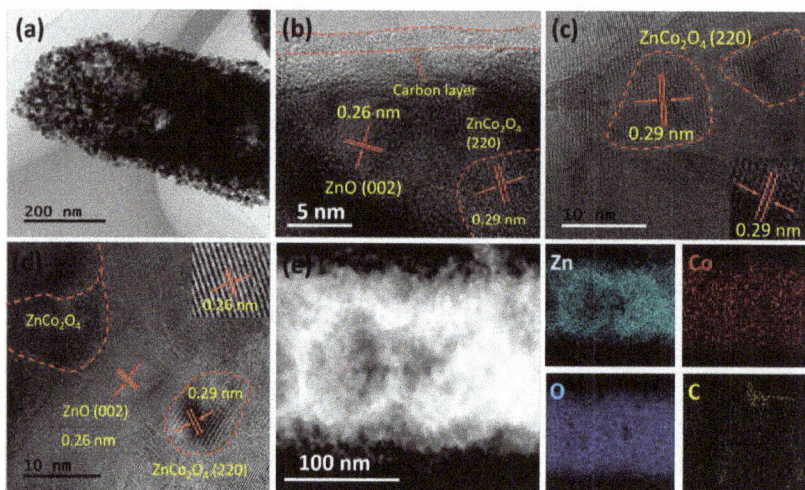

Figure 10. Micrographs of TEM (a) and HRTEM (b-d), in the latter the crystal lattices of the materials are shown, in addition, the elemental mapping of O, Zn, Co and C in the C-ZnCo$_2$O$_4$-ZnO is also presented. Reproduced with permission of Springer.

6. Role of N-doping

One of the main changes that are made to nanocarbons is dopING with nitrogen. This is due to the properties that you get from these materials when they are doped with nitrogen, among which electric conductivity highlights and facilitates charges transfers. These materials can be manufactured in atmospheres such as Ar and NH$_3$ using a high carbonization temperature [35]. In addition, metals such as Se can be used to obtain particles from the micropores of the material by melting-diffusion and infiltration methods.

Continuing the mentioned above, the bimetallic carbon materials doped with nitrogen have numerous reports due to the increase in their properties due to their textural characteristics and the addition of nitrogen. Thus, the bimetallic carbon of CoNi-MOFs doped with nitrogen was reported (Fig. 11) to be used for the adsorption of antibiotics (specifically norfloxacin NOR was studied) [36]. The material after pyrolysis (Fig. 12) had a high specific surface (871 m^{2}g^{-1}) and a high pore volume (0.75 cm^{3}g^{-1}). In addition, it possesses magnetic properties due to the Cobalt species which facilitates its recovery. The maximum experimental NOR adsorption capacity of the material was 55.12 mg g^{-1} at 298.15 K and a pH of 6.0. The revision of the kinetic data shows that the experimental data obtained from the NOR estimate versus time, reveals that the reaction matches a pseudo-second order model. The analysis of isotherma data can be adjusted Freundlich's model. Based on the results of the adsorption in a wide range of conditions, it was found that the dominant adsorption mechanisms are the filling of pores, the electrostatic interaction and the H-bond.

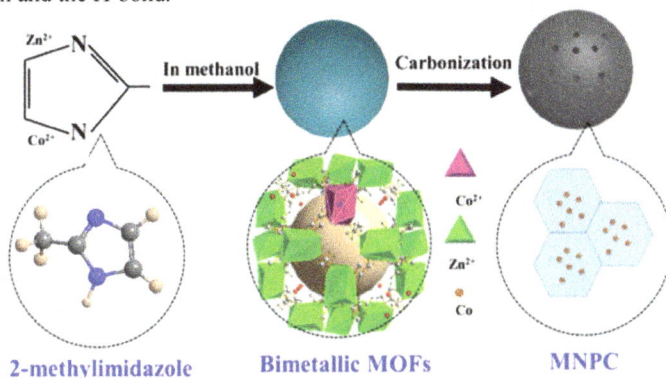

Figure 11. Scheme of magnetic nitrogen-doped porous carbon synthesis. Reproduced with permission of the MDPI.

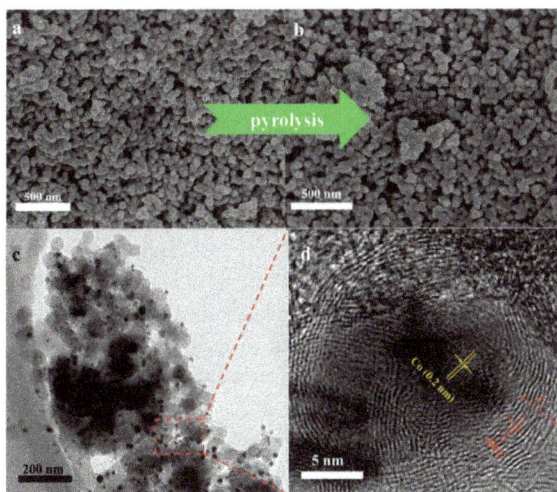

Figure 12. SEM micrographs of (a) bimetallic MOF before pyrolysis; (B) after the pyrolysis and (c) (d) micrographs after the pyrolysis of TEM and HRTEM respectively. Reproduced with permission of the MDPI.

7. Main applications of MOF-derived nanocarbons

In this chapter, we have noted the applications of the carbons derived from MOF, among which are catalysis and adsorption of pollutants, however, there are still other applications that have not been mentioned as is the case of gas capture such as CO_2 [37], electrolyte/membrane materials for fuel cells, batteries, adsorbents for removal of toxic species from drinking water, fabrication of pseudo-capacitance electrodes, drug delivery carriers, etc. [38]. The ZIF-8 nano-derived carbon is generally distinguished in applications related to the storage of electrochemical energy due to the metal particles on its surface that give it a high conductivity and high surface area, as it is observed in the composites of GO with MOF that contain cobalt, these composites can be called Co-MOFs@GO was prepared through a hydrothermal process [39], in addition to the previously mentioned, they have a potential application as hydrogen gas sensors after improving them by sputtering platinum (Pt) acting as a catalyst. This hybrid material can function as an alternative in the industry to detect hydrogen gas, in addition this material was synthesized from cobalt salts and BDC, so it can also be used for enantiomer detection in chiral drugs [40].

There are some other examples that are mentioned, where other strategies are used such as the controllable synthesis of 3D MOF hybrid matrices using semiconductor nanostructures as personal sacrifice templates [41]. This material can be applied directly in anodes and cathodes for the division of water, this material has characteristics of porosity similar to other carbons derived from MOF, however, in this case the pores obtained are well aligned and their diameter is approximately 50 nm [42]. It should also be mentioned that the carbons derived from MOF-8, present biocompativilidad, for which they can be applied as drug delivery carriers in medicine. In addition, the carbon derivative of isoreticular MOF-8 ($Zn_4O(ndc)_3$, ndc = naphthaline-2,6-dicarboxilate) was subjected to an exfoliation treatment with solvent, which resulted in a highly porous carbon material [43]. The treatment that was given to the material, allowed it to be more facts to disperse and gave it a high surface area ($1854 \ m^2g^{-1}$), also, it was used as a vitreous carbon electrode for the chloramphenicol detection. Furthermore, a simple strategy was offered for sodium storage improvement by phosphorus confining in the microporous carbon matrix doped with nitrogen derived from P@N-MPC (ZIF-8) [44]. It should also be noted that carbon doped with nitrogen and with pores smaller than 1 nm, facilitates the diffusion of ions in organic electrolytes, which makes this material promising in ORR (oxygen reduction reactions).

As in the previous case, where the use of monometallic carbon materials for the production of hydrogen has shown excellent results, bimetallic carbons have also been investigated, such as the ZIF-67 bucket, which is made from $CoSe_2$, and can be applied both in the reaction of the evolution of hydrogen (HER) in the cathode and the reaction of the evolution of hydrogen (OER) at the anode. A water electrolyzer equipped with two $CoSe_2/CF$ electrodes provided a water division current of 10 mA·cm^{-2} [45]. At a current of 20 mA·cm^{-2}, it can operate without degradation for 30 h. This study offers a cost-effective solution for water splitting.

In addition, there are also some MOF-derived trimetallic nanocarbons, as in the case of CuMnCo-MOF nanofibers, that are sought to be applied as nano-catalysts in the evolution reaction of the electrolytic oxygen in water. The preparation method comprises mainly the following steps: mix an alkaline solution of aspartic acid and a solution of copper nitrate, manganese nitrate and cobalt nitrate at room temperature, perform a suction filtration and dry to prepare a charge of nanofiber Cu-MOF Co(II) and Mn(II) ionic nanofiber, namely a CuMnCo-MOF nanofiber, and heating the CuMnCo-MOF nanofiber in air to prepare the composite carbon and the oxide nano-catalyst of the transition element. The catalyst showed to have an electrocatalytic activity and favorable electrochemical stability for the evolution of oxygen.

Other applications for materials synthesized from the pyrolysis of iron MOFs with large surface area are in the fields of removal and recovery of environmental pollutants (oil/hydrocarbon and dye/phenol). This is possible, since these nanocarbons possess hydrophobic behavior due to the absence of hydrophilic groups on their surface [46]. Their behavior is also highlighted as a magnetically separable and recyclable superadsorbent. In addition to its applications for water remediation, there are some reports of molybdenum-based pyrolyzed MOFs generating mesoporous molybdenum carbide nano-octahedrons composed of ultra-fine nanocristals, which can be applied in the electrochemical division of water, these materials show an electrocatalytic performance remarkable for the production of hydrogen from both acidic and basic solutions, which allows the production of clean and sustainable hydrogen fuel [47].

Conclusions and further outlook

In this chapter, we reviewed the great capacity of MOFs to be used as carbon precursors. The most studied MOFs are generally those containing Zn, Co and Cu; however, the number of reported MOFs and their applications continuously grow. Bi- and trimetallic nanocarbons, derived from MOFs, have found to reveal catalytic properties, which are sometimes better in comparison with their monometallic analogues. MOF-derived nanocarbons can be useful in the water splitting, treatment of pollutants in contaminated water (by adsorption and *in situ* destruction), as catalysts in the HER and ORR processes, as materials for supercapacitors, among many other uses. The area of the MOF-derived nanocarbons is relatively new research field, so many advances and discoveries are expected in the next decade.

References

[1] H. Wang, Q. Zhu, R. Zou, X. Qiang, Metal-Organic Frameworks for Energy-Related Applications, *Chem.*, 4.1 (2017), 52–80.

[2] K. Shen, X. Chen, J. Chen,Y. Li, Development of MOF-Derived Carbon-Based Nanomaterials for Efficient Catalysis, *ACS Catalysis*, 6.9 (2016), 5887–5903. https://doi.org/10.1021/acscatal.6b01222

[3] P.C. Rae, Y. S. Jae, K.R. Patent 20,130,031,926 (A), (2013).

[4] Q. Ren, H. Wang, X. Lu, Y.X. Tong, G.R. Li, Recent Progress on MOF-Derived Heteroatom-Doped Carbon-Based Electrocatalysts for Oxygen Reduction Reaction, *Advanced Science*, 5.3 (2018), 1700515. https://doi.org/10.1002/advs.201700515

[5] L. Lux, K. Williams, S. Ma, Heat-Treatment of Metal-Organic Frameworks for Green Energy Applications, *CrystEngComm*, 17.1 (2015), 10–22. https://doi.org/10.1039/c4ce01499e

[6] M. Yang, X. Hu, Z. Fang, L. Sun, Z. Yuan, S. Wang, W. Hong, X. Chen, D. Yu, Bifunctional MOF-Derived Carbon Photonic Crystal Architectures for Advanced Zn–Air and Li–S Batteries: Highly Exposed Graphitic Nitrogen Matters, *Advanced Functional Materials*, 27.36 (2017), 1–9. https://doi.org/10.1002/adfm.201770210

[7] X. Li, J. Zhang, Y. Han, M. Zhu, S. Shang, W. Li, MOF-Derived Various Morphologies of N-Doped Carbon Composites for Acetylene Hydrochlorination, *Journal of Materials Science*, 53.7 (2018), 4913–26. https://doi.org/10.1007/s10853-017-1951-3

[8] B. Chen, G. Ma, D. Kong, Y. Zhu, Y. Xia, Atomically Homogeneous Dispersed ZnO/N-Doped Nanoporous Carbon Composites with Enhanced CO_2 uptake Capacities and High Efficient Organic Pollutants Removal from Water, *Carbon*, 95 (2015), 113–24. https://doi.org/10.1016/j.carbon.2015.08.015

[9] W. Zhang, Z.Y. Wu, H.L. Jiang, S.H. Yu, Nanowire-Directed Templating Synthesis of Metal-Organic Framework Nanofibers and Their Derived Porous Doped Carbon Nanofibers for Enhanced Electrocatalysis, *Journal of the American Chemical Society*, 136.41 (2014), 14385-14388. https://doi.org/10.1021/ja5084128

Metal-Organic Framework Composites - Volume I
Materials Research Foundations 53 (2019) 235-257

Materials Research Forum LLC
https://doi.org/10.21741/9781644900291-11

[10] B. Ding, J. Wang, Z. Chang, G. Xu, X. Hao, L. Shen, H. Dou, X. Zhang, Self-Sacrificial Template Synthesis of Mixed-Valence-State Cobalt Nanomaterials with High Catalytic Activities for Colorimetric Detection of Glutathione, *Sensors and Actuators, B: Chemical*, 254 (2018), 329–36. https://doi.org/10.1016/j.snb.2017.07.104

[11] R. Chen, T. Zhao, T. Tian, S. Cao, P. R. Coxon, K. Xi, D. J. Fairen, R. Vasant, R. Cheetham, K. Anthony, Graphene-Wrapped Sulfur/Metal Organic Framework-Derived Microporous Carbon Composite for Lithium Sulfur Batteries, *APL Materials*, 2.12 (2014), 124109. https://doi.org/10.1063/1.4901751

[12] H. Wu, S. Wei, L. Zhang, R. Xu, H. Hng, X. Lou, Embedding Sulfur in MOF-Derived Microporous Carbon Polyhedrons for Lithium-Sulfur Batteries, *Chemistry - A European Journal*, 19.33 (2013), 10804–8. https://doi.org/10.1002/chem.201301689

[13] A. Banerjee, K. Upadhyay, D. Puthusseri, V. Aravindan, S. Madhavi, S. Ogale, MOF-Derived Crumpled-Sheet-Assembled Perforated Carbon Cuboids as Highly Effective Cathode Active Materials for Ultra-High Energy Density Li-Ion Hybrid Electrochemical Capacitors (Li-HECs), *Nanoscale*, 6.8 (2014), 4387–94. https://doi.org/10.1039/c4nr00025k

[14] K. Cendrowski, W. Kukulka, T. Kedzierski, S. Zhang, Poly (Vinylidene Fluoride) and Carbon Derivative Structures from Eco-Friendly MOF-5 for Supercapacitor Electrode Preparation with Improved Electrochemical Performance, *Nanomaterials*, 8.11, (2018), 890. https://doi.org/10.3390/nano8110890

[15] T. Segakweng, N. Musyoka, J. Ren, P. Crouse, H. Langmi, , Comparison of MOF-5- and Cr-MOF-Derived Carbons for Hydrogen Storage Application, *Research on Chemical Intermediates*, 42.5 (2016), 4951–61. https://doi.org/10.1007/s11164-015-2338-1

[16] A. Li, Y. Tong, B. Cao, H. Song, Z. Li, X. Chen, J. Zhou, G. Chen, H. Luo,, MOF-Derived Multifractal Porous Carbon with Ultrahigh Lithium-Ion Storage Performance, *Scientific Reports*, 7 (2017), 1–8. https://doi.org/10.1038/srep40574

[17] C.Y. Lee, U.S. Patent 20,180,214,849. (2018).

[18] P. Haijun, L. Xiaoming, N. Jiliang, C. Yuepeng, W.O. Patent 107,492,637. (2017).

[19] H. Xu, S. Zhou, L. Xiao, H. Wang, S. Li, Q. Yuan, Fabrication of a Nitrogen-Doped Graphene Quantum Dot from MOF-Derived Porous Carbon and Its Application for

Highly Selective Fluorescence Detection of Fe3+, *Journal of Materials Chemistry C*, 3.2 (2015), 291–97. https://doi.org/10.1039/c4tc01991a

[20] F. Zheng, G. Xia, Y. Yang, O. Chen, MOF-Derived Ultrafine MnO Nanocrystals Embedded in a Porous Carbon Matrix as High-Performance Anodes for Lithium-Ion Batteries, *Nanoscale*, 7.21 (2015), 9637–45. https://doi.org/10.1039/c5nr00528k

[21] S. Liu, J. Zhou, Z. Cai, G. Fang, Y. Cai, A. Pan, S. Liang, Nb2O5 Quantum Dots Embedded in MOF Derived Nitrogen-Doped Porous Carbon for Advanced Hybrid Supercapacitor Applications, *Journal of Materials Chemistry A*, 4.45 (2016), 17838–47. https://doi.org/10.1039/c6ta07856g

[22] H. Li, L. Chi, C. Yang, L. Zhang, F. Yue, J. Wang, MOF Derived Porous Co@C Hexagonal-Shaped Prisms with High Catalytic Performance, *Journal of Materials Research*, 31.19 (2016), 3069–77. https://doi.org/10.1557/jmr.2016.314

[23] H. Sung, M. Arumugam, Self-Templated Synthesis of Co- and N-Doped Carbon Microtubes Composed of Hollow Nanospheres and Nanotubes for Efficient Oxygen Reduction Reaction, *Small*, 13.11 (2017), 1–8. https://doi.org/10.1002/smll.201603437

[24] Y. Zhou, Y. Chen, L. Cao, J. Lu, H. Jiang, Conversion of a Metal-Organic Framework to N-Doped Porous Carbon Incorporating Co and CoO Nanoparticles: Direct Oxidation of Alcohols to Esters, *Chemical Communications*, 51.39 (2015), 8292–95. https://doi.org/10.1039/c5cc01588j

[25] L. Andrew, C. Hsuan, J. Bo, Multi-Functional MOF-Derived Magnetic Carbon Sponge, *Journal of Materials Chemistry A*, 4.35 (2016), 13611–25.

[26] N. Torad, M. Hu, S. Ishihara, H. Sukegawa, A. Belik, M. Imura, K. Ariga, Y. Sakka, Direct Synthesis of MOF-Derived Nanoporous Carbon with Magnetic Co Nanoparticles toward Efficient Water Treatment, *Small*, 10.10 (2014), 2096–2107. https://doi.org/10.1002/smll.201302910

[27] L. Torad, M. Hu, S. Ishihara, H. Sukegawa, A. Belik, M. Imura, K. Ariga, Y. Sakka, Y. Yamauchi, A Novel Metal – Organic Framework Route to Embed Co Nanoparticles into Multi-Walled Carbon Nanotubes for Effective Oxygen Reduction in Alkaline Media. *Catalysts*, 7.12, (2017), 364. https://doi.org/10.3390/catal7120364

[28] X. Liu, X. Quan, Fe-MOF Derived Ferrous Hierarchically Porous Carbon Used as EF Cathode for PFOA Degradation, *2017 International Conference on Environmental Pollution and Public Health, EPPH 2017*, (2017), 9–14. https://doi.org/10.4236/gep.2017.56002

[29] E. C. Walter, T. Beetz, M. Y. Sfeir, L. E. Brus, Crystalline Graphite from an Organometallic Solution-Phase Reaction, *Journal of the American Chemical Society*, 128.49 (2006), 15590–91. https://doi.org/10.1021/ja0666203

[30] Z. Hasan, D. Cho, I. Nam, C. Chon, H. Song, Preparation of calcined zirconia-carbon composite from metal organic frameworks and its application to adsorption of crystal violet and salicylic acid. *Materials*, 9.4, (2016),261. https://doi.org/10.3390/ma9040261

[31] S. Liang, W. Qian, M. Zhonglei, L. Ying, X. Juan, W.O. Patent 107,578,927. (2018).

[32] S. Wu, Y. Zhu, Y. Huo, Y. Luo, L. Zhang, Y. Wan, B. Nan, L. Cao, Bimetallic Organic Frameworks Derived CuNi/Carbon Nanocomposites as Efficient Electrocatalysts for Oxygen Reduction Reaction, *Science China Materials*, 60.7 (2017), 654–63. https://doi.org/10.1007/s40843-017-9041-0

[33] H. Sung, J. Michael, M. Arumugam, 1D Co- and N-Doped Hierarchically Porous Carbon Nanotubes Derived from Bimetallic Metal Organic Framework for Efficient Oxygen and Tri-Iodide Reduction Reactions, *Advanced Energy Materials*, 7.7 (2017), 1–9. https://doi.org/10.1002/aenm.201601979

[34] Q. Gan, K. Zhao, S. Liu, Z. He, MOF-Derived Carbon Coating on Self-Supported $ZnCo_2O_4$–ZnO Nanorod Arrays as High-Performance Anode for Lithium-Ion Batteries, *Journal of Materials Science*, 52.13 (2017), 7768–80. https://doi.org/10.1007/s10853-017-1043-4

[35] Z. Li, L. Yin, MOF-Derived, N-Doped, Hierarchically Porous Carbon Sponges as Immobilizers to Confine Selenium as Cathodes for Li-Se Batteries with Superior Storage Capacity and Perfect Cycling Stability, *Nanoscale*, 7.21 (2015), 9597–9606. https://doi.org/10.1039/c5nr00903k

[36] H. Wang, X. Zhang, Y. Wang, G. Quan, X. Han, J. Yan, Facile Synthesis of Magnetic Nitrogen-Doped Porous Carbon from Bimetallic Metal–Organic Frameworks for Efficient Norfloxacin Removal, *Nanomaterials*, 8.9 (2018), 664. https://doi.org/10.3390/nano8090664

[37] C. Watcharop, A. Katsuhiko, Y. Yusuke, A New Family of Carbon Materials: Synthesis of MOF-Derived Nanoporous Carbons and Their Promising Applications, *Journal of Materials Chemistry A*, 1.1 (2013), 14–19. https://doi.org/10.1039/c2ta00278g

[38] Y. Min, F. Kam, Z. George, Synthesis and Applications of MOF-Derived Porous Nanostructures, *Green Energy & Environment*, 2.3 (2017), 218–45.

[39] S. Fardindoost, S. Hatamie, A. Zad, F. Astaraei, Hydrogen Sensing Properties of Nanocomposite Graphene Oxide/Co-Based Metal Organic Frameworks (Co-MOFs@GO), *Nanotechnology*, 29.1 (2017), 7. https://doi.org/10.1088/1361-6528/aa9829

[40] W. Zhiling, C. Yu, Y. Xiaofeng, L. Zhilian, Z. Luyi, W.O. Patent 107,576,714. (2018).

[41] G. Cai, W. Zhang, L. Jiao, S. Yu, H. Jiang, Template-Directed Growth of Well-Aligned MOF Arrays and Derived Self-Supporting Electrodes for Water Splitting, *Chem*, 2.6 (2017), 791–802. https://doi.org/10.1016/j.chempr.2017.04.016

[42] N. Torad, Y. Li, S. Ishihara, K. Ariga, Y. Kamachi, H. Lian, H. Hamoudi, Y. Sakka, W. Chaikittisilp, K. Wu, Y. Yamauchi, MOF-Derived Nanoporous Carbon as Intracellular Drug Delivery Carriers, *Chemistry Letters*, 43.5 (2014), 717–19. https://doi.org/10.1246/cl.131174

[43] L. Xiao, R. Xu, Q. Yuan, F. Wang, Highly Sensitive Electrochemical Sensor for Chloramphenicol Based on MOF Derived Exfoliated Porous Carbon, *Talanta*, 167. January (2017), 39–43. https://doi.org/10.1016/j.talanta.2017.01.078

[44] W. Li, S. Hu, X. Luo, Z. Li, X. Sun, Confined Amorphous Red Phosphorus in MOF-Derived N-Doped Microporous Carbon as a Superior Anode for Sodium-Ion Battery, *Advanced Materials*, 29.16 (2017), 1605820. https://doi.org/10.1002/adma.201605820

[45] C. Sun, Q. Dong, J. Yang, Z. Dai, J. Lin, P. Chen, Metal–organic Framework Derived CoSe2 nanoparticles Anchored on Carbon Fibers as Bifunctional Electrocatalysts for Efficient Overall Water Splitting, *Nano Research*, 9.8 (2016), 34–43. https://doi.org/10.1007/s12274-016-1110-1

[46] A. Banerjee, R. Gokhale, S. Bhatnagar, J. Jog, M. Bhardwaj, MOF Derived Porous Carbon-Fe3O4 Nanocomposite as a High Performance, Recyclable Environmental Superadsorbent, *Journal of Materials Chemistry*, 22.37 (2012), 19694–99. https://doi.org/10.1039/c2jm33798c

[47] H. Wu, B. Xia, L. Yu, X. Yu, X. Lou, Porous Molybdenum Carbide Nano-Octahedrons Synthesized via Confined Carburization in Metal-Organic Frameworks for Efficient Hydrogen Production, *Nature Communications*, 6 (2015), 6512. https://doi.org/10.1038/ncomms7512

Metal-Organic Framework Composites - Volume I
Materials Research Foundations **53** (2019) 257-275

Materials Research Forum LLC
https://doi.org/10.21741/9781644900291-12

Chapter 12

Polyoxometalate-Based Metal-Organic Framework Composites

Yunpeng He, Shuijin Yang[*]

College of Chemistry and Chemical Engineering, Hubei Key Laboratory of Pollutant Analysis & Reuse Technology, Hubei Normal University, Huangshi Hubei 435002, China

yangshuijin@163.com

Abstract

A series of polyoxometalates(POMs)-based Metal-Organic Frameworks (MOFs) were synthesized through direct incorporation between POMs and MOFs. In these compounds, the catalytically active polyanions as noncoordinating guests in the metal-organic frameworks host matrix. To overcome the difficulties of lack of sufficient thermal and chemical stability, a lot of research was done to introduce POMs to different MOFs for constructing various POM supporting MOFs with desired properties. POM based MOFs as a new type of functional materials developed rapidly, and is used in electrocatalysts, photocatalysis, oxidative desulfurization, lithium storage and adsorption.

Keywords

Polyoxometalates (POMs), Metal-Organic Frameworks (MOFs), Catalysts, Oxidative Desulfurization, Adsorption

Contents

1.   Introduction

At present, one of the development directions of inorganic chemistry is to combine with materials science. Inorganic functional materials were prepared by using the idea of crystal engineering. For example, organic and inorganic hybrid, composite, self-assembly and enhanced functional materials have received a lot of attention. Researching the principles of nanochemistry is strongly dependent on the successful construction of complex nanosystems in a large scale in a highly accurate assembling manner [1-3].

Metal organic frameworks (MOFs) have attracted the attention of scientists all around the world during the last years. The combination of organic and inorganic subunits has led to a vast chemical versatility, giving rise to more than ten thousand MOF structures [4].

The assembly of metal-organic frameworks (MOFs) constructed from metal ions as connectors and ligands as linkers have emerged as a class of porous molecular solid materials with designable, adjustable structure and chemical functionality [5-7]. In this research field, the scope for employment of functional organic linkers in MOFs and the encapsulation of different active entities, such as metal or metal oxide nanoparticles or polyoxometalates (POMs) and medicine molecules, within their channels have led to the development of MOFs for a wide variety of uses, such as catalysis, luminescence, separation, adsorption, gas storage, molecular detection and drug delivery, and so on [8-14].

Polyoxometalates (POMs) bear unique properties and exhibit a diverse compositional range. Since 1929, six basic configurations of polyoxometalates have been identified: Keggin, Dawson, Anderson, Waugh, Silverton and Lindqvist [15-21], in this case, the more common structure is Keggin and Dawson. Polyoxometalates (POMs) with controllable shape, size, and large negative charges possess potential multiple coordination sites and the ability [22]. However, their applications are limited by their low specific surface area, low stability, the conglomeration of POM particles and high solubility in aqueous solution. One of the ways to overcome these drawbacks is encapsulating them within metal-organic frameworks [23].

POMs are either directly part of the frameworks of MOFs or encapsulated within the cavities of MOFs, we called these kinds of materials POM-based MOF materials. Thus, such materials can combine the advantages of both POMs and MOFs. Introducing POMs into MOF systems to obtain POM-based MOF (POM@MOF) composite materials is a hot topic in this field of desulfurization, catalysis, adsorption, storage, and so on.

The topic of this review focuses on the recent progress in POM-based MOF materials, porous POM-based MOF materials including their electrocatalysts, photocatalysis, oxidative desulfurization, lithium storage and adsorption.

Metal-Organic Framework Composites - Volume I   Materials Research Forum LLC
Materials Research Foundations **53** (2019) 257-275   https://doi.org/10.21741/9781644900291-12

2. Application of POM-based MOF in electrocatalysts

POM-based MOF electrocatalysts, owing to their ability to mediate electron, proton and/or oxygen atom transfer reactions, which make them very attractive for energy-related electrocatalytic applications. The general methodologies used in the preparation of POM-based MOF composites and their different immobilization on electrodes is briefly described.

Basu et al. [24] have reported CoII-based Keggin $K_6[CoW_{12}O_{40}]$ as a guest in the cages of the ZIF-8 metal-organic framework (MOF) host material (Fig 1).

Faradaic efficiency of POM@ZIF-8 was evaluated by determining the amount of O_2 evolved in constant current coulometry (at j=1 mAcm^{-2}) and was found to be 95.7%. Turnover frequencies (TOF) were calculated from the quantitative oxygen evolution (TOF = 12.5 s^{-1}) and also from electrochemical measurements (TOF = 10.19 s^{-1}) (using Tafel plot and calculated surface coverage value). Such high TOF is extremely rare among all electrocatalytic WOC reported. Thus, it can be said that the requirement of very high overpotential is the most serious drawback of POM@ZIF-8 as an electrochemical WOC, while the other factors, i.e., Faradic efficiency, turnover frequency, recyclability, robustness, etc., prove the potential of such POM encapsulated MOF systems to perform electrocatalytic water oxidation.

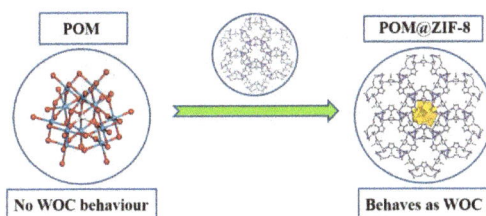

Fig 1. Activation of Keggin ($K_6[CoW_{12}O_{40}]$) toward electrochemical water oxidation byencapsulation in ZIF-8.

Nohra et al. [25] have reported the high electroactivity of POM-based metal organic frameworks (POM@MOFs). (TBA)$_3$[PMoV$_8$MoV$_{14}$O$_{36}$(OH)$_4$Zn$_4$][C$_6$H$_3$(COO)$_3$]$_{4/3}$·6H$_2$O (ε(trim)$_{4/3}$) is a 3D open-framework (Fig 2). These hybrids built of molecular Keggin units are connected by 1,3,5-benzene tricarboxylate linkers, with channels occupied by tetrabutylammonium (TBA) counterions. A TOF value of 6.7 s^{-1} at η = 0.20 V for HER in aqueous medium has been achieved with the POM@MOF-based electrodes. This robust performance was due to the POM@MOF structure and to confinement effects.

Metal-Organic Framework Composites - Volume I
Materials Research Foundations **53** (2019) 257-275

Materials Research Forum LLC
https://doi.org/10.21741/9781644900291-12

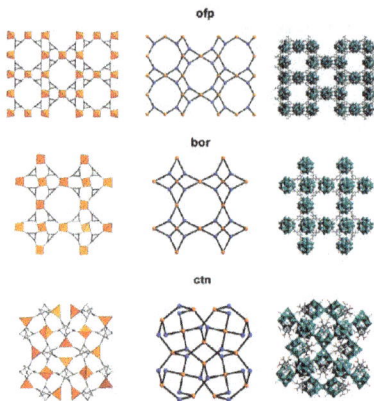

Fig 2. Crystal structures of the three ε(trim)$_{4/3}$ POMOFs described in the simulation section

In 2015, Qin et al. [26] have also designed two novels POM@MOFs and assessed their electrocatalytic activity towards HER. POM@MOFs and the organic links were benzene tribenzoate (H$_3$BTB) and [1, 1'-biphenyl]-3, 4', 5- tricarboxylate (H$_3$BPT) fragments (Fig 3a and b), for NENU-500 and NENU-501, respectively. NENU-500 presented the highest activity in acidic conditions (Fig 3c and d), with η10 = 0.237 V vs. RHE and was able to maintain the HER activity after 2000 cycles.

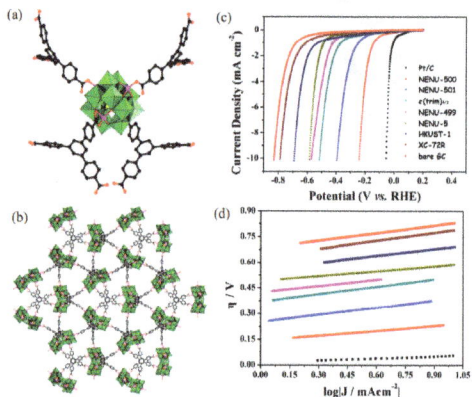

Fig 3. Structure of NENU-500: (a) connection mode between Zn-ε-Keggin and BTB3-fragments, (b) 3D (3,4)-connected framework. Electrochemical characterization of the prepared catalysts: (c) LSV polarization curves in 0.5 M H$_2$SO$_4$ aqueous solution and (d) the corresponding Tafel plots

This unique feature of POMOFs to allow the stabilization of electroactive POMs in MOF-type scaffold opens up promising perspectives for the design of more efficient catalysts.

3.   Application of POM-based MOF in photocatalysis

Employing the cages or channels with tunable sizes in MOFs, POMs can be encapsulated in MOFs as guests yielding POM@MOFs crystalline hybrids, in which POMs are dispersed at the molecular level, the hydrolytic stability of both POMs and MOFs can be greatly enhanced, and the catalytic functions of POMs may be combined with the porosity and selective sorption properties of MOFs.

Shah et al. [27] reported two cobalt containing polyoxometalate anions $K_7[Co^{II}Co^{III}W_{11}O_{39}(H_2O)]$ and $Na_{10}[Co_4(PW_9O_{34})_2(H_2O)_2]$ were encapsulated in MIL-100 (Fe) to form leaching free PMOF composites was prepared by hydrothermal synthesis (Fig 4).

Photocatalytic water oxidation activities for $K_7[Co^{II}Co^{III}W_{11}O_{39}(H_2O)]$ and $Na_{10}[Co_4(PW_9O_{34})_2(H_2O)_2]$ were examined in borate buffer solutions at pH = 9 and 8 for respectively. $Na_2S_2O_8$ was used as sacrificial electron acceptor and $Ru(bpy)_3^{2+}$ as photosensitizer under visible light (> 420 nm).

Compare bare $K_7[Co^{II}Co^{III}W_{11}O_{39}(H_2O)]$ and $Na_{10}[Co_4(PW_9O_{34})_2(H_2O)_2]$, O_2 yield in case of $[CoIICoIIIW_{11}O_{39}(H_2O)]^{-7}@$ MIL-100(Fe) in first 10 minutes increased from 30% to 41% while turn over frequency (TOF) value in the first two minutes jumped from 0.48 s^{-1} to 0.53 s^{-1}. The TOF value for $[Co_4(PW_9O_{34})_2(H_2O)_2]^{-10}@MIL-101(Fe)$ is calculated to be 9.2×10^{-3} s^{-1} in first 20 minutes and 72% yield of produced O_2 is achieved in 70 minutes of irradiation. These values are higher than TOF and O_2 yield of POM@MIL-101(Cr) (66% and 7.3 x 10^{-3} s^{-1}) reported previously. The results showed that the catalytic activity was significantly enhanced. Both of the composites also showed significant structural stability and can be recycled. This enhanced activity of composite catalysts is attributed to synergic effect between components.

Zhang et al. [28] developed a simple and effective charge-assisted self-assembly process to encapsulate a polyoxometalate (POM) inside a metal-organic framework (MOF) built from $[Ru(bpy)_3]^{2+}$-derived dicarboxylate ligands and $Zr_6(\mu_3-O)_4(\mu_3-OH)_4$ secondary building units(SBUs) (Fig 5a).

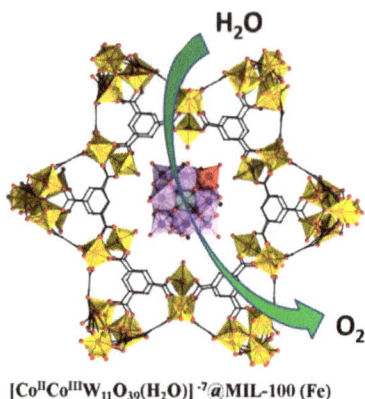

$[Co^{II}Co^{III}W_{11}O_{39}(H_2O)]^{-7}$@MIL-100 (Fe)

Fig 4. Crystal structures of $[Co^{II}Co^{III}W_{11}O_{39} (H_2O)]^{-7}$@MIL-100(Fe)

Hierarchical organization of photosensitizing and catalytic proton reduction components in such a POM@MOF assembly enables fast multi-electron injection from the photoactive framework to the encapsulated redox-active POMs upon photoexcitation, leading to efficient visible-light driven hydrogen production (Fig 5b).

The modular and tunable nature of this synthetic strategy should allow the design of multifunctional MOF materials for other applications.

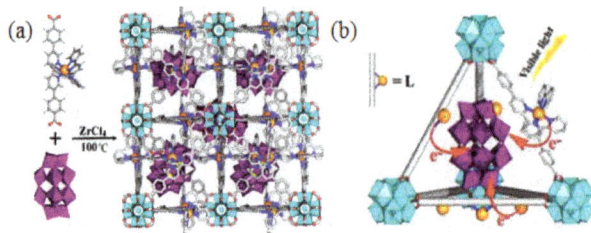

Fig 5. (a) One-pot synthesis of the POM@UiO system via charge assisted self-assembly; (b) schematic showing synergistic visible light excitation of the UiO framework and multi-electron injection into the encapsulated POMs

Li et al. [29] have developed a simple, general, and efficient method for constructing photocatalytic active metal-organic framework (MOF)-based composite materials for

Metal-Organic Framework Composites - Volume I   Materials Research Forum LLC
Materials Research Foundations **53** (2019) 257-275   https://doi.org/10.21741/9781644900291-12

visible light-driven hydrogen production. Here, several transition metal-substituted Wells-Dawson-type polyoxometalates (POMs) were successfully immobilized into a Cr-MOF of the MIL-101 structure, resulting in a series of POM@MOF composite materials [POM=$K_8HP_2W_{15}V_3O_{62}\cdot9H_2O(P_2W_{15}V_3)$,$K_8P_2W_{17}(NiOH_2)O_{61}\cdot17H_2O(P_2W_{17}Ni)$, $K_8P_2W_{17}(CoOH_2)O_{61}\cdot16H_2O$ $(P_2W_{17}Co)$].

These composite materials accommodate and enrich cationic photosensitizer (PS) ruthenium (II) tris(bipyridyl) ($Ru(bpy)_3^{2+}$) from the solution, allowing the PSs to surround the POM proton reduction catalysts, this greatly improves their ability to adsorb cationic PSs. The introduction of POMs into the MOF framework not only causes the polyoxoanions to arrange themselves uniformly in the 3D matrix, but also enriches the PSs, allowing them to surround around and make effective contact with catalysts. The hydrogen production rate can reach up to 25578 $\mu molh^{-1}g^{-1}$ by enriching both the anionic Mo_2S_{12} clusters and cationic PSs in the porous MOF platform (Fig 6).

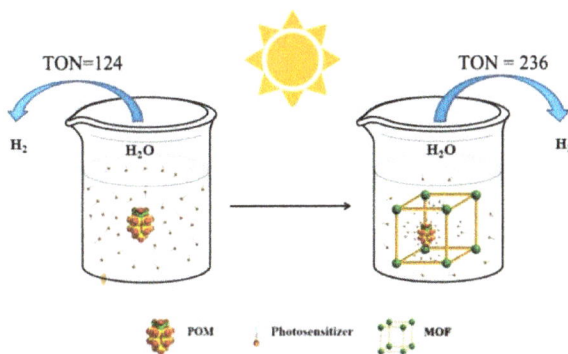

Fig 6. Schematic diagram of hydrogen production by POM@MOF photocatalysis

Zhong et al. [30] encapsulated PW_{12} into MOFs HKUST-1 through facile liquidassisted grinding method, and applied the obtained $[Cu_2(BTC)_{4/3}(H_2O)_2]_6[HPW_{12}O_{40}]$ nanocrystals (NENU-3N) as catalysts in the degradation of phenol. It is the first time of Liquidassisted grinding was applied in the preparation of nanocrystalline POM@MOFs hybrids. The NENU-3N nanocrystals catalyzed the degradation of phenol on the basis of both MOF and POM catalytic activities, representing the first example of POM@MOFs catalyst boosting catalytic oxidation reaction with double actives sites (Fig 7).

Strikingly, up to 97% conversion and 88% mineralization have successfully realized by perfect cooperative catalysis between POMs and Cu (II) nodes in MOFs at 35 ℃.

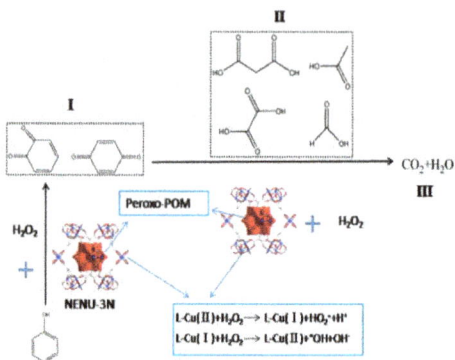

Fig 7. Speculation for the degradation processes of phenol in the CWPO of phenol with NENU-3N as catalyst

4.   Application of POM-based MOF in oxidative desulfurization

Fuel desulfurization as a crucial subject has aroused people's attention with the environmental problems becoming more and more important. Oxidative desulfurization (ODS) is one of the most promising methods due to its high efficiency, low cost, easy operation and mild reaction condition. The ODS process comprises two steps: (i) oxidation of the organosulfur compounds to the corresponding sulfoxides and/or sulfones, and (ii) the removal of the oxidized compounds by extraction, adsorption or distillation [31-33].

Introducing polyoxometalates (POMs) into MOF systems to obtain POM-encapsulating MOF (POM@MOF) composite materials is one such hot topic in this field.

Gao et al. [34] with the loading of polyoxometalate (POM), a hybrid material (CNTs@MOF-199) supported catalysts (CNTs@MOF-199-POM) are designed and prepared using one-pot procedure.

The test results indicate that the catalyst CNTs@MOF-199-$Mo_{16}V_2$ possesses superior catalytic activity, with a desulfurization efficiency of up to 98.30 %. In addition, CNTs@MOF-199-$Mo_{16}V_2$ exhibits an excellent reusability, and the catalytic efficiency is only slightly reduced after recycling for 7 times (Fig 8a). Besides, the kinetic studies indicate that the desulfurization process belongs to apparent first-order kinetic reaction (Fig 8b). And the apparent activation energy is 12.89 kJ / mol.

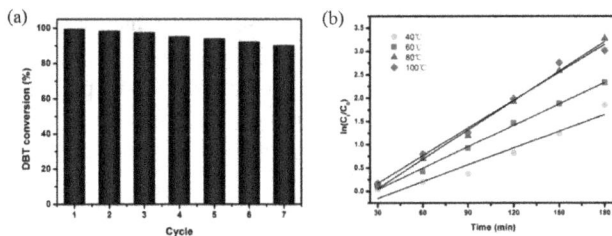

Fig 8. (a) Stability test results of the catalyst; (b) pseudo-first-order kinetics for the oxidation of DBT

Hao et al. [35] has synthesized a new cationic triazole-based metal-organic framework encapsulating Keggin-type polyoxometalates, with the molecular formula $[Co(BBPTZ)_3][HPMo_{12}O_{40}] \cdot 24H_2O$ (BBPTZ=4,4'-bis(1,2,4-triazol-1-ylmethyl)biphenyl) by hydrothermal method.

The structure of $[Co(BBPTZ)_3][HPMo_{12}O_{40}] \cdot 24H_2O$ contains a non-interpenetrated 3D $CdSO_4$ (cds)-type framework with two types of channels that are interconnected with each other; straight channels that are occupied by the Keggin-type POM anions, and wavelike channels that contain lattice water molecules (Fig 9). $[Co(BBPTZ)_3][HPMo_{12}O_{40}] \cdot 24H_2O$ represents a new POM@MOF hybrid example, which is composed of a cationic porous MOF based on N-donor ligands and Keggin-type polyoxoanions. Moreover, the catalytic activities of such POM@MOF composites are not only dependent on the POM moieties but also the pore size and volume of the MOF units. The catalytic activity of $[Co(BBPTZ)_3][HPMo_{12}O_{40}] \cdot 24H_2O$ in the oxidative desulfurization reaction indicates that it is not only an effective and size-selective heterogeneous catalyst, but it also exhibits distinct structural stability in the catalytic reaction system.

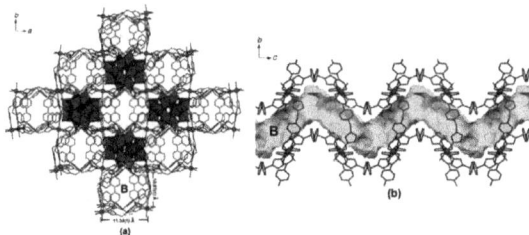

Fig 9. (a) Ball-and-stick and polyhedral view of the POM@MOF structure of composite viewed along c axis; (b) the wavelike channel B in the POM@MOF of composite viewed along an axis.

An aluminum 2-aminoterephthalate based metal-organic framework (MOF) material was applied for the first time to prepare highly efficient heterogeneous catalysts in desulfurization processes [36]. Sandwich-type $[Eu(PW_{11}O_{39})_2]^{11-}$ polyoxometalate (POM) was supported on Al(III) and Cr(III) MIL-type MOFs, NH_2-MIL-53(Al) and MIL-101(Cr), and extensive characterization confirmed the incorporation of the POM on the two supports.

Both composite materials have shown to be active and robust heterogeneous catalysts for the efficient removal of sulfur-containing compounds from model diesel, and the influence of the solid support on the catalytic performance of the active species was further assessed. The POM/MIL(Al) revealed notable catalytic performance, since complete desulfurization was obtained after 2 h of reaction (Fig 10). Furthermore, this remarkable heterogeneous catalyst revealed to be stable and recyclable for various catalytic cycles.

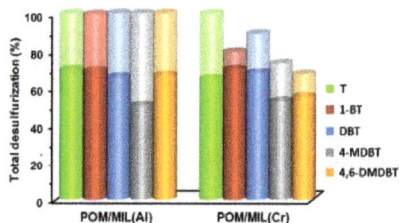

Fig 10. Percentage of each sulfur compound removed from the model diesel in the presence of the heterogeneous POM/MIL(Al) and POM/MIL(Cr) catalysts after 10 min (darker part) and 2 h (entire bars) of reaction

4.   Application of POM-based MOF in lithium storage

Lithium-ion batteries (LIBs) have become the most important mobile storage technology [37-38]. However, there is still a great challenge to improve the performance of LIBs, for example, by further improving the energy density, rate performance, and cycle life polyoxometalates (POMs) generally possess redox properties, which have great potential applications for electrochemical energy storage [39-40].

However, when POMs are explored for anode materials, they can be easily dissolved in the electrolyte, resulting in significant capacity degradation. To overcome this difficulty, POMs are usually combined with MOFs to fabricate polyoxometalate-based metal-organic frameworks (POMOFs). POMOFs combine the advantages of POMs and MOFs and are a perfect candidate for anode materials.

Metal-Organic Framework Composites - Volume I   Materials Research Forum LLC
Materials Research Foundations **53** (2019) 257-275   https://doi.org/10.21741/9781644900291-12

Wei et al. [41] report a novel nanocomposites based on polyoxometalates-based metal–organic frameworks (POMOFs)/reduced graphene oxide (RGO) for lithium-ion batteries. It demonstrates the advantages of polyoxometalates (POMs), metal–organic frameworks (MOFs) and RGO, thus shows the hybrid behavior of battery and supercapacitor. A reversible capacity of 1075 mAh g^{-1} was maintained after 100 cycles, and the capacity retentions are nearly 100 % both at 2000 and 3000 mA g^{-1} for over 400 cycles.

This study demonstrates the utilization of PMG nanocomposites as anode materials for rechargeable LIBs for the first time (Fig 11). It shows the cooperative capacity of battery-supercapacitor behavior for superior lithium storage by combine the advantages of the POMs, MOFs and RGO together, such as multi-electron redox properties of POMs, high electronic conductivity of RGO and open frameworks of MOFs which can buffer the volume expansion and ensure good cycling stabilities.

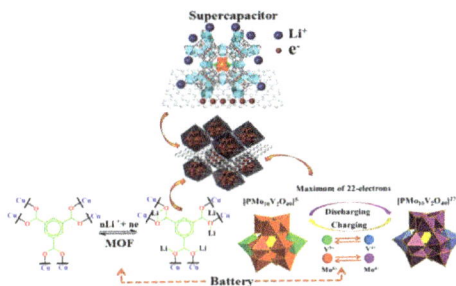

Fig 11. The schematic diagram of the possible mechanism is for the cooperative capacity of battery-supercapacitor for the PMG nanocomposites.

Dong et al. [42] found a simple hot-pressing method to incorporate polyoxometalate (POM)-based metal-organic frameworks (MOFs) onto three-dimensionally structured carbon cloth (CC), denoted as HP-NENU-5/CC (Fig 12), which immobilizes POMs into the MOFs avoiding the leaching of POMs and employs HP-NENU-5/CC as a flexible, conductive, and porous anode material.

The integrated HP-NENU-5/CC electrode is the best for lithium storage, which displays the largest capacity of 1723 mAh g^{-1} at a current density 200 mA g^{-1} for 100 cycles. Importantly, nearly 99.8% of the capacity is retained at 1000 mA g-1 for 400 cycles. It shows the outstanding electrochemical performances for lithium storage, which are attributed to the redox properties of POMs, the porous property of MOFs, and the excellent electronic conductivity of CC. This new strategy effectively improves the electrochemical performance by addressing the dissolution of POMs and the low electronic conductivity of MOFs, which can be applied in the new design for the development of MOFs, POMs, and POMOF-based materials for energy storage.

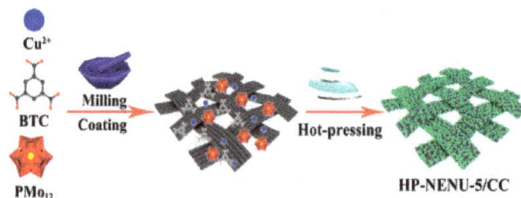

Fig 12. Synthesis of HP-NENU-5/CC

Zhang et al. [43] skillfully encapsulate ionic liquids (ILs) into polyoxometalate-based metal-organic frameworks (POMOFs) to fabricate a series of ILs functionalized POMOFs crystals (denoted as POMs-ILs@MOFs) (Fig 13), which immobilize POMs in the cage of MOFs avoiding the leaching of POMs and obtain an enhanced conductivity by the modification of the ILs. One of POMs-ILs@MOFs ($PMo_{10}V_2$-ILs@MIL-100) crystals show superior cycling stability and high rate capability when they are used as anode materials, which are the best amongst all the reported MOFs, POMs and POMOFs crystals materials. The outstanding performances are attributed to the hybrid behavior of a battery-supercapacitor, which is a synergistic effect among ILs, POMs, and MOFs.

When $PMo_{10}V_2$-ILs@MIL-100 crystals are used as anode materials for LIBs, it delivers a high capacity of 1248 mAh g^{-1} at 0.1 A g^{-1} after 100 cycles. The battery performances of PMo10V2-ILs@MIL-100 crystals, including capacity, cycling stability and rate behavior, are significantly improved compared with the MIL-100, $PMo_{10}V_2$, $PMo_{10}V_2$@MIL-100 and ILs@MIL-100 crystals, and are the best amongst all the reported MOFs, POMs and POMOFs-based crystals materials.

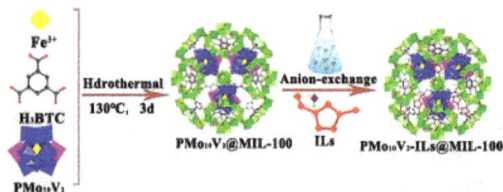

Fig 13. Schematic diagram of the synthesis process of $PMo_{10}V_2$-ILs@MIL-100 crystals

Metal-Organic Framework Composites - Volume I   Materials Research Forum LLC
Materials Research Foundations **53** (2019) 257-275   https://doi.org/10.21741/9781644900291-12

5. Application of POM-based MOF in adsorption

Yan et al. [44] have adopted a simple one-pot method to synthesize the POM@MOF composites. A mesoporous MIL-101 was used as the platform to encapsulate different POMs affording three new POM@MOF composites, POM@MIL-101 (POM=$PW_{11}V$, PW_{12}, SiW_{12}). The adsorption efficiency of MIL-101 towards MB in aqueous solution was significantly improved by immobilizing of hydrophilic POMs into its cages. Notably, these composites possess a fast adsorption rate and high uptake capacity of the dye molecules (Fig 14a). They could capture the dye molecules (such as MB) from the mixture of the dyes (such as MB and MO with similar sizes) with high selectivity. Also, the POM@MIL-101 materials could be readily recycled and reused, and no POM leached during the dye adsorption process (Fig 14b). The results of adsorption and separation of dyes have also expanded the application of this kind of compound.

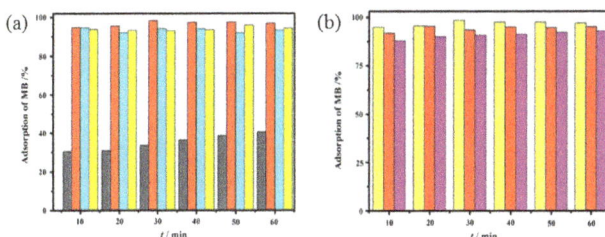

Fig 14. (a) Different POMs encapsulated in MIL-101 working on adsorption of MB: gray for MIL-101; red for $PW_{11}V$@MIL-101; azure for PW_{12}@MIL-101; yellow for SiW_{12}@MIL-101; (b) The reusability of $PW_{11}V$@MIL-101: light-yellow for the first cycle; red for the second cycle; magenta for the third cycle.

Ghahramaninezhad et al. [45] successful interaction of zeolitic imidazolate framework-8 (ZIF-8) with a water-soluble Keplerate-type polyoxometalate ($\{Mo_{132}\}$) is accomplished and a POM/ZIF-8 nano-composite material (ZIF-8@$\{Mo_{132}\}$) is achieved. The POM/ZIF-8 nano-composite enables the uptake of lithium functional groups (Li^+ cations) due to the existence of $\{Mo_{132}\}$ in its structure. As a result of interactions between ZIF-8@$\{Mo_{132}\}$ and an organolithium compound, a Li-trapped POM/ZIF-8 nano-composite material (ZIF-8@$\{Mo_{132}\}$@Li^+) is readily obtained (Fig 15). In particular, the ZIF-8@$\{Mo_{132}\}$@Li^+ nano-composite displays a surprising enhancement in CO_2 adsorption capacity, the most significant greenhouse gas. The CO_2 uptake of ZIF-8@$\{Mo_{132}\}$@Li^+ is around 26.40 wt%, while the capacity of pristine ZIF-8 is only around 4 wt%. The presence of lithium cations inside and on the surface of the as-synthesized nano-composite plays a key role in the enhancement of CO_2 storage。

Notably, the ZIF-8@{Mo_{132}}@Li^+ nano-composite also exhibits high stability under real conditions, where the sample was exposed to humid air for 7 months and no changes in CO_2 capacity were observed.

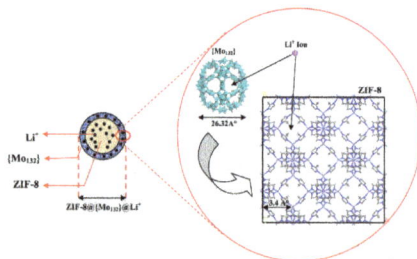

Fig 15. Schematic diagram of the as-synthesized nano-composite of
ZIF-8@{Mo_{132}}@Li^+

A novel environmental friendly adsorbent $H_6P_2W_{18}O_{62}$/MOF-5 was synthesized by a simple one-step reaction under solvothermal conditions [46]. The removal rate of $H_6P_2W_{18}O_{62}$/MOF-5 was quite greater (85%) than that of MOF-5 (almost zero), showing that the adsorption performance of porous MOF-5 can be improved through the modification of $H_6P_2W_{18}O_{62}$.

Further study revealed that $H_6P_2W_{18}O_{62}$/MOF-5 exhibited a fast adsorption rate and selective adsorption ability towards the cationic dyes in aqueous solution (Fig 16). The removal rate was up to 97% for cationic dyes methylene blue (MB) and 68% for rhodamine B (RhB) within 10 min. However, anionicdye methyl orange (MO) can only reach to 10%.

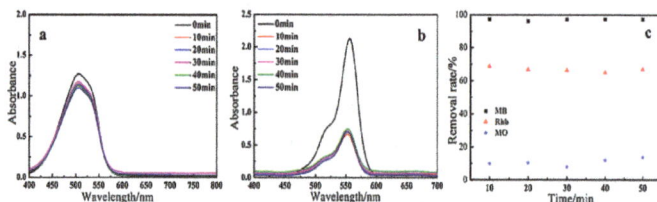

Fig 16. The adsorption capability of $H_6P_2W_{18}O_{62}$/MOF-5 toward MO (a), Rhb (b) and the
removal rate of Rhb, MO and MB (c)

Concluding Remarks

In this review, the application of POMs-based MOFs composites for various fields was reviewed. A series of outstanding work has been done to encapsulate POMs within MOFs. The resulting composites have been applied to electrocatalysts, photocatalysis, oxidative desulfurization, lithium storage, adsorption, and so on.

Undoubtedly, huge progress is being made in the field of POMs-based MOFs for many fields. The use of experimental and computational methods, as well as the increasing knowledge regarding the combination of different types of materials, has led to the improvement of science and technology. Nevertheless, there is still considerable work to be done in order to achieve a set of more stable POMs-based MOFs materials.

Reference

[1] K. Suzuki, S. Sato, M. Fujita, Template synthesis of precisely monodisperse silica nanoparticles within self-assembled organometallic spheres, Nat. Chem. 2(2010) 25-29. https://doi.org/10.1038/nchem.446

[2] Z. H. Kang, E. B. Wang, B. D. Mao, et al, Controllable fabrication of carbon nanotube and nanobelt with a polyoxometalate-assisted mild hydrothermal process, J. Am. Chem. Soc. 127(2005) 6534-6535. https://doi.org/10.1021/ja051228v

[3] Z. H. Kang, Y. Liu, S. T. Lee, Small-sized silicon nanoparticles: new nanolights and nanocatalysts, Nanoscale. 3(2011) 777-791. https://doi.org/10.1039/c0nr00559b

[4] J. R. Long, O. M. Yaghi, The pervasive chemistry of metal–organic frameworks, Chem. Soc. Rev. 38(2009) 1213–1214. https://doi.org/10.1039/b903811f

[5] H. K. Chae, D. Y. Siberio-Perez, J. Kim, et al, A route to high surface area, porosity and inclusion of large molecules in crystals, Nature. 427(2004) 523-527. https://doi.org/10.1038/nature02311

[6] K. T. Holman, Molecule-constructed microporous materials: long under our noses, increasingly on our tongues, and now in our bellies, Angew. Chem., Int. Ed. 50(2011) 1228-1230. https://doi.org/10.1002/anie.201006783

[7] J. Q. Sha, J. W. Sun, C. Wang, et al, Syntheses study of Keggin POM supporting MOFs system, Cryst. Growth Des. 12(2012) 2242-2250. https://doi.org/10.1021/cg201478y

[8] A. Aijaz, T. Akita, N. Tsumori, et al, Metal-organic framework-immobilized polyhedral metal nanocrystals: reduction at solid-gas interface, metal segregation,

core-shell structure, and high catalytic activity, J. Am. Chem. Soc. 135(2013) 16356-16359. https://doi.org/10.1021/ja4093055

[9] Y. Yuan, F. X. Sun, L. N. Li, et al, Porous aromatic frameworks with anion-templated pore apertures serving as polymeric sieves, Nat. Commun. 5(2014) 4260-4268. https://doi.org/10.1038/ncomms5260

[10] K. Manna, T. Zhang, W. B. Lin, Stable porphyrin Zr and Hf metal–organic frameworks featuring 2.5 nm cages: high surface areas, SCSC transformations and catalyses, J. Am. Chem. Soc. 136(2014) 6566-6569. https://doi.org/10.1039/c5sc00213c

[11] M. J. Dong, M. Zhao, S. Ou, et al, A luminescent dye@ MOF platform: emission fingerprint relationships of volatile organic molecules, Angew. Chem. Int. Ed. 53(2014) 1575-1579. https://doi.org/10.1002/anie.201307331

[12] K. Mo, Y. H. Yang, Y. Cui, A homochiral metal-organic framework as an effective asymmetric catalyst for cyanohydrin synthesis, J. Am. Chem. Soc. 136(2014) 1746-1749. https://doi.org/10.1021/ja411887c

[13] D. Tian, Q. Chen, Y. Li, et al, A mixed molecular building block strategy for the design of nested polyhedron metal-organic frameworks, Angew. Chem. Int. Ed. 53(2014) 837-841. https://doi.org/10.1002/anie.201307681

[14] W. W. Zhan, Q. Kuang, J. Z. Zhou, et al, Semiconductor@metal-organic framework core-shellHeterostructures: a case of ZnO@ZIF-8 nanorods with selective photoelectrochemical response, J. Am. Chem. Soc. 135(2013) 1926-1933. https://doi.org/10.1021/ja311085e

[15] D. Y. Du, J. S. Qin, S. L. Li, et al. ChemInform abstract: recent advances in porous polyoxometalate-based metal-organic framework materials, Chem. Soc. Rev. 43(2014) 4615-4632. https://doi.org/10.1039/c3cs60404g

[16] A. M. Khenkin, L. Weiner, Y. Wang, et al, Electron and oxygen transfer in polyoxometalate, H(5)PV(2)Mo(10)O(40), catalyzed oxidation of aromatic and alkyl aromatic compounds: evidence for aerobic Mars-van Krevelen-type reactions in the liquid homogeneous phase, J. Am. Chem. Soc. 123(2001) 8531-8542. https://doi.org/10.1021/ja004163z

[17] N. M. Okun, T. M. Anderson, K. I. Hardcastle, et al, Cupric Decamolybdodivanadophosphate. a coordination polymer heterogeneous catalyst for rapid, high conversion, high selectivity sulfoxidation using the ambient environment, Inorg. Chem. 42(2003) 6610-6612. https://doi.org/10.1021/ic0348953

[18] T. Rüther, V. M. Hultgren, B. P. Timko, et al, Electrochemical investigation of photooxidation processes promoted by sulfo-polyoxometalates: coupling of photochemical and electrochemical processes into an effective catalytic cycle, J. Am. Chem. Soc. 125(2003) 10133-10143. https://doi.org/10.1021/ja029348f

[19] C. L. Hill. Stable, Self-Assembling, Equilibrating catalysts for green chemistry, Angew. Chem. Int. Ed. (2004) 402-404. https://doi.org/10.1002/anie.200301701

[20] J. M. Clemente-Juan, E. Coronado. Coord, Magnetic clusters from polyoxometalate complexes, Chem. Rev. 193-195(1999) 361-394. https://doi.org/10.1016/s0010-8545(99)00170-8

[21] T. Yamase, P. V. Prokop, Photochemical formation of tire-shaped molybdenum blues: topology of a defect anion, [Mo(142)O(432)H(28)(H(2)O)(58)](12-), Angew. Chem. Int. Ed. 41(2002) 466-469. https://doi.org/10.1002/1521-3773(20020201)41:3<466::aid-anie466>3.0.co;2-w

[22] J. J. Gong, W. S. Zhang, Y. Liu, et al, Keggin polyanion and copper cluster based coordination polymer towards model for complex nanosystem, Dalton Trans. 41(2012) 5468-5471. https://doi.org/10.1039/c2dt30284e

[23] R. Canioni, C. Roch-Marchal, F. Sécheresse, et al, Stable polyoxometalate insertion within the mesoporous metal organic framework MIL-100(Fe), J. Mater. Chem. 24(2011) 1226-1233. https://doi.org/10.1039/c0jm02381g

[24] O. Basu, S. Mukhopadhyay, S. K. Das, Cobalt based functional inorganic materials: electrocatalytic water oxidation, J. Chem. Sci. 130(2018) 93-108. https://doi.org/10.1007/s12039-018-1494-4

[25] B. Nohra, H. El Moll, L. M. Rodriguez Albelo, et al, Polyoxometalate- based metal organic frameworks (POMOFs): structural trends, energetics, and high electrocatalytic efficiency for hydrogen evolution reaction, J. Am. Chem. Soc. 133(2011) 13363-13374. https://doi.org/10.1021/ja201165c

[26] J. S. Qin, D. Y. Du, W. Guan, et al, Ultrastable polymolybdate-based metal-organic frameworks as highly active electrocatalysts for hydrogen generation from water, J. Am. Chem. Soc. 137(2015) 7169-7177. https://doi.org/10.1021/jacs.5b02688

[27] W. A. Shah, A. Waseem, M. A. Nadeem, et al, Leaching-free encapsulation of cobalt-polyoxotungstates in MIL-100 (Fe) for highly reproducible photocatalytic water oxidation, Appl. Catal., A. 567(2018) 132-138. https://doi.org/10.1016/j.apcata.2018.08.002

[28] Z. M. Zhang, T. Zhang, C. Wang, et al, Photosensitizing metal-organic framework enabling visible-light-driven proton reduction by a Wells-Dawson-type polyoxometalate, J. Am. Chem. Soc. 137 (2015) 3197-3200. https://doi.org/10.1021/jacs.5b00075

[29] H. Li, S. Yao, H. L. Wu, et al, Charge-regulated sequential adsorption of anionic catalysts and cationic photosensitizers into metal-organic frameworks enhances photocatalytic proton reduction, Appl. Catal. B. 224(2018) 46-52. https://doi.org/10.1016/j.apcatb.2017.10.031

[30] X. H. Zhong, Y. Lu, F. Luo, et al, A nanocrystalline POM@MOFs catalyst for the degradation of phenol: effective cooperative catalysis by metal nodes and POM guests, Chem. Eur. J. 224(2018) 3045-3051. https://doi.org/10.1002/chem.201705677

[31] J. M. Campos-Martin, M. C. Capel-Sanchez, P. Perez-Presas, et al, Oxidative processes of desulfurization of liquid fuels, J. Chem. Technol. Biotechnol. 85(2010) 879-890. https://doi.org/10.1002/jctb.2371

[32] X. Ma, A. Zhou, C. Song, A novel method for oxidative desulfurization of liquid hydrocarbon fuels based on catalytic oxidation using molecular oxygen coupled with selective adsorption, Catal. Today. 123(2007) 276-284. https://doi.org/10.1016/j.cattod.2007.02.036

[33] E. Ito, J. A. R. Van Veen, On novel processes for removing sulphur from refinery streams, Catal. Today. 116(2006) 446-460. https://doi.org/10.1016/j.cattod.2006.06.040

[34] Y. Gao, Z. Lv, R. Gao, et al, Oxidative desulfurization process of model fuel under molecular oxygen by polyoxometalate loaded in hybrid material CNTs@MOF-199 as catalyst, J. Hazard. Mater. 359(2018) 258-265. https://doi.org/10.1016/j.jhazmat.2018.07.008

[35] X. L. Hao, Y. Y. Ma, H. Y. Zang, et al, A polyoxometalate-encapsulating ctionic metal-organic framework as a heterogeneous catalyst for desulfurization, Chem. Eur. J. 21(2015) 3778-3784. https://doi.org/10.1002/chem.201405825

[36] C. M. Granadeiro, L. S. Nogueira, D. Julião, et al, Influence of porous MOF support in the catalytic performance of Eu-polyoxometalate based materials: desulfurization of model diesel, Catal. Sci. Technol. 6(2016) 1515-1522. https://doi.org/10.1039/c5cy01110h

[37] Y. Zhao, X. F. Li, B. Yan, et al, Recent developments and understanding of novel mixed transition-metal oxides as anodes in lithium ion batteries, Adv. Energy Mater. 6(2016) 1502175. https://doi.org/10.1002/aenm.201502175

[38] J. Xu, J. Mahmood, Y. Dou, et al, 2D frameworks of C2N and C3N as new anode materials for lithium-ion batteries, Adv. Mater. 29(2017) 1702007. https://doi.org/10.1002/adma.201702007

[39] J. J. Chen, M. D. Symes, S. C. Fan, et al, High-performance polyoxometalate-based cathode materials for rechargeable lithium-ion batteries, Adv. Mater. 27(2015) 4649-4654. https://doi.org/10.1002/adma.201501088

[40] H. Wang, S. Hamanaka, Y. Nishimoto, et al, Control of the grafting of hybrid polyoxometalates on metal and carbon surfaces: toward submonolayers, J. Am. Chem. Soc. 134(2012) 4918-4924.

[41] T. Wei, M. Zhang, P. Wu, et al, POM-based metal-organic framework reduced graphene oxide nanocomposites with hybrid behavior of battery-supercapacitor for superior lithium storage, Nano Energy, 34(2017) 205-214. https://doi.org/10.1016/j.nanoen.2017.02.028

[42] A. M. Zhang, M. Zhang, D. Lan, et al, Polyoxometalate-based metal-organic framework on carbon cloth with a hot-pressing method for high-performance lithium-ion batteries, Inorg. Chem. 57(2018) 11726-11731. https://doi.org/10.1021/acs.inorgchem.8b01860

[43] M. Zhang, A. M. Zhang, X. X. Wang, et al, Encapsulating ionic liquids into POM-based MOFs to improve their conductivity for superior lithium storage, J. Mater. Chem. A. 6(2018) 8735-8741. https://doi.org/10.1039/c8ta01062e

[44] A. X. Yan, S. Yao, Y. G. Li, et al, Incorporating polyoxometalates into a porous MOF greatly improves its selective adsorption of cationic dyes, Chem. Eur. J. 20(2014) 6927-6933. https://doi.org/10.1002/chem.201400175

[45] M. Ghahramaninezhad, B. Soleimani, M. N. Shahrak, A simple and novel protocol for Li-trapping with a POM/MOF nano-composite as a new adsorbent for CO2 uptake, New J. Chem. 42(2018) 4639-4645. https://doi.org/10.1039/c8nj00274f

[46] X. X. Liu, W. P. Gong, J. Luo, et al, Selective adsorption of cationic dyes from aqueous solution by polyoxometalate-based metal–organic framework composite, Appl. Surf. Sci. 362(2016) 517-524. https://doi.org/10.1016/j.apsusc.2015.11.151

Keyword Index

About the Editors

Dr. Anish Khan is currently working as Assistant Professor, at Chemistry Department, Centre of Excellence for Advanced Materials Research (CEAMR), Faculty of Science, King Abdulaziz University, Jeddah, Saudi Arabia. **Ph.D.** Completed from Aligarh Muslim University, India in 2010. He has 13 years research experience of working in the field of organic-inorganic electrically conducting nano-composites and its applications in making chemical sonsor. He completed Postdoctoral from School of Chemical Sciences, University Sains Malaysia (USM) on electroanalytical chemistry for one year. More than 115 research articles have been published in refereed international journals. More than 10 international conferences/ workshop and 7 books published and 15 Book chapters. Around 20 research project completed. Member of American Nano Society, Field of specialization is polymer nanocomposite/cation-exchanger/chemical sensor/micro biosensor/nanotechnology, application of nanomaterials in electroanalytical chemistry, material chemistry, ion-exchange chromatography and electro-analytical chemistry, dealing with the synthesis, characterization (using different analytical techniques) and derivatization of inorganic ion-exchanger by the incorporation of electrically conducting polymers. Preparation and characterization of hybrid nano composite materials and their applications, Polymeric inorganic cation –exchange materials, Electrically conducting polymeric, materials, Composite material use as Sensors, Green chemistry by remediation of pollution, Heavy metal ion-selective membrane electrode, Biosensor on neurotransmitter.

Prof. Bahaa M. Abu-Zied studied Chemistry in at Assiut University (Egypt) from 1985 to 1989. In 1994 he received his Master degree in Surface Chemistry and Catalysis from the same university. After obtaining his PhD from Assiut University in 1997, he was appointed as Assistant Professor at the Chemistry Department, Faculty of Science, Assiut University. At the same year he joined the 33rd International Seminar at the University of Karlsruhe (Germany) for 16 months. During that period, he joined the group of Prof. Thomas Turek at the "Institute für Chemische Verfahrenstechnik" at the same University. In 2004 he was promoted as Associate Professor and in 2010 as full Professor at Assiut University. During the period 2002-2011, he joined the research group of Prof. Wilhelm Schwieger at the "Institute für Chemische Reaktionstechnik" Friedrich-Alexander-University, Erlangen-Nürnberg (Germany) for a short research visits (3-6 months). Since 2012, he is on leave to King Abdulaziz University in Jeddah (Saudi Arabia).

Dr. Mahmoud A. Hussein is an Associate professor of Polymer Chemistry, polymer chemistry Lab, Chemistry Department, Faculty of Science, Assiut University (AU), Egypt. He got a new position within Asiri group at Chemistry Department, King Abdulaziz University (KAU), Jeddah, Saudi Arabia. He obtained BSc in Chemistry and PhD in Organic Polymer Synthesis from Assiut University, Egypt. He got a post doctoral position in the University of Nice Sophia

Antipolis, France and University Sains Malaysia, Malaysia. He visited school of Industrial technology, University Sains Malaysia and Faculty of Engineering, University of Porto (UP) as visiting researcher two times for each. His research interests in the area of polymer synthesis, characterization and applications for different fields, polymer composites materials, and other smart materials. Dr. Hussein published +70 research articles in ISI journal and attend several conferences and workshops in the field of polymer chemistry and martial science. He present 17 poster and ppt in international conferences. In addition he trained in more than 20 training courses which are related to the field of study.

Prof. Abdullah Mohammed Ahmed Asiri is Professor in Chemistry Department – Faculty of Science -King Abdulaziz University. **Ph.D.** (1995) From University of Walls College of Cardiff, U.K. on Tribochromic compounds and their applications. More than 1000 Research articles and 20 books published. The chairman of the Chemistry Department, King Abdulaziz University currently and also the director of the center of Excellence for Advanced Materials Research. Director of Education Affair Unit–Deanship of Community services. Member of Advisory committee for advancing materials, (National Technology Plan, (King Abdul Aziz City of Science and Technology, Riyadh, Saudi Arabia). Color chemistry. Synthesis of novel photochromic and thermochromic systems,Synthesis of novel colorants and coloration of textiles and plastics, Molecular Modeling, Applications of organic materials into optics such as OEDS, High performance organic Dyes and pigments. New applications of organic photochromic compounds in new novelty. Organic synthesis of heterocyclic compounds as precursor for dyes. Synthesis of polymers functionalized with organic dyes. Preparation of some coating formulations for different applications. Photodynamic thereby using Organic Dyes and Pigments Virtual Labs and Experimental Simulations. He is member of Editorial board of Journal of Saudi Chemical Society, Journal of King Abdul Aziz University,Pigment and Resin Technology Journal, Organic Chemistry Insights, Libertas Academica, Recent Patents on Materials Science, Bentham Science Publishers Ltd. Beside that he has professional membership of International and National Society and Professional bodies.

Dr. Mohammad Azam is currently an Associate Professor of Chemistry at the King Saud University, Riyadh Saudi Arabia. He earned his Ph.D. degree in 2009 from the Aligarh Muslim University, Aligarh, India. He is working in the area of coordination chemistry. His research interests focus mainly on the construction of coordination compounds and their properties. He has published over 50 research articles in various journals of international repute.